Sino–US Energy Triangles

The remarkable performance of the Chinese economy in the last three decades has placed China at the centre of the world stage. In 1993, China became a net importer of energy, although it was not until the early 2000s that the world began to pay more attention to China's energy needs and its potential impact on the world. With China's energy search occurring within a hegemonic global structure dominated by the United States, the US watches with interest as China enhances its ties with energy-rich states.

The book examines this triangular relationship and questions whether the US and China are in competition regarding access to the energy of a third state, within the context of a potential power transition. It includes case studies on China's energy relationship with countries such as Canada, Australia, Saudi Arabia, Nigeria, Brazil, Kazakhstan, Iran, Sudan and Venezuela and aims to understand the ways a rising power interacts with an existing leading power and the possible outcomes of this competition. The analytical framework employed helps the reader to understand not only the nature and pattern of triangles among the US, China and the resource rich states under 'resource diplomacy', but also the salient features of US–China competition around the world.

Making an impressive contribution to the literature in fields such as US–China Relations, International Relations, Chinese Foreign Policy and Global Energy Geopolitics, this book will appeal to students and scholars of these subjects.

David Zweig is Chair Professor, Division of Social Science, Hong Kong University of Science and Technology, and Director of the Center on China's Transnational Relations. He is Senior Research Fellow of the Asia–Pacific Foundation, Vancouver, Canada, and Associate Chairman of the Center on China and Globalization, Beijing, China. His widely known work includes *Internationalizing China: Domestic Interests and Global Linkages* (2002).

Yufan Hao is Chair Professor of Political Science and Dean of the Faculty of Social Sciences at the University of Macau. He has written and edited 24 books and monographs, and authored more than 50 journal articles and book chapters on Chinese Politics, Chinese Foreign Relations, US–China Relations and Macau Studies.

Politics in Asia series

Sino–US Energy Triangles

Resource diplomacy under hegemony

Edited by David Zweig and Yufan Hao

Routledge
Taylor & Francis Group

LONDON AND NEW YORK

First published 2016
by Routledge
2 Park Square, Milton Park, Abingdon, Oxfordshire OX14 4RN

and by Routledge
711 Third Avenue, New York, NY 10017

First issued in paperback 2016

*Routledge is an imprint of the Taylor & Francis Group,
an informa business*

British Library Cataloguing in Publication Data
A catalogue record for this book is available from the British Library

Library of Congress Cataloging in Publication Data
Sino-U.S. energy triangles : resource diplomacy under hegemony / edited by
David Zweig and Yufan Hao.
    pages cm. – (Politics in Asia series)
    Includes bibliographical references and index.
    1. Energy consumption–Political aspects–China. 2. Energy development–
Political aspects–China. 3. Petroleum industry and trade–China. 4. China–
Foreign economic relations. 5. China–Foreign relations. 6. China–Foreign
relations–United States. 7. United States–Foreign relations–China.
    I. Zweig, David. II. Hao, Yufan.
    HD9502.C62S565 2015
    333.790951–dc23   2015000970

ISBN 13: 978-1-138-62902-8 (pbk)
ISBN 13: 978-1-138-77808-5 (hbk)

Typeset in Times New Roman
by Sunrise Setting Ltd, Paignton, UK

Contents

List of illustrations

Figures

Tables

Contributors

Philip Andrews-Speed is a Principal Fellow at the Energy Studies Institute of the National University of Singapore. Recent books include *China, Oil and Global Politics* with Roland Dannreuther (Routledge, 2011) and *The Governance of Energy in China: Transition to a Low-Carbon Economy* (Palgrave Macmillan, 2012).

John W. Garver teaches in the Sam Nunn School of International Affairs at Georgia Institute of Technology. He has written and studied China's international relations since becoming intrigued by this topic during the Vietnam War when he discovered, low and behold, that the US was fighting in Vietnam to contain China. Since then he has written a bunch of books and articles trying to understand China's relations with various countries. He has visited Iran twice, once in late 1978 as the revolutionary movement was gaining momentum and he had the distinct honor of being chased by a mob, and again in 2004 during the presidency of Mohammad Khatami. That time the mob was missing.

Yufan Hao is Chair Professor of Political Science and Dean of Faculty of Social Sciences at University of Macau. He obtained his PhD from the Johns Hopkins University (SAIS) in 1989, and taught at Colgate University from 1990 to 2005. He has written and edited 24 books and monographs, and authored more than 50 journal articles and book chapters on Chinese politics, Chinese foreign relations, US–China relations, and Macao Studies.

Mikkal Herberg is Research Director on Asian energy security at The National Bureau of Asian Research (NBR). He is also a Senior Lecturer on international energy and politics at the Graduate School of International Relations and Pacific Studies at the University of California, San Diego. Previously he spent 20 years in the oil industry in strategic planning roles for ARCO (now BP), where from 1997–2000 he was Director for Global Energy and Economics, responsible for worldwide energy, economic, and political analysis. He also headed country risk management and held positions including Director of Portfolio Risk Management and Director for Emerging Markets.

Wenran Jiang has been the Director of Canada–China Energy and Environment Forum since 2004. He has written extensively on the rise of China and its impact on the Chinese political economy and the rest of the world, with a major focus on the shifting balance of power in the global economy, international finance, and energy and resource sectors. He is currently completing a book on energy security and Chinese foreign policy.

Susana Moreira is an Extractive Industries Specialist at the World Bank. She received a PhD degree from the Johns Hopkins University School of Advanced International Studies, focusing on Chinese national oil companies' investment strategies in Latin America and Sub-Saharan Africa. She also serves as a sub-Saharan Africa analyst for Freedom in the World.

Sébastien Peyrouse, PhD, is a Research Professor at the Central Asia Program in the Institute for European, Russian and Eurasian Studies (George Washington University). His main areas of expertise are political systems in Central Asia, economic and social issues, Islam and religious minorities, and Central Asia's geopolitical positioning toward China, India and South Asia.

Sonja Regler has studied political science and philosophy at universities in Germany, France and China. She completed her PhD research on the topic 'Struggle over Darfur – the Development of Chinese Security Politics in the Darfur crisis' in September 2012 and now works as a Program Manager with EULEX in Kosovo.

Yitzhak Shichor, gained his PhD from London School of Economics and Political Science and is Lipson Chair Professor Emeritus of Political Science and Asian Studies at the Hebrew University of Jerusalem and the University of Haifa. He is the former Head of the Tel-Hai Academic College, and Fellow at the Institute of National Security Studies, Tel Aviv.

Ian Taylor is Professor in International Relations and African Politics at the University of St Andrews and also Chair Professor in the School of International Studies, Renmin University of China. He is also Professor Extraordinary at the University of Stellenbosch, South Africa; an Honorary Professor at the Institute of African Studies, China; and a Visiting Scholar at Mbarara University, Uganda. Focusing largely on Africa he has authored eight academic books, edited another eight and has published over 60 peer-reviewed scholarly articles, over 60 chapters in books and numerous working papers, reports, etc. He holds a DPhil from the University of Stellenbosch and an MPhil from the University of Hong Kong. He also holds a teaching certificate from the International Language Institute, Cairo, Egypt. Prior to joining St Andrews he taught African Politics for 4 years at the University of Botswana and has conducted research in and/or visited 38 African countries.

Nicholas Thomas is an Associate Professor in the Department of Asian and International Studies, City University of Hong Kong. He has published widely on Chinese foreign policy, East Asian regionalism and non-traditional security challenges. He is currently working on a project exploring the subnational dimensions of China's Northeast Asian politics.

Alex Vines is head of the Africa Programme at Chatham House and a senior lecturer and Co-Director of the African Studies Centre at Coventry University. He regularly publishes on Angola and sits on the boards of the Journal of Southern African Studies, the South African Journal of International Affairs and Africa Review (the journal of the African Studies Association of India).

Cynthia Watson serves on the faculty at the National War College, concentrating on China's evolving international security posture, particularly in Latin America. She earned a doctorate from the University of Notre Dame, a master's from the London School of Economics, and an honours undergraduate degree from the University of Missouri Kansas City. She is a member of numerous professional associations and has published extensively in national security fora.

David Zweig is Chair Professor, Division of Social Science, and Director, Center on Environment, Energy and Resource Policy at The Hong Kong University of Science and Technology. He is also a Senior Research Fellow for the Asia–Pacific Foundation of Canada and Associate Chairman for the Center on China and Globalization, Beijing.

Acknowledgements

Editing books is no easy task. Editors are highly dependent on other people: on funders to supply funds for the conference; on authors to write their original drafts for the conference and to make the revisions in a timely fashion; on co-editors to help with the heavy load; on administrators who manage the process, who often make the in-text editing after the editor had made changes on hard copy, and who follow up with delinquent authors; reviewers for presses who need to find something meaningful in your book and to recommend publication; and on editors at the publishing house to agree to send the manuscript out for review, to help edit it, to keep pressure on you to complete the project and to get the final effort accomplished. For that reason, I always advise junior colleagues to eschew this mode of publication until they have tenure.

In this case, my thanks go especially to Stephanie Rogers with whom I had worked previously, and Hannah Mack, both at Routledge, for believing in the project. I also thank the editors at another press who politely waited my decision of where I was going to publish.

The book benefited from an opportunity provided by the Asia Society, which funded a travelling seminar with myself, three authors (Thomas, Garver and Moreira), and Xu Xiaojie of the Chinese Academy of Social Sciences to present our ideas in both Houston and New York where we received excellent comments. The ideas behind this project were first laid out in an online publication, '"Resource Diplomacy" Under Hegemony: Foreign Policy "Triangularism" and Sino-American Energy Competition in the 21st Century', in Mikkal E. Herberg and David Zweig's *China's 'Energy Rise', The US, and the Geopolitics of Energy*, published by the Pacific Council on International Policy in April 2010. My thanks go to David Karl for his help on that publication.

This book is no different to many other books, which takes time, effort and cooperation to succeed. I sometimes wonder why I have been involved in so many edited or co-edited projects, given the amount of time and effort involved. In this case, the task of writing so many well-developed case studies to test my ideas was simply too great. For that reason, I am deeply grateful to a talented, thoughtful and knowledgeable group of academics who helped me cover such a large field and for helping me test my own model. I appreciate that they actually took my ideas on the 'triangular' nature of Sino–American energy relations seriously.

I am also thankful to the reviewers at Routledge who made an enormous number of very valuable comments, particularly the comment that this study is valuable, worthy of publication and will be widely read. Xu Xiaojie, of the Chinese Academy of Social Sciences, and his team gave important comments at several points also along the way.

Funding for the conference, which led to this book, was supplied the Lee Hysan Foundation (Hong Kong), The China Institute (University of Alberta) and the School of Humanities and Social Sciences at The Hong Kong University of Science and Technology. Funding for some of the research used in the papers on Canada, Venezuela and Australia came from a grant from the Hong Kong Research Grants Council, entitled 'Resource Diplomacy under Hegemony'.

Particular thanks to Yufan Hao, Dean of the School of Social Science and Humanities at the University of Macau, who co-sponsored the conference and without whose support the project would not have been possible. I also greatly appreciate his role as co-editor. Thanks to Karen Greening at Sunrise Setting Ltd for successfully guiding the production process to completion with only limited friction. A truly remarkable feat!

My deep thanks also to my Executive Officer, Ms. Meggy Wan, who has been a great help to me for many years. And to my wife, Joy, and my children Rachel and Aaron.

Abbreviations

AD	Acción Democrática
ADIZ	Air Defence Identification Zone
AFP	Agence France Presse
ANZUS	Australia, New Zealand, United States Security Treaty
AOSC	Athabasca Oil Sands Corporation's
APEC	Asia–Pacific Economic Cooperation
APSCO	Asia Pacific Space Cooperation Organization
AU	African Union
AUSFTA	Australia–US Free Trade Agreement
BNDES	the Brazilian Development Bank,
BTC	Baku–Tiblisi–Ceyhan
CAFTA	China–Australia Free Trade Agreement
CAPP	Canadian Association of Petroleum Producers
CASS	Chinese Academy of Social Sciences
CBBC	China–Brazil Business Council
CCB	China Construction Bank
CCP	Chinese Communist Party
CDB	China Development Bank
CERC	US–China Clean Energy Research Center, the
CFIUS	Committee on Foreign Investment in the United States
CGNPC	Guangdong Nuclear Power Group
CIC	China Investment Corporation
CICIR	Chinese Institute for Contemporary International Relations
CIF	China International Fund Ltd.
CIS	Commonwealth of Independent States
CITIC	China International Trust and Investment Corporation
CNL	Chevron Nigeria Limited
CNOOC	China National Offshore Oil Corporation
CNPC	China National Petroleum Corporation
COPEI	Partido Social Cristiano de Venezuela
COSBAN	China–Brazil High-level Coordination and Cooperation Committee

COSTIND	Committee on Science, Technology, and Industry for National Defense
CPA	Comprehensive Peace Agreement
CPC	Caspian Pipeline Consortium
CRDF	Collective Rapid Deployment Force
CSCEC	China State Construction Engineering Corporation
CSIH	China Sonangol International Holding (Ltd.)
CSRC	China Securities Regulatory Commission
CSTO	Collective Security Treaty Organization
DFAT	Department of Foreign Affairs and Trade
ECOWAS	Economic Community of West African States
EIA	Energy Information Administration (US)
ELN	Ejército de Liberación Nacional
EPC	Engineering Procurement and Construction
EU	European Union
Ex–Im Bank	China's Export–Import Bank
FATA	Foreign Acquisitions and Takeovers Act
FDI	foreign direct investment
FIRB	Foreign Investment Review Board
FNLA	National Front for the Liberation of Angola
FOCAC	Forum of Chinese–African Cooperation
FTA	Free Trade Agreement
FTZ	Free Trade Zone
GASCAC	Sinopec's contract for the second phase of Gasene
GDP	gross domestic product
GRN	Gabinete de Reconstrução Nacional
IAEA	International Atomic Energy Agency
ICC	International Criminal Court
IEA	International Energy Agency
IMF	International Monetary Fund
IOCs	International oil companies
IOSC	International Oil Spill Conference
IPAP	Individual Plan of Action for the Partnership
IRI	Islamic Republic of Iran
JEM	Justice and Equality Movement
LHS	left-hand side
LNG	Liquefied natural gas
LTO	Light tight oil
MEND	Movement for the Emancipation of the Niger Delta
MOU	Memorandum of Understanding
MPLA	Popular Movement for the Liberation of Angola
MPNU	Mobil Producing Nigeria Unlimited
NAFTA	North American Free Trade Agreement
NAOC	Nigerian Agip Oil Company Limited

NATO	North Atlantic Treaty Organization
NDP	New Democratic Party
NDRC	National Development and Reform Commission
NEA	National Energy Administration
NNPC	Nigerian National Petroleum Corporation
NOC	National Oil Companies
NPD	Nigerian Petroleum Development
NPT	Non-Proliferation Treaty
NSA	National Security Agency (US)
NSW	New South Wales
ODI	Overseas direct investment
OECD	Economic Cooperation and Development
OML	Oil Mining Lease
ONGC	Oil and Natural Gas Corporation
OPEC	Organization of Petroleum Exporting Countries
OSCE	Organization for Security and Co-operation in Europe
PLA	People's Liberation Army
PLAN	People's Liberation Army Navy
PRC	People's Republic of China
RHS	right-hand side
RRCs	Resource Rich Countries
RRS	Resource Rich States
SABIC	Sinopec and Saudi Basic Industry Corporation
SASAC	State-owned Assets Supervision and Administration Commission
SCO	Shanghai Cooperation Organization
SDB	State Development Bank
SEC	Securities and Exchange Commission
Sinopec	China Petrochemical Corp
SIPRI	Stockholm International Peace Research Institute
SLA	Sudan Liberation Army
SLOCs	Sea lanes of communication
SMEs	Small and medium-sized enterprises
SOE	State-owned enterprises
SPDC	Shell Petroleum Development Company of Nigeria (Ltd.)
SPLM	Sudanese People's Liberation Army
SPR	Strategic Petroleum Reserves
SSI	Sonangol Sinopec International
SSTPC	Sinopec–SABIC Tianjin Petrochemical Company
SWF	Sovereign wealth fund
TPNL	Total Petroleum Nigeria Limited
TPP	Trans-Pacific Partnership
UAE	United Arab Emirates
UN	United Nations
Unipec	China International United Petroleum & Chemicals Co., Ltd.

UNITA	National Union for the Total Independence of Angola
Unocal	Union Oil Company of California
US	United States
US$	US Dollars
USAID	United States Agency for International Development
WECO	World Energy China Outlook
WTI	West Texas Intermediate
WTO	World Trade Organization
YASREF	Yanbu Aramco Sinopec Refinery

Introduction

Sino–US energy competition in resource rich states

Yufan Hao and David Zweig

The spectacular performance of the Chinese economy in the last three decades has put China on the centre of the world stage. With its gross domestic product (GDP) quadrupling in less than 16 years, China has become the world's second economic powerhouse. Every day brings astonishingly new figures from China: China has become the largest trading state; China's foreign reserves exceed the three trillion dollar mark; China has surpassed the United States (US) to become the world's largest automobile market; China will have its first stealth jetfighter and its first aircraft carrier, etc. As China's economic and military capabilities grow, people are debating the likely impact of China's rise. Some believe that China may be more dissatisfied about the existing international system as its power grows; others believe that China may be more cooperative as it becomes an affluent country with a population that is increasingly middle-income and an overall more pluralistic society.

The future direction of China will have great implications for the US. Reports speculate about the potential dangers posed by a rising China to the US. For example, the *New York Times* reported that the Chinese have reached the Caribbean beaches to build 'a brand new $35 million stadium' in the Bahamas (Archibold, 2012).'The tiny island nation of Dominica has received a grammar school, a renovated hospital and a sports stadium…. Antigua and Barbuda got a power plant and a cricket stadium, and a new school is on its way' (Etzioni, 2012). In June 2013, Chinese President Xi Jinping visited Trinidad and Tobago, Costa Rica and Mexico before he met with President Obama in Sunnylands, California, and extended China's influence within the US's backyard.

There is every reason to hope that US–China relations will follow a smooth and peaceful course, but neither history nor theory can assure us that it will be so. Although some suggest that US–China energy competition may lead inevitably to confrontation (Friedberg, 2005: 45), others see competition without conflict (Kirshner, 2008). Yet scholarly analysis by way of case studies is lacking. This volume fills that need by examining the real and potential competition between China and the US for influence over third countries that are rich in resources, as well as US and Chinese policy towards their major energy suppliers.

This study is a collective effort, led by David Zweig, to understand how a rising power interacts with the existing leading power and the possible outcome of this competition. First, we test the hypothesis that because of the structure of the international system, with the US as the hegemon and China as a rising challenger, the US and China should encounter each other in almost every corner of the world. Second, due to China's challenge to US global dominance, the US is likely to use China's need for energy resources to slow down its rise. No doubt the Chinese feel this way because they see the US everywhere, trying to 'contain' their development; however, although theory (as well as Chinese concerns) would predict that the US should interfere in China's effort to resolve its energy needs as a way to slow down its rise, is this true?

The theme of the book, as Zweig emphasizes in Chapter 1, is that this competition over resources generates a triangular relationship among the US, China and the resource rich state (RRS). According to one view, if two sides of a triangle gang up on the other, the outlier can make the relationship between the other two difficult to manage (Alterman, 2011: 28). Zweig's framework, however, seeks to discern conditions under which China's search for energy generates tension, conflict or cooperation with the US. Zweig divides China's energy partners, the RRSs, into three groups based on the how they are perceived by the US – pariahs, neutrals or allies – and hypothesizes that the US's response to China's engagement with these countries will be strongly influenced by the nature of its ties to the RRS. Similarly, Zweig suggests that the RRS can respond to China in several ways – by pushing back against China, manipulating both sides or drawing close to China and perhaps China's political system – again based partly on the nature of that state's ties to the US. In the end, these two dyadic ties may influence Sino–US interactions towards this third RRS.

This analytical framework helps us understand not only the nature and pattern of triangles among US, China and the RRSs under 'resource diplomacy', but also the salient features of US–China competition around the world. Some countries examined in this book, particularly US 'pariahs', take advantage of Chinese energy thirst by moving close to China in order to protect themselves from the US. Other countries may try to accommodate China, but push back when domestic or even foreign forces (i.e., the US) demand they do so. The citizens of democratic countries are particularly keen in resisting selling national companies to what they perceive as Chinese Communist Party-dominated state-owned enterprises (SOE). Finally, some RRS manipulate the Sino–US competition to maximize their economic gain from China (and the US), while enjoying a close military and political relationship with Washington.

Different energy strategies: the US and China

The US's energy security concerns and its global energy strategy are fundamentally two-fold: avoid large oil supply disruptions or oil transit disruptions, and avoid oil price shocks that may damage the US economy or that of its allies. In pursuit of these interests, the US has pushed for a global energy policy that emphasizes free and transparent market trade and the free flow of investment to drive global supply, demand and prices. According to Herberg (Chapter 3), the US thrust is 'market over mercantilism'. The US has pushed aggressively for the broadest possible market access to reserves for new exploration and development to boost global oil supply, and also collaborated with other oil consumers to manage the potential impact of oil supply disruptions, particularly through the International Energy Agency (IEA). At the same time, Washington, as the hegemon, protects key transit routes to ensure that the US and its allies have secure oil supply. As the hegemon, Washington is ready to take military action if necessary, making itself a key player normatively, diplomatically and militarily in all of the key oil and gas exporting regions of the world.

China, on its side, has developed a three-pronged strategy to secure external oil and gas supplies and satisfy its energy needs: (1) reduce domestic industrial consumption by increasing energy efficiency – industry consumes two-thirds of all of China's energy supplies (not households or automobiles); (2) invest in a range of new and renewable forms of energy to diversify energy sources, including unconventional gas reserves, nuclear power, wind and solar power, and shale (Chapter 2); and (3) most importantly, securing external energy supplies.

As China's net imports of both crude oil and oil products more than tripled between 2000 and 2010, China has tried to diversify the sources of these imports (see Chapter 14 for an assessment of these efforts). The Chinese government has given loans to the national oil companies (NOC) in various RRSs and has constructed new infrastructure to secure delivery of oil and gas to China. China's NOCs have also aggressively invested in overseas oil and gas assets believing that, in a global energy crisis, only NOCs will provide the needed services. According to Andrew-Speed (Chapter 2), China's NOCs have a stake in more than 200 projects in about 50 countries as a result of these Chinese overseas upstream investments. To enhance the transport of energy purchases, China has recently rapidly expanded its naval capabilities, creating a blue water navy that could be deployed to project its power beyond its coast. China is also building ports that can receive Chinese ships – cargo vessels and oil tankers – in Pakistan, Bangladesh, Myanmar and Sri Lanka. Finally, China has developed a significant tanker fleet that can transfer oil to China should a shipping crisis emerge.

These actions have profound implications for the US. American military planners are aware of China's growing military power. Policy analysts in Washington are concerned about China's energy strategy and overseas influence in three respects. First, China's increasing need for energy, and its efforts to secure energy supply through equity oil purchases, poses a potential threat to the market order that the US works to maintain, destabilizing energy markets and distorting prices. Some Americans particularly fear that China may try to 'lock up' global energy supplies, rather than open up new ones (Herberg, 2011).

Second, China has begun competing for oil in markets that the US has traditionally dominated, such as Canada, Saudi Arabia, Angola, Nigeria, Venezuela and Brazil. China's effectiveness is due in part to its willingness to work with the various NOCs in RRSs, although today, most oil majors have become more willing to do so.

Third, Washington is uncomfortable about China's involvement in countries that the US deems as 'pariahs', such as Iran or Sudan, because Chinese investment or loans may obstruct the US's efforts to pursue its strategic interests in those countries. In this respect, it is not competition over oil and gas, per se, that raises the eyebrows of Washington policy makers; rather, it is the different policy objectives that Beijing and Washington pursue with those 'pariah states'.

Pariahs move closer to China

As Zweig suggests in Chapter 1, some states that Washington policy makers consider to be pariahs offer to resolve China's energy thirst in return for protection from US or UN sanctions. In Iran and Sudan, China quickly filled the vacuum created when the US and its allies withdrew from both countries. In truth, China needs their oil, but they need China's protection.

In 2007 and 2009, Chinese energy firms entered the Iranian energy section as European and Japanese firms exited under the pressure of US sanctions (Chapter 11). By investing US$48–$50 billion in 2009 (35–40 per cent involved contracts signed and under execution), China became the largest investor in the upstream energy sector in Iran. In 2011, Iran was the third largest source of crude oil for China, and approximately 10 per cent of China's oil imports were from Iran (US Energy Information Administration, 2013). Approximately 80 per cent of China's total imports from Iran are oil and the rest are mineral and chemical products (Simpson, 2010). However, as tension between Washington and Tehran over Iran's nuclear program escalated in the 2000s, Washington repeatedly asked China to support sanctions against Iran. That request put China in a dilemma. China supports the principle of non-proliferation advocated by Washington and its allies, but Beijing worries that Washington's real intention towards Tehran is 'regime change' and broader US geostrategic interests in the Middle East. Undermining Iran's economy merely abets the US effort to overthrow the regime. Also, abandoning Iran in the face of US pressure would undermine China's relations with Iran and China's long-term efforts to gain a foothold in Iran's energy sector. However, ignoring US demands increases distrust between Washington and Beijing and potentially undermines efforts to improve Sino–US relations.

According to Garver (Chapter 11), Beijing has tried to balance US demands, and the probability of gains or losses in Sino–US ties with its own connections to Iran by at times acceding to or rejecting particular US requests. As Garver further points out, Beijing cooperates with the US enough to make Washington happy, but not enough to undermine China's economic cooperation with Iran, demonstrating pragmatism in dealing with both Washington and Tehran. By accepting Tehran's disclaimers that it was not seeking nuclear weapons, Beijing opposed sanctions on two grounds: first, sanctions, they argue, make resolving issues more difficult; and second, sanctions would interfere with China's economic relations with Iran. China's position with regard to Iran is that because it does not participate in US-led sanctions against Iran, China's importation of oil from Iran is based on economic need, and China will continue to trade with Iran without violating relevant resolutions of the UN Security Council or undermining interests from a third party or the international community.

According to Regler (Chapter 12), China's involvement in Sudan mirrors that in Iran. China's objectives in Sudan were economic and energy-oriented. Chinese NOCs quietly entered Sudan after Washington sanctioned it as a state sponsor of terrorism and for human rights abuses, allowing Chinese NOCs to become the biggest stakeholders in the country's energy sector. However, the Darfur crisis tested China's resource diplomacy because China's involvement in Sudan has been severely criticized by Western governments, media and human rights activists. Yet, as with Iran, China feared that the US's real goal was to change the regime and regain access to Sudanese oil.

Sudan, like Iran, stuck close to China in an effort to weaken US pressure. China opposed sanctions against the Sudanese government, but eventually abstained from voting for UN Resolution 1593, which referred the case of Darfur to the International Criminal Court in March 2005. That action upset Sudan's President, Omar al-Bashir, but pleased the Americans.

Some suggested that China's balancing acts in Iran and Sudan did not satisfy Washington, Tehran or the people of Sudan, and may erode Sino–Iranian relations, Sino–Sudanese relations and Sino–US trust. China, therefore, shifted from a passive response to active involvement in mediating the Iranian–US conflict in 2009, and in 2007 they actively worked to resolve the Darfur issue. In 2007, Hu Jintao publicly criticized the Sudanese president, a very rare act for a Chinese leader, and advised the Sudanese government to accept Resolution 1706 of the UN Security Council. As Regler points out, the 'China card' proved very effective in pressuring Khartoum to resolve the issue. Still, China's energy quest in these states has created tension between Washington and Beijing.

One state that could be placed in the pariah camp, although we debated this issue heatedly at the conference, is Venezuela. The leader, Hugo Chavez, who died in March 2013, had worked enthusiastically to undermine US influence in Latin American, while at the same time enhance China's influence in the US's 'backyard'. His propaganda was stridently anti-American and he opposed the 'Washington Consensus', reflecting very much a 'Beijing Consensus'. By allowing only 40 per cent foreign ownership of oil fields, he drove US majors out of the country. As he tried to end Venezuela's long-term dependence on the US, he offered Venezuelan oil to China, hoping that 'China will help us'.

China, however, kept some distance from Chavez and did not play up his nationalistic, anti-American policies. Still, China engaged more deeply in Venezuela's energy sector and built good bilateral ties. It helped Venezuela send a communication satellite into space and has trained military officers. Chinese NOCs invested heavily in Venezuela's energy sector and China has dramatically increased its energy imports from that country. Although in 2007 imports from Venezuela comprised only 2.6 per cent of China's total oil imports, by 2013 Venezuela supplied 6.5 per cent of China's oil imports, moving into 5th place.

Unlike in the cases of Sudan and Venezuela, the US neither imposed sanctions on Venezuela nor mounted any global coalition to force it to change its policies – although it reportedly had tried to assassinate Chavez earlier in his presidency. This response may be because Venezuela had not violated any international norms that could become the target of sanctions at the UN; it simply wanted distance from the US and its oil majors. Rather than confronting Chavez, both the US oil majors and the US State Department chose instead to withdraw from interactions with Venezuela, leaving it to sink or swim on its own. In fact, as Watson and Zweig (Chapter 13) show, Venezuela's oil sector has suffered significantly from the loss of US technology and management. More importantly, even as this major regional shift occurred, the US did not try to contain China's growing ties with Venezuela as one might have anticipated if the US was truly acting as a threatened hegemon determined to contain a rising China.

Allies, democracies and pushback

Australia shows how China's economic power can influence a US ally to support China's global interests. Due to Australia's dependence on resource exports for its economic security, China has become Australia's key foreign economic relationship, whilst squeezing its normative and strategic alliance with the US. After 2005, trade relations with China underwrote an unparalleled economic boom in Australia; however, as a close ally of the US, with whom it shares democratic values, Australia feels the pull from the US (and in some cases from its own concerns about the 'China threat') to maintain close strategic links with the hegemon. Moreover, a shift to a closer relationship with China will have important implications for other US allies, such as South Korea.

Balancing between the hegemon and the challenger has never been easy for Australia. As Thomas (Chapter 5) shows, US government officials have pressured Australia not to 'accommodate' China's strategic interests or national goals, even as Australia became increasingly dependent on China's growth for its own economic security. This difficult balance has been further complicated by the ongoing rebalancing of power and authority between the US and China as the US's 'pivot' to East Asia has intensified America's expectations about Australian security.

However, Australia's response is not solely driven by external forces, such as China's booming economy and the US's expanding military role in the region; in fact, domestic politics, including leadership and societal voices, play an important role in this democracy. New prime ministers usually try to readjust the resource triangle, while Australian citizens manifest a form of 'pushback' whereby, despite recognizing the importance of China to their personal wellbeing, they fear China's economic inroads into Australia's economy, particularly the sale of Australian companies to the rising China.

Australia's status as a 'middle power' (Cooper *et al.*, 1993) has also led it to try to extricate itself from its current dilemma by acting as a balance between its two partners (Metherell, 2005). Traditionally, Realist scholars would regard China's power transition process as a 'zero-sum game'; however, complementary interests may create accommodative and harmonious relations between the hegemon and the challenger with the help of a balancing power. Thus, Australia's close security alliance with the US, and its vital economic partnership with China, places Canberra in a nuanced position between the two powers, prompting many Australian commentators to ponder the implications of modifying this balancing policy between China and the US (He, 2014; White, 2012). Although China reminds Australia of its own ability to talk directly to the US, the Americans also rebuke Australia for accommodating China's military rise.

A second US ally, Canada, has a far more spotty experience in selling energy to China, making it far less dependent on China for resource exports and economic development than Australia. Despite close ties since the early 1970s, between 2005 and 2009, China proved unwilling to buy oil from a country – other than the US – that prioritized its human rights policies. Only after Prime Minister Harper disconnected trade and human rights in 2009 did the Chinese government allow Chinese companies to again purchase any Canadian companies. Since then, China has sought numerous energy deals, including its largest foreign direct investment (FDI) project to date – the US$16 billion purchase of Nexen oil.

Proximity also necessitates that Canada maintains very close economic and political ties to the US. Although the US remains rather confident in its alliance with Canada – not expecting any strategic concessions due to China's economic weight, as has occurred with Australia – the Americans do remind Canada that it cannot begin shipping large quantities of its oil to China without consulting the US government (Chapter 6). The Canadians, however, can be piqued when Americans try to remind them of North America's energy interdependence. But because some Canadian and US energy companies are intertwined – Nexen owned rigs in the Gulf of Mexico – approval from the US's Committee on Foreign Investment in the US (CFIUS) must take place before some Canadian companies can be sold to China. Nevertheless, Chinese capital is a welcome addition to Canadian companies, who are strapped for cash, to develop the Tar Sands in Alberta and build pipelines to the west coast of Canada.

But again, as in democratic Australia, Canadian politicians – particularly the Opposition New Democratic Party, the Canadian Security Services, think tanks, as well as many Canadian citizens – doubt the wisdom of selling Canadian companies to Chinese SOEs whom Canadians distrust. Most recently, China's ongoing anti-corruption campaign, and alleged misdeeds by Chinese managers working for PetroChina in Canada, has reinforced concerns about the non-transparent nature of Chinese companies and their willingness to ignore Canadian business norms.

One of the US's closest allies, Saudi Arabia, drew close to China at the bequest of the US, making the Saudi case different from Canada and Australia. As a staunchly anti-communist country, Saudi Arabia should have abjured from selling oil to 'communist' China. Nevertheless, Saudi oil exports to China have boomed. In 2009, Saudi Arabia exported more crude oil to China than to the US, comprising over one-fifth of China's total energy imports, turning Saudi Arabia into China's leading oil supplier. Numerous Sino–Saudi joint ventures in downstream production play a major role in developing the delivery of energy to Chinese consumers.

Ironically, Sino–Saudi energy relations developed with the blessing, if not instigation, of the US, which saw Sino–Saudi energy relations as a way to block Iran from using its oil and gas to pressure Beijing to protect Iran from US-organized sanctions. In fact, Washington may have facilitated a Sino–Saudi entente and China's entry into the Saudi marketplace when it encouraged Sino–Saudi collaboration against the Mujahidin in Afghanistan (Ali, 2005: 175–6), and turned a blind eye to the Sino–Saudi 1988 missile deal of which it had undoubtedly been aware (Shichor, 1989). In 2010, therefore, the US encouraged King Abdullah to offer China long-term oil supply guarantees in return for supporting sanctions against Iran (Zambellis, 2010: 5). In 2011, Khalid al-Falih, Aramco's CEO, said that the bilateral relationship was not based solely on the selling of oil to China; rather, the two countries were 'strategic partners whose many relationships…are founded on mutual respect, independence and mutual benefit' (Dipaola, 2011).

The oil, however, has come with conditions. As the Chinese expand and modify their refineries to adapt to the high-sulphur ('sour') Saudi crude (Meidan, 2013: 211–14), they appear reconciled to having the Saudis as China's leading oil supplier. Given its future oil needs and Saudi Arabia's reliability, Beijing has no choice. Also, reflecting the triangularity of the links between these three countries, Washington has benefited greatly from the Sino–Saudi oil connection because China set aside its traditional non-interventionist policy, and the US's sidestepping of the UN, to assist Saudi Arabia by accepting the two US wars against Iraq in 1991 and 2003.

During his visit to Saudi Arabia in January 2010 the Chinese Foreign Minister was 'pressed hard' to adopt a more active Chinese role against Iran's nuclear threat to Saudi security (WikiLeaks, 2010a). In return, Deputy Foreign Minister Prince Torki al-Saud al-Kabir announced that Saudi Arabia 'is willing to take actions to address those concerns [i.e. sign long-term contracts], but must have Chinese cooperation in stopping Iran's development of nuclear weapons as a quid pro quo' (WikiLeaks, 2010b). The Saudi Foreign Minister also remarked that, 'China is perfectly aware of the scope of its responsibilities and its obligations, including the position it holds on the international stage and as a permanent member of the [UN] Security Council' (WikiLeaks, 2010b).

Neutral countries engage in manipulation

An increasing number of 'neutral' countries try to take advantage of US–China resource competition to advance their own national goals. Brazil, Angola, Nigeria and Kazakhstan – the 'neutral' countries analyzed in this book – show that these neutral states were able to manipulate China and the US and interact with both major powers with greater flexibility than that demonstrated by allies or pariahs.

As Moreira (Chapter 10) demonstrates, a country with strong leadership, such as Brazil, can benefit from Sino–US competition over resources. Both Chinese and US companies seek to be involved in Brazil's new pre-salt findings, yet although Brazil needs FDI and sophisticated technology to develop the pre-salt fields, Brazil created a new oil regime that allows the government and its NOC, Petrobras, to develop the energy sector on their own terms. With its USD$10 billion loan to Petrobras, China achieved a foothold in the pre-salt region, but at an increasing cost because competitive US companies also weighed in. Increasing US and Chinese competition, therefore, helped maximize Brazil's gains. In Brazil, as in Nigeria, a change of president led to a major shift in the influence of the US and China.

Angola, another neutral state, has been quite pragmatic in its foreign ties, proving quite willing to shift between China and the US as the need arises. According to Vines (Chapter 7), Angola desperately needed new partners and new sources of aid and FDI to reconstruct the country after its disastrous 20-year civil war. International financial institutions, however, demanded greater transparency and structural readjustment before such funds could be forthcoming. But China's model of cooperation, based on credit for oil, lower interest rates, longer repayment time and, most importantly, the attachment of no conditions, was deeply appreciated by an administration that steals large amounts of the nation's energy wealth. Also, China proved willing to invest in many important infrastructure projects that Western institutions would not support. In its early post-civil war years, Angola, therefore, leaned towards and relied heavily upon China, which proved quite willing to supply large amounts of funds. The US's concern about its own ties to Angola increased during this period as China became deeply involved in that country. Washington began to worry that 'the dramatic increase in Angolan exports to China could eventually come at the expense of its longstanding priority market in the US' (Eurasia Group, 2011).

Angola's financial needs were far greater than what one donor, China, which itself was a developing state, could supply. Angola, therefore, decided that globalization offered an opportunity 'to diversify international relations and to accept the principle of competition, which has, in a dynamic world, replaced the petrified concept of zones of influence that used to characterize the world' (Zhao, 2011: 6). Angola therefore began to strengthen ties with the West and other developing states around 2007. Moreover, by 2011, several knowledgeable analysts of Sino–Angolan relations were remarking that, despite the significant investments made by China and its willingness to loan billions of dollars to help Angolan development, China seemed to be losing influence as Angola sought to diversify its economic partners and avoid dependence on China (Corkin, 2011; Horta, 2011).

The situation in Kazakhstan differs in major ways from our other cases because the US is not a member of the dominant triangle, which is based on the dynamic relations between China, Russia and Kazakhstan. According to Peyrouse:

> insofar as the United States has also to be included, one must speak of a ménage à quatre, or else refer to the existence of several triangles: the main, driving one being the Russia–China–Kazakhstan triangle; the Russia–USA–Kazakhstan being second in importance; and the USA–China–Kazakhstan one being of less relevance. Indeed, there is no triangular relationship China–USA–Kazakhstan that does not also involve Russia, which is omnipresent in the agendas of all three countries.

> (Peyrouse, Chapter 9, this volume)

The main story here is Russia's resurgence into the region since the rise of President Putin, and China's rapidly growing economic and energy ties, a process that makes the Russia–China–Kazakh triangle the most important one for all three players. China's trade with Kazakhstan surpasses that of Russia, the EU and the US, and is on a sharply upward trajectory. China seeks access to oil and gas resources partly through building pipelines, while not antagonizing Russia, which sees Kazakhstan as part of its traditional sphere of influence where it hopes to maintain its monopoly on pipelines established during the Soviet era. Moreover, although China can ship oil from this country, its NOCs have failed to gain investment or development rights in major Kazakh oil fields, a sign that Sino–Kazakh ties face serious constraints.

For the US, pipeline politics, as well as strategic ties, remain dominant because trade is secondary. Russia's re-emergence, and Putin's ability to control the direction of oil shipments, remains the key focus of the US. The US leverage emerges because its energy firms continue to dominate oil production. But the US effort to ship oil westward through Turkey to Europe also challenges Chinese interests because a major goal of China is to ship Kazakh oil eastward. In fact, the US pipeline effort in this region of the world is one of the few examples where there is real Sino–US competition over oil, which to a certain extent can be seen as a 'zero-sum game'.

The final neutral case, Nigeria, is rapidly becoming one of China's largest trading partners in Africa, with trade jumping from US$384 million in 1998 to US$10.7 billion in 2012 (CEIC China, 2014). In January 2006, Abuja and Beijing signed a memorandum of understanding (MoU) to establish a 'Strategic Partnership', the first African country to do so. China's first scheduled direct flight to Africa was inaugurated in December 2006 from Beijing to Lago via Dubai. As Taylor argues (Chapter 8), Nigeria and China had excellent ties, particularly during the presidency of President Olusegun Obasanjo, who was a keen supporter of 'oil for infrastructure' projects. The change of presidency, however, led to reconsideration of many contracts and a dramatic decline in China's influence in Nigeria, and although the Nigerians accepted Chinese investment in their energy sector, they did so with caution and only when Western firms showed a lack of interest.

On the other hand, Nigeria is the US's most important partner in sub-Saharan Africa. The two countries have extensive trade links, Nigeria is a major player in the Economic Community Of West African States (ECOWAS), which carries out peacekeeping in this unstable region of the world, and Nigeria is also a major producer of oil. Approximately two million Nigerians live in the US and, according to Johnnie Carson, US Assistant Secretary for African Affairs in 2010, America maintains enormous influence in Nigeria.

The US is not particularly worried about Chinese inroads into Nigeria. Quoting Carson, Taylor (Chapter 8) argues that the US has set 'trip wires', which should China cross, the Americans might get anxious. As of 2010, however, the Chinese had not crossed those lines. This means that unlike some other neutral states, Nigeria has only limited opportunities to play one side off against the other.

Nigeria's opportunities are also limited by the unpredictability and secondary costs of doing business with such a corrupt and inept government, and where oil production occurs in a very dangerous region of the country. Chinese companies are quite risk averse and feel quite uncomfortable sending their people into the Niger River Delta, which is the site of a major rebellion. Also, today much Western scholarship emphasizes the importance of 'good governance' that is separate from the issue of democratic development. In this case, Nigeria is a prototype of 'bad' governance, where elites steal enormous amounts of oil (worth perhaps US$1.4 billion annually) even as companies have to pay exorbitant bribes. Although Nigerians assert that Chinese firms were quick to engage in such bribery to seal their deals, the Chinese, along with everyone else, find doing business in Nigeria extremely difficult.

Managing US–China resource competition

Several points can be made related to Sino–US cooperation and competition. The US's need for energy imports has been declining rapidly due to the shale revolution, which dampens Sino–US competition for oil and gas in the world. As a result, the US government is far less concerned about energy supplies in 2014 than it was in 2011 when we held our conference, thereby ameliorating most of the political dynamics over oil supplies from neutral states, such as Nigeria and Angola. In fact, as Taylor (Chapter 8) shows, West African crude exports to the US are down as much as two-thirds, giving China a much more privileged position in these countries' market place. The case studies reflect these changes because the writers have updated their papers, but the chapters also reflect the reality of the past ten years. Still, the hypothesis that China's energy relationships are largely triangular not bilateral remains, in our minds, worthy of testing, largely because this view reflects Chinese leadership concerns. So, what have we discovered about Sino–US energy competition?

First and foremost, we find that the US has not tried to deny China access to resources as a strategy to slow down its rise. In most countries, the competition tends to be commercial in nature and more businesslike. The only clear attempt by the US to deny China's energy access was when the US blocked the sale of Union Oil Company of California (Unocal) to the Chinese National Offshore Oil Corporation (CNOOC). One Congressman admitted to opposing the sale because he thought it would slow down China's challenge to the US, but this case involved an American oil company and American technology, not competition in a third state. Western oil companies also tried to keep Chinese firms out of Nigeria, but for commercial rather than strategic reasons. However, the US has not blocked China's growing investment in US or Canadian oil and gas companies nor have they publicly opposed China's inroads in Venezuela, a country with whom the US has an important energy relationship. In fact, as of 2010, China's investment in the energy sector in the US topped US$17 billion (Dezember and Areddy, 2012).

Second, there are few political or ideological elements in Beijing's involvement in Iran, Sudan or Venezuela, although some people in Washington suspect a well-coordinated strategy aimed at undermining US domination, and eventually displacing the US as Number 1 (Lieberthal and Wang, 2012). This book tests the hypothesis that dominates China's worldview that the US will use its hegemonic status and influence over oil production, shipments and markets to slow China's rise. In the minds of most Chinese officials, a rational US cannot but utilize this opportunity to protect its global dominance.

Third, the US and other Western countries are concerned about China's loans to developing countries because China generally does little to combat corruption, foster transparency or raise environmental and social standards. China is, however, becoming more concerned about the quality of economic policy making in the recipient countries, a position in line with US interests. In 2010, when China and Venezuela were negotiating a US$20.6 billion credit facility, Chinese consultants spent 18 days in Venezuela drafting a plan for Caracas to reform and grow its economy. As Erica Downs observed:

> They provided roadmaps for Caracas to achieve price stability within three years and to improve the climate for foreign investment and economic development with six years. The Chinese also recommended that the Venezuelan focus on developing four pillar industries – agriculture, mining, real estate and consumer goods into the engine of economic growth.
>
> (Downs, 2011: 90)

Third, given the many resource triangles in the world, the US and China should both think about trilateral relationships rather than bilateral ones when dealing with RRSs. China must consider the US factor in its bilateral relations with these oil-supplying countries, particularly if the US insists forcefully, pressing the challenging power cooperate with the dominant one. The US, for its part, needs to be sensitive to China's energy 'anxiety' (see Chapter 14).

China therefore cooperates with the US over certain issues, even though it is often not willing to do so. In the case of Sudan, China was initially convinced that the US sought a regime change that would undermine Chinese energy supplies. Still, according to Regler (Chapter 12), China helped end the killing in Darfur once it received US assurances that human rights, not energy, was motivating the US. Similar changes have occurred with Iran. In May 2009, Iran was China's number one source of oil, supplying more than 200,000 barrels/day, but due to US pressure, China slowed the implementation of existing deals with Iran and reduced oil imports. By mid-2010, Iran had dropped from second to eighth in the list of oil suppliers to China, with only 189,000 barrels/day. As Garver shows in Chapter 11, China values its relationship with Washington more than its relationship with energy suppliers because it has larger strategic interests in mind.

As for the hegemon, the US prefers a more cooperative resource partnership with China beyond Iran. According to Garver (Chapter 11), Washington seeks to demonstrate to Beijing that the US does not intend to slow China's rise by choking its resource imports. In Afghanistan, Iran, Saudi Arabia and Sudan, Washington has over the last several years demonstrated a more understanding attitude toward China's efforts to get the raw materials it needs; however, Washington will still sanction China's banks and oil firms if they invest in Iran's energy sector.

Despite this, China has increased its capability – politically, economically and even militarily – to resist US pressure and can even restrict US unilateral initiatives, as Shichor (Chapter 4) demonstrates. Although China cares about US concerns, there is a limit to China's willingness to compromise to US pressure because China has to pursue its core interest to drive the Chinese economy. Competition has its structural roots in the global system and has the potential to lead to a head-on confrontation if not properly managed. We can only hope that today's commercial competition will not turn into a geostrategic rivalry tomorrow.

Conclusion

As the case studies will show, the triangular framework employed in this book helps explain the pattern and nature of some of the relationships between the US, China and those RRSs within the global balance of power and the current political economy.

A number of observations can be made. First, resource-rich 'pariahs' tend to bandwagon with China but strong US pressure, and China's desire to maintain relations with the US and enhance its global image, can limit Beijing's support for the pariahs. In the case of Venezuela, however, the US preferred to let 'sleeping dogs lie', in that it did not challenge China's entrance into the Western hemisphere.

Second, the US's traditional allies, particularly the democracies, are sensitive to US pressure, although they may resent efforts by the US to limit their resource deals with the People's Republic of China. Domestic politics, however, creates 'pushback' against China's buying efforts because popular opinion mistrusts China and Chinese state-owned industries. Opposition parties will also use close business ties with China as a weapon with which to challenge the ruling party. In the case of Saudi Arabia, the US had other strategic interests – undermining Sino–Iranian ties – that led it to support its ally's links with China.

Finally, some neutral RRSs (such as Angola or Brazil) shift loyalties between China and the US, and perhaps manipulate Sino–US competition to their own advantage. When Angola, therefore, needed China's help to defer pressure from international financial institutions, they accepted Beijing's money, but after almost a decade, once they worried about becoming too dependent on China, the Angolan government again sought the West's involvement in its national development. Nigeria, too, has shifted back and forth, first as part of the Western camp, and then as a new 'strategic partner' of the Chinese. Nevertheless, it will be interesting to see whether China's leverage will increase following dramatic cuts in US energy imports from West Africa, which leaves China as a key importer to replace the US's decreasing demands for imports.

References

Ali, S.M., 2005. *US–China Cold War Collaboration, 1971–1989.* London: Routledge.

Alterman, J.B., 2011. The vital triangle. *In:* B. Wakefield and S.L. Levenstein, eds. *China and the Persian Gulf: Implications for the United States.* Washington, DC: Woodrow Wilson International Center for Scholars, 27–37.

Archibold, R.C., 2012. China buys inroads in the Caribbean, catching US notice. *New York Times,* April 7.

CEIC China, 2014. China's Foreign Trade Statistics, various years.

Cooper, A., Higgott, R. and Nossal, H., 1993. *Relocating Middle Powers: Australia and Canada in a Changing World Order.* Vancouver, BC: University of British Columbia Press.

Corkin, L., 2011. China and Angola: strategic partnership or marriage of convenience? *CMI Angola Brief,* 1 (1), 1–4.

Dezember, R. and Areddy, J.T., 2012. China's money trail: China foothold in US energy. *The Wall Street Journal,* March 6.

Dipaola, A., 2011. Aramco, CNPC to construct refinery in Yunnan Province. *China Daily,* March 22, p. 6.

Downs, E., 2011. *Inside China, Inc: China Development Bank's Cross-Border Energy Deals.* Washington, DC: John L. Thornton China Center at Brookings.

Etzioni, A., 2012. The Chinese Are Coming! *The National Interest,* May 7.

Eurasia Group, 2011. African Security in Strategic Perspective. *Eurasia Group,* December 22.

Friedberg, A.L., 2005. The future of US–China relations: is conflict inevitable? *International Security,* 30 (2), 7–45.

He, B., 2014. Collaborative and conflictive trilateralism: perspectives from Australia, China, and America. *Asian Survey,* 54 (2), 247–72.

Herberg, M., 2011. China's energy rise and the future of US–China energy relations. *New America Foundation,* June 21. Available online from: http://newamerica.net/publications/policy/china_s_energy_rise_and_the_future_of_us_china_energy_relations (accessed 4 April 2015).

Horta, L., 2011. China's waning influence in Angola. *Diplomatic Courier,* 26 August.

Kirshner, J., 2008. The consequences of China's economic rise for Sino–US relations: rivalry, political conflict and (not) war. *In:* R.S. Ross and F. Zhu, eds. *China's Ascent: Power, Security, and the Future of International Politics.* Ithaca, NY: Cornell University Press, 238–59.

Lieberthal, K. and Wang, J., 2012. *Addressing US–China Strategic Distrust.* Washington, DC: The Bookings Institution.

Meidan, M., 2013. Muddling through with Chinese characteristics: Beijing's energy policy and diplomacy in West Asia and North Africa. Unpublished doctoral dissertation. Science Po, Paris.

Metherell, M., 2005. PM's vision: Australia as honest broker. *Sydney Morning Herald,* April 1.

Shichor, Y., 1989. *East wind over Arabia: origins and implications of the Sino–Saudi missile deal.* Center for Chinese Studies, China Research Monograph 35, Berkeley, CA.

Simpson Jr, G.L., 2010. Russian and Chinese support for Tehran. *Middle East Quarterly,* Spring 63–72.

US Energy Information Administration, 2013. China: overview. Available online at: www.eia.gov/countries/cab.cfm?fips=CH (accessed 4 April 2015).

White, H., 2102. *The China Choice: Why America Should Share Power*. Collingwood, Australia: Black Inc.

WikiLeaks, 2010a. FM Sand: China needs to more actively counter Iranian nukes. 10RIYADH123, January 27.

WikiLeaks, 2010b. Cablegate: Saudi Foreign Ministry pressing China to stop Iranian proliferation. 10RIYADH, January 26.

Zambelis, C., 2010. Shifting sands in the Gulf: the Iran calculus in China–Saudi Arabia relations. *China Brief 10,* May 14. Available online at: www.jamestown.org/single/?no_cache=1&tx_ttnews%5Btt_news%5D=36371&tx_ttnews%5BbackPid%5D=7&cHash=305f 844510.

Zhao, S., 2011. The China–Angola Partnership: A Case Study of China's Oil Relations in Africa. *China Briefing*, May 25, p. 6.

Part I
Conceptual frameworks

1 Modelling 'resource diplomacy' under hegemony

The triangular nature of Sino–US energy relations[1]

David Zweig

Introduction

Today's international power structure can be characterized as unipolar with a somewhat declining US hegemon facing a rapidly rising China. To maintain its rapid growth, China must engage in 'resource diplomacy' and seek resources and energy from resource rich states (RRS) around the globe. The hegemon, however, has interests in all corners of the world where China's resource search is ongoing, turning China's bilateral ties into triangular ones, with the US as the third partner.

This paper proposes that a triangular model composed of the US, China and the RRS reflects an excellent framework through which to analyze China's global search for energy and the US's response. It helps us understand the US and China's efforts to enhance their own energy security and the response of the RRS to that competition, and to project whether energy competition will lead to cooperation, competition or even conflict between the US and China.

Some Chinese academics and government officials argue that the US, looking to contain China and maintain its hegemony, uses China's energy need to slow its development (Lu *et al.,* 2011; Zhao, 2006), yet no one has analyzed this issue in depth. The goal of this paper is to lay out a framework for evaluating this hypothesis, which is then applied to the case studies in Part 2 of the book where each author assesses the extent to which efforts by China to gain oil in many RRS gains the US's attention and whether the US tries to influence the outcome of China's efforts.

China's and the US's search for energy security

Before discussing the triangularity of energy ties between the US, China and the RRS, we must first clarify the strategies of both China and the US to energy security. These strategies are somewhat fluid, particularly for the US, which is shifting from being an energy importer to an important energy exporter. Both countries look internally and externally in their search for enhanced energy security.

Key aspects of China's energy strategy include energy efficiency, diversifying supplying countries, developing oil substitutes, securing oil delivery channels, building a strategic energy reserve and increasing overall supply. China's 11th and 12th Five Year Plans call on the country to constrain the growth of energy demand and reduce energy intensity (Andrews-Speed, 2009). In November 2003, Hu Jintao called on officials to 'give equal weight to economizing and resource exploitation...and actively developing oil substitutes' (Wen, 2004).

The deal that the Chinese Communist Party (CCP) has struck with China's enlarging middle class – to supply them automobiles, air conditioning and a better quality of life in return for acquiescing to one-party rule – forces the CCP to increase significantly supplies of oil and gas. Similarly, with growth and employment still dependent on exports, China needs more energy to remain the 'factory to the world'. In 2008, 50 per cent of China's energy imports were used to manufacture goods for the rest of the world (International Crisis Group, 2008).

This effort to access more oil internationally is coordinated between China's foreign establishment, the central leadership and the national oil companies (NOC), particularly China National Petroleum Corporation (CNPC), Sinopec and China National Offshore Oil Corporation (CNOOC), who have been encouraged by the State Council, the China Development Bank and the Export–Import Bank to purchase equity oil abroad. In 2004, the National Development and Reform Commission (NDRC) and the Chinese State Council promulgated a list of resources and countries from which imports would be subsidized through favourable loans (Ministry of Commerce, 2004), with a follow-up list in 2007. However, some wise Chinese analysts recognize that because 'peak oil' makes oil competition appear as a zero-sum game, China is accused of taking other people's oil, and so they advocate cutting China's energy imports (Wang, 2011).

China's NOCs have significantly expanded the range of countries, regions and territories from which they import energy resources (Table 1.1). The number of countries surpassed 35 in 2013, although within 10 years, some of these countries will stop exporting to meet their own growing domestic demands. In many cases, China's top leaders visit RRS and they promise various development programs in return for supporting bids for oil fields by Chinese NOCs. Coordinated loan programs in the form of 'infrastructure for oil' (roads, rails, ports and schools), mostly by the China Development Bank, which is a practice used in Africa, have helped China's bid to secure oil supplies (Brautigam, 2009; Downs, 2011).

Think tanks inform China's leaders of the risks to China's energy security posed by the global system. One meeting in March 2002, run by the Chinese Institute of International Studies, emphasized that China's energy security faced an unfavorable shift; the West, particularly the US, could contain China's efforts to exploit overseas oil and gas resources; and, violent fluctuations of oil prices and energy supply could affect China's energy security. These concerns strike at the heart of the three components of energy security: supply, delivery and price (Yong, 2002: 59).

Purchasing oil and gas abroad forms only one component of energy security; one must also guarantee the safe delivery of energy to China. When General Xiong Guangkai was Deputy Chief of Staff he opined that the 'contest for energy sources and secure supply channels among major powers and powerful nations

Table 1.1 Sino–US oil imports by category of US ties, 2007–2012

	By China (%)			By US (%)		
	2007	*2010*	*2012*	*2007*	*2010*	*2012*
Pariah						
Iran	12.9	8.9	8.1	–	–	–
Sudan	6.5	5.3	0.9[1]	–	–	–
Venezuela	2.6	3.2	5.6	11.5	9.9	10.7
Subtotal	*22.0*	*17.3*	*14.7*	*11.5*	*9.9*	*10.7*
Neutral						
Angola	15.8	16.5	14.8	5.0	4.2	2.6
Brazil	1.5	3.4	2.2	1.7	2.8	2.2
Congo	3.0	2.1	2.0	0.6	0.8	0.3
Equatorial Guinea	2.1	0.3	0.7	0.6	0.5	0.5
Kazakhstan	3.8	4.2	3.9	0.1	0.2	–
Libya	1.8	3.1	2.7	0.8	0.5	0.7
Nigeria	1.0	0.5	0.3	10.8	10.7	4.8
Russia	9.2	6.4	9.0	1.1	2.9	1.2
Vietnam	0.3	0.3	0.3	0.3	0.1	0.1
Subtotal	*38.4*	*36.8*	*36.0*	*21.0*	*22.6*	*12.4*
US ally						
Australia	0.0	1.2	1.4	0.0	0.1	0.1
Canada	0.0	0.1	0.2	18.6	21.4	28.4
Indonesia	1.4	0.6	0.2	0.2	0.4	0.1
Iraq	0.9	4.7	5.8	4.8	4.5	5.6
Kuwait	2.3	4.1	3.9	1.8	2.1	3.6
Mexico	–	0.5	0.4	14.1	12.5	11.4
Norway	–	–	0.0	0.6	0.3	0.3
Oman	8.6	6.6	7.2	0.3	0.1	0.1
Saudi Arabia	16.6	18.7	19.9	14.5	11.7	16.0
UAE	2.3	2.2	3.2	0.1	0.0	n.a.
Yemen	2.0	1.7	1.3	0.1	n.a.	n.a.
Sub-total	*35.9*	*41.0*	*44.2*	*55.5*	*53.6*	*66.0*
Total	*96.3*	*95.0*	*94.9*	*88.0*	*86.1*	*89.1*

Source: EIA (various years), China Customs Statistics.

Note:

1 Due to conflict between Sudan and South Sudan over the transfer fee of oil through Sudan's pipelines, South Sudan closed down all oil shipments in 2012.

has become fiercer', so the nation's energy problem 'needs to be seriously taken into account and dealt with strategically' (Xiong, 2004: 5). In 2003, President Hu Jintao, referring to a 'Malacca dilemma', commented that 'certain powers have all along encroached on and tried to control the navigation through the Straits' (Wen, 2004). In summer 2006, Zhang Wenmu from the University of Aeronautics and Astronautics described China's energy vulnerability as

> rapidly changing from a relationship of relative dependence to one of abso-
> lute dependence…China is almost helpless to protect its overseas oil import
> routes…The most crucial conduit connecting China with the region and the rest
> of the world is the sea lanes, and therefore China must have a powerful navy.
> (Zhang, 2006: 19–20)

A US military report in November 2008 quotes a Chinese naval strategist saying that 'The straits of Malacca are akin to breathing itself…to life itself' (American Foreign Policy Council, 2009). In 2008, China, therefore, announced that it was building an aircraft carrier, which is now completed with a second one on the way.

Finally, China has established a network of oil and gas pipelines that circumvent the high seas. In 2006, China's first oil pipeline from Central Asia was opened for business and in 2013 a new oil pipeline from Russia was linked to Daqing, a major refinery centre in northeast China. Gas pipelines have also been completed from central Asia and from Myanmar, limiting China's dependence on the Malacca Straits.

The US's search for energy security and its response to China

The US has both domestic and global strategies to enhance energy security. Domestic goals include increasing output of oil and gas within the US, cutting demand, particularly in the transport sector (Herberg, Chapter 3, this volume), enhancing energy efficiency and dealing with environmental issues, such as pol-lution and climate change. External efforts occur in the near and far. The US supports policies to expand and accelerate cross-border energy investment, oil and gas pipelines and electricity grid connections with Mexico and Canada.

Globally, the Bush Administration announced a policy to support US energy firms competing internationally by negotiating with RRS through multilateral and bilateral institutions to open their markets to foreign investment and trade. It also sought to 'promote the transparency, timeliness, and accuracy of the data that guide global oil markets' (National Energy Policy Development Group, 2001) so as to avoid price shocks to the global economy and to allow for more oil extraction. It also planned to expand the sources and types of global energy sup-plies, strengthen the capacity to respond to disruptions of oil supplies and increase reserves being held by major consumers of oil (National Energy Policy Develop-ment Group, 2001).

The US military's job is to keep oil flowing to its own shores and those of its allies, despite having to pass through major bottlenecks around the globe (National Energy Policy Development Group, 2001). The US worries about the stability of supplies on which its allies and the world economy depend because disruptions to the global market that cause higher prices affect the US economy as well.

Ironically, in the brief period since the authors in this volume began their work (circa 2010) and its completion, the energy situation in the US has changed dramatically. For decades the idea of the US gaining energy independence was 'the preserve of quixotic rhetoric' (Yergin, 2012), but in 2013 the International Energy Agency (IEA) projected that the US could overtake Saudi Arabia as the world's biggest oil producer by 2020. During this period, the US will rapidly become far less dependent on oil imports and, in fact, it has already become an exporter of liquefied natural gas. In 2008, the expectation was for a decline in US oil production and an increase in imports, heightening the importance of energy security. Instead, US oil output has risen 25 per cent since 2008 and the IEA estimates that it will increase a further 30 per cent by 2020 to 11.1 million barrels/day. US petroleum imports have fallen from 60 per cent of consumption in 2005 to 42 per cent in 2013. Within a decade, with shale gas production rising from 2 per cent of US natural gas production to 37 per cent, the US will overtake Russia as the world's largest natural gas producer. Clearly, the US concept of energy security still emphasizes ensuring supplies for its allies, such as US-produced liquefied natural gas (LNG) for Japan in the wake of Fukishima, continuing to decrease its dependence on foreign oil, developing new technologies to increase efficiency and cutting domestic demand. As of 2013, the US was ranked fifth in terms of energy security among major developed countries (Goode, 2013).

Is China's resource diplomacy really triangular?

This chapter and volume argue that a triangular relationship exists between the US, China and RRS and as a rising China expands its energy relationships, it intersects with the economic, strategic, political or moral ties that the US has with the RRS around the world. China's bilateral relations – an apparent '*tête à tête*' with RRS – are therefore often triangular – a '*ménage à trois*' – where the third partner in the relationship is the US.

Chinese and Americans both recognize the 'triangularity' of their ties. In 2008, former US Deputy Assistant Secretary of State for East Asia, Tom Christensen, remarked that, 'The US–China relationship is much more about third areas in the world than it has been in the past' (Christensen, 2008). A researcher at CNPC admitted that, 'The research team at CNPC considers the US interests in all their analysis and recommendations. We always believe that the US will want to block the deal as part of its strategy to contain China' (CNPC interview, March 2011). Australian leaders constantly confront this triangle. In 2004, Prime Minister Howard observed that, 'one of the great successes of this country's foreign relations is that we have simultaneously been able to strengthen our longstanding ties with the United States of America, yet at the same time continue to build a very close relationship with China' (Howard, 2004). Finally, Xu Xiaojie, a leading analyst of China's energy policy sees Sino–US competition in third states as an important trend of the recent era (Xu, 2013: 16).

For triangularity to be invoked, each party must take the third party into consideration as it deals with its bilateral partner. According to Dittmer (1981), states are profit maximizers, trying to gain the 'pivot' position which has good ties with the other two states even as they are hostile towards each other, granting the pivot maximum flexibility (Dittmer, 1981). However, in traditional triangular theory, states simply try to ensure their security (personal communication with Iain Johnston, 2010). The extent of intervention or response by the various members of the triangle may vary if the issues at stake involve national security rather than energy extraction. One may therefore hypothesize that the US as the hegemon is more likely to flex its muscles in a triangular setting when the issue affects US security, the security of its allies or moral issues, such as nuclear proliferation (which can influence US national security) or human rights, than when the issue is just oil. As expected, the US pressures China on Iran and Sudan, but when Venezuela tries to strengthen its energy ties with China to the detriment of the US, the US redirects its energy supply (author's interview in US State Department, April 2010). Similarly, the US may respond with different intensity when China seeks energy or resources in RRS that are US allies, such as Canada, Australia or Saudi Arabia. Many of our cases involve political economy and the stakes are, therefore, lower than they would be if the threats were to US national security. This lowers the likelihood of intervention by the US because the US must see energy competition with China as sufficient cause to intervene in China's ties with a RRS. In the case of China, however, one wonders if Chinese NOCs or Chinese officials consider the US's perception of its ties with its energy suppliers as it deepens that engagement. Finally, we must test to see if the RRS considers the attitudes of one side when dealing with the other state.

A brief discussion about the actors is required. The most important actor is the Executive branch of government – prime ministers, presidents and general secretaries – who make policy, and their government ministers, who implement it. Some lead full-blown democracies, whilst others are populists running illiberal democracies. Many RRS are led by autocrats, who care little about popular views. Other important actors include the international energy companies (IOCs) or NOCs, whose degree of independence from national governments is variable, particularly in China and the RRS (Ma and Andrews-Speed, 2006). In China, financial institutions such as the China Development Bank finance much of China's overseas energy deals, although their level of independence from the Executive is uncertain (Downs, 2011). Parliaments in democratic RRS and the US raise concerns about Chinese investments and the morality of China's energy security strategy and may pass laws constraining Chinese investment. Society, also an important actor, influences China's drive for energy security through public opinion, think tanks and the media, and often generates or reports on the resistance to Chinese efforts in democratic RRS.

Figure 1.1 shows the triangular nature of China's seemingly bilateral energy ties. As China expands resource ties with each state, it finds that the US already has a relationship with each target of China's foreign policy. When President Chavez of Venezuela promised to deliver oil to 'the Chinese fatherland', the US

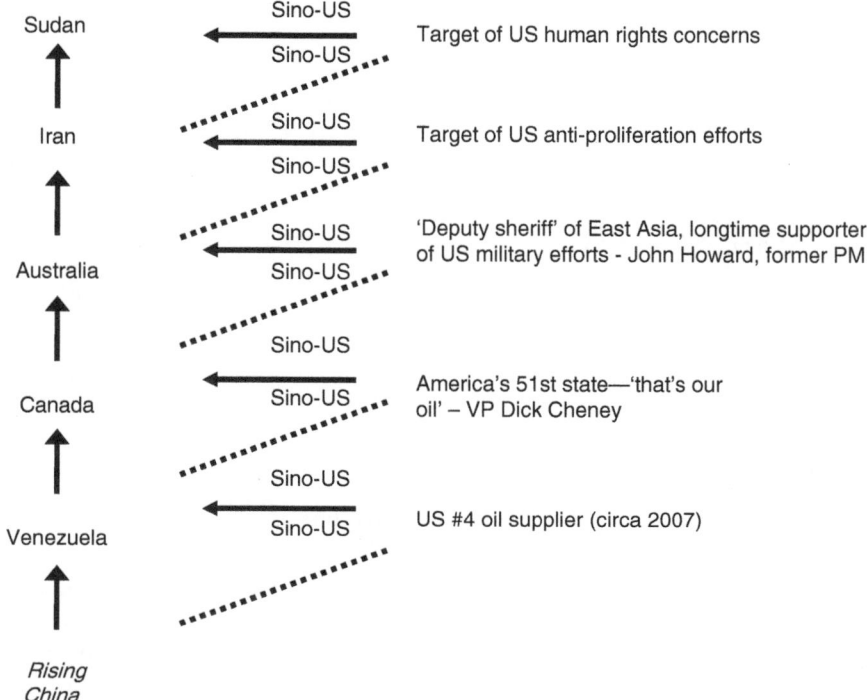

Figure 1.1 US interests in China's relations with the RRS.

Congress instructed the Congressional Research Service to study the impact that a total freeze on Venezuelan oil shipments could have on the US, given that Venezuela was one of the top four oil suppliers to the US. A 2001 report on US energy security directed by VP Dick Cheney labelled Canada's Tar Sands 'a pillar of sustained North American energy and economic security', (National Energy Policy Development Group, 2001: 8-8) and in 2012, when China's CNPC tried to buy Canadian energy company Nexus, many American officials, past and present, raised concerns about the deal (Chapter 6, this volume).

A more sophisticated model of this triangularism (see Figure 1.2) recognizes that the nature of each of the three bilateral relationships varies due to certain conditions. The following section suggests categories through which to analyze the bilateral ties in light of the third side of the triangle.

The nature of the ties between the US and the RRS (A–B) is a dominant factor in this analysis affecting the intensity of US concerns about China's ties to the particular RRS (C–B), and perhaps the degree of caution China adopts in dealing with the same RRS. Regarding the nature of the ties between China and the RRS, the nature of the RRS's ties to the US can affect its ties to China (B–C), which may also be influenced by the type of government or 'regime type' in the RRS. In democracies, such as Canada, Australia or Zambia, Chinese companies often confront domestic resistance or 'pushback' from citizens, the media or opposition

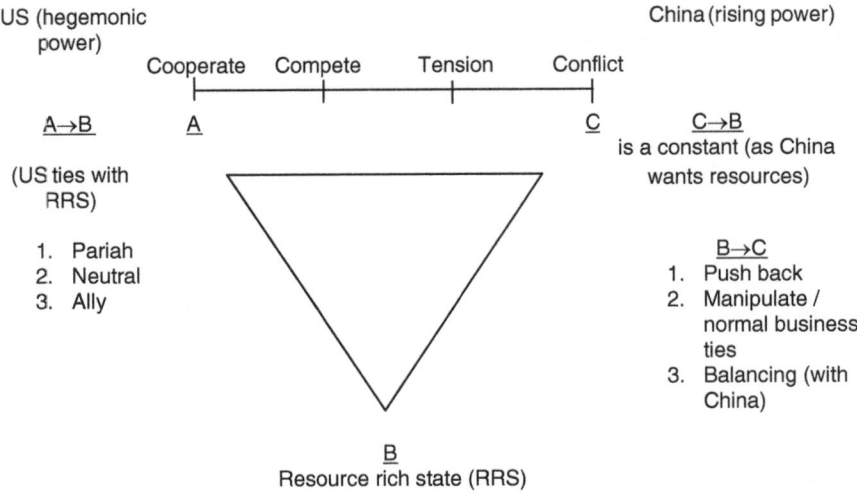

Figure 1.2 'Triangularizing' Sino–US resource competition.

parties as they negotiate their China deals. Stable authoritarian regimes, such as Angola, face little such pressure. Finally, the nature of US–RRS ties, as well as China's ties with the RRS, could also have some impact on Sino–American relations.

Pariahs, neutrals and allies: US perceptions of China's energy suppliers

China's energy partners are divided into three types based upon the US's perception of those states: pariahs, US allies and neutrals.

Pariahs: the US's moral dilemma and energy suppliers

Pariahs are states whose international or domestic behaviour is deemed unacceptable by global standards as determined by the US. Examples include Iran's nuclear proliferation activities, Myanmar's previous violations of its citizens' political rights, Venezuela's anti-US rhetoric and foreign policy, or the Sudanese government's complicity in violence against refugees in Darfur. Some are RRS from which the US has withdrawn, such as Sudan, or where US domestic laws forbids US companies from doing business, such as in Iran. Although these states are often subject to Western sanctions, they reap the benefit of China's homage to national sovereignty because China downplays or ignores their domestic or regional policies, making it hesitate to pressure these countries to conform to global norms. It may also protect them from UN sanctions and continue to do business despite calls for embargoes. Although the US is guilty of selective moral outrage – it never withdrew from Iran despite the Shah torturing prisoners – China's agnosticism about some international norms exacerbates moral disagreements and tensions with the US. Roger Robinson, vice-chairman of the US–China

Economic and Security Review Commission, said that China's support for pariahs to cater to its energy needs 'shows a stunning indifference to human rights' (Fisher-Thompson, 2005). Although the commission is not a government institution, it is mandated and funded by Congress whose views it can influence.

Pariah states often support China bilaterally and in multilateral organizations. After China and Iraq went their separate ways in the 1990s, Iran was China's only friend in the Persian Gulf (Wu, 2011). Iran abjured from the international isolation of China that followed the 1989 Tiananmen crackdown. According to researchers in the Chinese Institute for Contemporary International Relations (CICIR), just because the US calls for sanctions, must China reject a friend who is developing a nuclear capability to defend itself against the US? In fact, they saw no examples where sanctions stopped a state from developing nuclear weapons once it had decided to do so (Zweig interview at CICIR, 2005).[2] Why should China complicate relations with Iran, which has some of the largest oil and gas supplies in the world, when that effort is likely to fail? They also noted that the Bush Administration gave India nuclear technology, despite India's flagrant misuse of Canadian technology to build a bomb.

The amount of energy China receives from pariah RRS encourages it to maintain these links. Iran has been a major supplier of oil and natural gas to China, and in 2007, with Sudan and Venezuela, pariah states supplied 22 per cent of China's oil imports (Table 1.1); however, due to the triangular nature of Sino–Iranian–US ties, China has cut its oil imports from Iran in response to US pressure. As of 2010, Iran supplied 8.9 per cent of China's oil imports because Chinese firms who seek to work with US companies cannot trade with Iran (Jaipuriyar, 2010). By 2012, this was was down to 8.1 per cent (Table 1.1).

Pariahs come in different forms. The loss of US influence in Venezuela may be more acute than in regions outside the US's 'backyard'. The shift of Venezuelan oil to China could embolden leftist leaders in Latin America who want to nationalize their oil fields and turn access over to China (Kurlantzick, 2007: 210). Unlike Sudan and Iran, whose oil the US abandoned long ago, most Venezuelan oil is still refined in the US, undermining Venezuela's anti-Americanism. The US, however, has still decreased its dependence on Venezuela. In 2004, US imports from Venezuela were 1.55 million barrels per day; by 2008, it was 1.19 million barrels per day, and by 2013 it was only 0.797 million barrels per day, down 49 per cent in ten years (US Energy, 2014). Moreover, oil from western Canada's Tar Sands allowed the US to distance itself from Chavez (interview at US State Department, April 2010). By 2012, the US imported less oil from Venezuela than at any other time since 1984, but that drop may have been due to factors other than a conscious strategy to abjure from importing oil from a pariah (Blas, 2012).

Allies: Canada, Australia and Saudi Arabia

US allies who are RRS, for example Australia, Canada, Mexico and Saudi Arabia, comprise a second category. Despite deep military and economic ties with the US, and despite the fact that states prefer to trade with their allies, (Gowa, 1994) these

RRS feel some pull from China's expanding energy purchases, foreign invest-ment and economic cooperation. But how much space does the senior partner in the alliance grant to its junior partners? Does it use its power dominance to influ-ence their economic and strategic ties to China? According to Walt (2009), allies of the dominant power in a unipolar world lack leverage because the dominant power needs them less than he would in a competitive bipolar system. But he also recognizes that states with 'unique assets', such as RRS, have more bargaining power than other medium-sized allies in a unipolar world (Walt, 2009: 98).

Allies of the US do engage with China. The Saudis are China's top oil supplier, building refineries in China and developing a strong relationship. As of 2009, they exported more oil to China than to the US (Shichor, Chapter 4, this volume). However, the US encouraged the Saudis to sell oil to China to disengage it from Iran. As for the Australians and the Canadians, the former seems to have much more leeway to move closer to China than the latter.

Two patterns are possible with the US's allies: first, the US could worry more about China's resource diplomacy if the RRS's economic dependency on China generates some strategic shift, as occurred in 2004 when Australian Foreign Minister Downer tried to argue that under the ANZUS (Australia, New Zealand, United States Security Treaty), Australia was not committed to fight alongside the US if it was to get involved in a war with China over Taiwan. As Wolfers (1962: 29) informed us, 'most dangerous to the amity between peacetime allies are sus-picions concerning the reliability of allied pledges of future assistance'. Alterna-tively, the security component in the US–RRS alliance structure could overpower the pull of strong economic ties between the RRS and China.

Neutrals: competing on relatively equal terms

Although most analysts of international relations would apply the term 'neutral' to states that do not take sides in a conflict and 'neutrality' to the legal status of states that choose this position, neutral states in my analysis refer to states that are neither US-declared pariahs nor US allies. Some, such as Nigeria, may lean closer to the US, whilst others, such as Angola, were closer to China for more than a decade. This may seem like an overly large category, but the number of states remains limited because we are only interested in 'neutral' states that are resource rich (Table 1.1).

Neutral states possess characteristics pertinent to our analysis. First, they hold a significant amount of the world's oil, comprising a good percentage of the RRS, particularly in Africa, Latin America and Central Asia. By 2020, just 15 oil-producing countries in the world will account for 75 per cent of the net growth of oil supply capacity (van Geuns, 2008). Second, many neutral states are run by powerful dictators or are illiberal democracies dominated by populist leaders. Although some have no ideological predilection to oppose the US or capitalism, some populist leaders in Latin America oppose the capitalist West. Many leaders are Friedman's 'petro-dictators' whose power increases along with the price of oil (Friedman, 2008), but they generally do not treat their citizens with sufficient contempt to become a focal point of hegemonic concerns over human rights. Also, as neutrals, they are likely to possess greater flexibility within the triangular structure. While pariahs need China's protection and allies are constrained by their strategic ties with the US, neutrals can deal with whoever contributes more to their state coffers or offers less intrusive investment terms.

As illiberal democracies or authoritarian states, they still worry about US efforts to promote democracy and financial transparency around the globe. China's comparative advantage in these states, relative to the West, is its willingness to ignore their bad behaviour – particularly their corruption – even as China makes major loans to their governments. In 2012, the US, therefore, bought only 18 per cent of its oil from neutrals, whilst China bought 41.8 per cent.

In Angola, President Dos Santos has reinvested his ill-gotten gains overseas, making him the third richest person in Brazil (Ghazvinian, 2007). In 2003, the International Monetary Fund (IMF) was using Angola's need for financial support to rebuild the economy after the civil war as leverage to extract greater transparency in Angola's use of oil revenues. But China's US$5 billion loan let the regime off the hook. Arieh Neier, Head of George Soros' Open Society Institute that promotes an 'Extractive Industries Transparency Initiative' targeting African leaders who enrich themselves, not their people, through non-transparent energy deals and financial allocations, believes that Chinese loans to resource-rich African states, such as Angola, significantly undermine this effort (interview with the author in Hong Kong, 2008). To this extent, the US would probably prefer that the Angolan government change its policies, but it is far more interested in ensuring continued access to oil for US oil majors.

An important neutral RRS is Russia, which is energy-rich and militarily capable, and whose ruling group has little love for the US. Sino–Russian energy cooperation still runs hot and cold, however, because Russia plays China as it plays the rest of the world (Chow, 2005), as demonstrated during its negotiations with China and Japan over the Angarsk–Nakhodka pipeline. Although the US does not rely on Russian energy, Russia's supplies to Europe are critical and Sino–Russian energy cooperation is an important part of China's strategy to diversify its sources.

China and the resource rich states: side two of the triangle

RRS deal with China largely in three ways: soft balancing, pushback or engaging in normal business relations.

Pariahs facing international sanctions need China, and China, late to the energy game, has difficulty saying no to pariah RRS seeking support, particularly versus the US. By soft balancing, these states use diplomatic, rather than military, means 'to obtain outcomes contrary to US preferences, outcomes that could not be gained if the balancers did not give each other some degree of mutual support' (Walt, 2009: 104). China's rise and need for energy and resources, therefore, allow states such as Venezuela, Ecuador or Bolivia to distance themselves from the US, whilst helping states such as the Democratic Republic of the Congo to survive

(Global Witness, 2011). Despite the limited flexibility afforded to pariah states under these conditions, Myanmar's shift towards the West shows that China must be careful not to treat the pariah as a client state.

Democratic RRS who are mainly US allies may respond in two ways. On the one hand, economic engagement creates constituencies within democracies that may lobby for closer ties (Milner, 1988). Thus, Australians generally support a Free Trade Agreement with China and the business community is viewed as rather pro-China. On the other hand, citizens in democratic RRS worry about selling their country's own resource extraction companies to state-owned enterprises, which they see as extensions of the Chinese government. The resultant 'pushback' complicates China's resource diplomacy. Although exports to China have driven Australia's economy since 2005, an Australian government bill insisted that sales of Australian firms to all foreign state-owned firms must undergo review by the Foreign Investment Review Board (FIRB). Similarly, when China's Minimetals tried to buy Canada's Noranda, popular hostility among the Canadian populace and concerns within the Canadian government scuttled the deal (Litvak, 2006).

Changes in national leaders may affect ties to China as well, particularly in democracies. When the Liberal Party under Jean Chretien governed Canada (1993–2003), China was seen as a viable investor in, and consumer of, oil from the Tar Sands in Alberta. But Prime Minister Harper, a staunch conservative and human rights advocate, let ties languish for five years (2005–10), and only in 2011 did he open the door to Chinese investment. Similarly, a change of leader in Nigeria ended the close business ties between the two countries as the new president rejected the 'infrastructure for oil' arrangement of his predecessor (Taylor, Chapter 8, this volume).

Although some RRS are greatly influenced by external forces, such as whether it is being sanctioned by the US, EU or UN, the case studies in this book show that RRS have a degree of 'agency' and can manoeuvre between China and the US. This may be particularly true for neutral states. We must, therefore, focus on the domestic qualities of the RRS state, the regime type (whether it is a democracy or an authoritarian regime), the level of regime stability and its level of transparency. Also of critical importance is its level of economic dependency on Chinese investment and trade. We must be sensitive to any soft balancing by the RRS with China, or any weakening of the RRS's alliance with the US, which could lead the US to try to influence this side of the triangle. Finally, neutral states seem likely to adopt 'normal business relations', meaning that they negotiate with Chinese and Western oil majors to get the best deal, but authoritarian neutrals prefer China's 'Beijing Consensus' over the US's 'Washington Consensus' because China rarely politicizes the trade relationship.

In summary, we find that triangularity is at work among the RRS, the US and China. Some leaders in RRS, particularly US allies that are democracies, take a nuanced posture towards China out of deference to the US and often 'pushback' when difficulties emerge with China, or when China's 'resource diplomacy' becomes too assertive. At the other extreme, 'pariahs' have little choice but to balance with China for self-preservation in the face of US threats.

Closing the triangle: Sino–US relations and the search for energy

Finally, the framework proposed calls on us to explore the extent to which China's ties with the RRS and the nature of the RRS's ties to the US could influence the level of cooperation, competition, tension or conflict between the US and China. Klare (2001) has been arguing that energy could trigger more conflicts in the future, while Blumenthal and Lin (2006) saw a growing risk that China and the US would 'clash over sources of fossil fuels in the Middle East and other oil-patch states that are not models of stability or representative governance' (Blumenthal and Lin, 2006). How could such incidents emerge? The most likely scenario could result if China was to attack a US oil rig drilling in contested territorial waters, such as off the coast of Vietnam, triggering an American response. Such a confrontation, however, was more likely before China's leaders created the National Security Council, which brings the naval forces of five government agencies under central control. A conflict on the land seems less likely because the two countries' sources of supply differ, with significant overlap occurring only in Brazil, Venezuela, Kuwait, Iraq, Angola and Saudi Arabia. In fact, some confrontation in Venezuela might have been possible, but when the Venezuela government nationalized US oilfields, the Americans simply pulled out. Even if the US or Israel should decide to carry out a surgical strike on Iran's nuclear plants, China will not intervene militarily. Also, the regional overlap is not great (Pollack, 2008). As of 2012, over 50 per cent of US oil imports came from Mexico and Canada. Surprisingly, the US imports only 18 per cent of its oil from the Middle East compared to China, which imports 50 per cent from the Middle East.

Tension will persist long into the future. Kirshner (2008), who does not see war on the horizon, does believe that 'much politics, short of war, and the scramble for energy will provide yet another source of tension between the two states, deriving from geopolitical wariness, divergent interests, and Hirschmanesque affects that accompany China's relentless economic rise' (Kirshner, 2008: 255). The US dislikes China's energy security strategy, including China's emphasis on buying equity oil, which reduces the amount of oil on the global market, China's willingness to buy oil from US-declared pariah states, such as Iran and Sudan, and its willingness to protect them from international sanctions. According to former Assistant Secretary of State for East Asia, Chris Hill, the US and its allies must ensure that in its search for resources, 'China does not underwrite the continuation of regimes that pursue policies seeking to undermine, rather than sustain, the security and stability of the international community' (Hill, 2005). Interviews by the author in Washington with senior staff of the Senate Foreign Relations Committee and a Congressional Aide in the House of Representatives from an oil-producing state confirmed that the US's primary concern is that China helps pariahs evade Western sanctions. Chinese resistance to such pressures creates ill will within the US Congress (Garver, Chapter 11, this volume), but is unlikely to trigger direct conflict. Moreover, China cut its oil purchases from Iran (Table 1.1) in line with US requests, defusing some of the tension between the two powers.

Diplomatic battles for influence in RRS occur not infrequently. US officials worry about Chinese inroads into Latin America. On 2 March 2006, the US Congress, House International Relations Committee, Subcommittee on Western Hemisphere, held a hearing entitled 'Energy Security in the Western Hemisphere'. At that session, Dan Burton, Chairman of the House Subcommittee on Western Hemisphere claimed that, 'We should always look at Latin America in relation to the Monroe Doctrine, we have concerns: Chavez, Castro, Ortega, Morales in Bolivia, and their connections with communist China' (Hawksley, 2006). Former Secretary of State Condoleezza Rice, in a 2006 Senate testimony, said she was shocked how the 'politics of energy' was 'warping diplomacy around the world', as growing states, such as China and India, were engaged 'in an all-out search for energy…that is, really sending them into parts of the world where they've not been seen before, and challenging, I think, for our diplomacy' (Mufson, 2006). Some of this tension arises over the direction of new pipelines – either heading east or west – from countries around the Caspian Sea (Yergin, 2011).

Competition over oil fields, which takes place mostly among companies based in the two countries, can also generate some tension. In the US's opinion, Chinese government support for its NOCs, by supplying them cheap money, allows them to compete unfairly in the international market. This issue was part of the reason why the US Congress blocked the sale of Unocal to CNOOC in 2005, an event that made Chinese companies quite cynical about the US demands that China adopt free market principles (interview with CNPC analyst, 2010). According to a different Congressman, he worked against the Unocal deal because he believed that denying China access to this company and its technology would slow down China's rise.

Much of the competition simply occurs among IOCs and NOCs in the RRS, where neutrals play one side against the other in order to maximize their own interests; however, this competition among energy companies within neutral states generates less state-to-state tensions than for pariahs. Government loans, aid and development projects enhance China's oil companies' competition with other energy companies, resulting in its mode of engagement in 'normal business relations' often being scrutinized by Western governments and their parliaments.

Chinese firms still compete effectively with US IOCs (and other IOCs) because they accept minority shareholder status in oil fields whose majority shareholder is the NOC of the RRS. China, as a developing state, sympathizes with aspirations for resource sovereignty in ex-colonial states. The moment an RRS nationalizes its oil fields or demands that an IOC accept minority equity status, causing the Western IOC to withdraw entirely, the opportunity arises for China to become the NOC's partner, thereby gaining access to new sources of oil.

Finally, Sino–US energy cooperation is now a hot topic generating scholarly analysis (Hu, 2008; Hu and Zha, 2007), collaborative energy projects, joint research, new organizations and government interactions. In fact, a US Department of Energy report in 2006 concluded that China's purchases of energy assets were not harmful to the US and could in fact be beneficial because they 'will enlarge the total global oil supply' (Andrews-Speed and Dannreuther, 2011).

To promote further production, Conoco and CNOOC have a joint project in the Penglai oilfield in the Bohai Gulf, while in 2009, Sinopec launched a program to expand energy development with the US in third countries, partly to maintain low energy prices and the security of supply (China Petroleum and Chemical Cooperation, 2009).

Most Sino–US cooperation targets clean energy. During visits to China, David Sandalow, then Assistant Secretary of Energy for Policy and International Affairs, focused far more on clean energy and environment rather than on competition for resources (interview with the author, April 2010), as does the 'US–China Ten Year Energy and Environment Cooperation Framework' (US Department of Energy, 2011). Some of the joint research organizations and research centres include the US–China Energy Cooperation Project (founded by the American Chamber of Commerce in Beijing), the US–China Clean Energy Forum and the US–China Clean Energy Research Center (CERC). Some of these organizations see the shale revolution as a means to deepen Sino–US energy cooperation because Chinese companies need US knowhow to develop their enormous but isolated shale supplies. In 2009, the US Secretary of State, in a memo leaked by Wikileaks, asked embassies in Beijing, Delhi and Canberra, amongst others, to 'assess the state of shale gas development and/or potential for development in their host country and report their findings to Washington', turning shale gas development into an important component of Sino–American cooperation (Brennan, 2013).

China and the US cooperate over the security of supply lines, and although sea-lane security deeply worries China's leaders, it enhances military-to-military relations. Joint anti-piracy efforts in the Gulf of Aden have proven relatively successful and a new organization, the Commission on Energy and Geopolitics, which was established in January 2014 and is led by retired US generals, admirals and intelligence directors, seeks, amongst other aspects of enhancing US energy security, to 'involve China in maritime security operations to protect oil shipping' (Secure Energy, 2014).

Finally, the US and China can cooperate in bringing peace to internal wars in RRS because they share the goal of increasing the world's oil supply. Sudan is one example where, since the new conflict broke out between the governments in the north and the south, neither country has been unable to put its oil onto the world market (Table 1.1). China, once perceived in the West as a source of the problem, is now seen as an honest broker that can bring the civil war to a close (Regler, Chapter 12, this volume).

Conclusion

This chapter argues that many of China's bilateral energy relationships are really triangular because China's resource diplomacy occurs under a unipolar system with a US hegemon. Particularly in the case of US defined pariahs, the Americans should press China not to let its energy hunger override its moral responsibilities to reign in bad behaviour; yet in these cases, China needs to keep the oil flowing, setting up a complicated triangular relationship.

In the case of US allies, the US may worry that China's economic clout could weaken the security ties that bind them to their strategic partners; yet China's investment helps US allies grow at a time when US demand is weak and the country is not flush with cash. The US may tolerate an RRS becoming economically dependent on China as long as no strategic shift follows. If the ally is a major supplier of energy to the US, we should anticipate a more cautious US approach. The Chinese should also be circumscribed in their willingness to risk the wrath of the US by drawing too close to a long-term US ally.

Finally, US neutrals should have the greatest flexibility. They are neither dependent on China's protection from international pressure, as are pariahs, nor do they have formal (or informal) military alliances with the US, which the US might guard jealously. When authoritarian leaders in Africa reject financial transparency or Latin American leaders seek greater autonomy from US influences, they can still use their oil to draw China's attention and support. Balancing with China, however, has its risks, both for the RRS and for China, and brings scrutiny from the US Congress. In neutral states, such as Nigeria and Angola, the US and China may both fill their oil tankers without bumping into each other and even cooperate in extracting oil from current or new fields.

Just how much Sino–American competition, cooperation, tension or conflict emerges from these triangular ties awaits the findings of the case studies. I have speculated somewhat and been enlightened in part by having read all the papers more than once. Nevertheless, we will leave it to the readers to draw their own conclusions about the impact China's 'resource diplomacy under hegemony' has on Sino–US relations.

Notes

1   Funding came from 'Resource Diplomacy under Hegemony', Research Grants Council (HK), Project no: 646010, 2010–2013. Thanks for comments from Zhang Baohui, readers from Routledge and University of Kentucky Press, and research by Kang Siqin.
2   Discussions between the author and several analysts at CICIR took place in May 2006.

References

American Foreign Policy Council, 2009. *China Reform Monitor*. No. 738, 1 February. Available online at: www.afpc.org/publication_listings/viewBulletin/551(accessed 1 April 2015).

Andrews-Speed, P., 2009. China's ongoing energy efficiency drive: Origins, progress and prospects. *Energy Policy,* 37 (4): 1331–44.

Blas, Javier, 2012. Washington's reliance on Caracas for oil declines: US net imports of Venezuelan oil hit 30-year low. *Financial Times*, 13 December.

Blumenthal, Daniel and Lin, Joseph, 2006. Oil obsession: energy appetite fuels Beijing's plans to protect vital sea lines. *Armed Forces Journal*, 1 June.

Brautigam, Deborah, 2009. *The Dragon's Gift: The Real Story of China in Africa*. Oxford: Oxford University Press.

Brennan, Eliot, 2013. Shale Gas: The Key in the US' Asia Pivot? China–US Focus, 8 March. Available online at: www.chinausfocus.com/energy-environment/shale-gas-the-key-in-the-us-asia-pivot/ (accessed 1 April 2015).

China Petroleum and Chemical Cooperation, 2009. *Win–Win! Unveiling the New Era for Sino–American Oil Industry Cooperation.* 28 September. Available online at: www.uschinaogf.org/forum9/pdfs/lt-11-SINOPEC-HouHongbin_english.pdf (accessed 1 April 2015).

Chow, Edward, 2005. On China's Energy Relations with Russia and Central Asia. *Conference on China's Search for Energy Security and Implications for the US.* National Defense University, Washington, DC, 27–28 September.

Christensen, Thomas, 2008. Comments by Thomas Christensen. The US–China Institute, University of Southern California, March.

Andrews-Speed, P. and Dannreuther, R., 2011. *China, Oil and Global Politics.* London: Routledge.

Dittmer, Lowell, 1981. The strategic triangle: an elementary game–theoretical analysis. *World Politics,* 33 (4), 485–515.

Downs, Erica, 2011. *Inside China, Inc.: China Development Bank's Cross-Border Energy Deals.* Brookings Institution, John L. Thornton China Center Monograph Series, Washington, DC. No. 3, March.

EIA, various years. For 2007–2009 see: www.iea.org/publications/freepublications/publication/china_2012.pdf and for 2010–2012 see: www.iea.org/media/freepublications/security/EnergySupplySecurity2014_China.pdf (accessed 20 May 2015).

Fisher-Thompson, Jim, 2005. China no threat to United States in Africa, US official says. *IIP digital,* 28 July.

Friedman, Thomas L., 2008. *Hot, Flat and Crowded: Why The World Needs a Green Revolution – and How We Can Renew Our Global Future.* London: Allen Lane.

Ghazvinian, John, 2007. *Untapped: The Scramble for Africa's Oil.* New York, NY: Harcourt.

Global Witness, 2011. *China and Congo: Friends in Need.* London: Global Witness.

Goode, Darren, 2013. Energy security improves but US still at risk of oil disruptions. *Politico.* Available online at: www.politico.com/story/2013/10/us-energy-security-oil-disruptions-98298.html (accessed 1 April 2015).

Gowa, Joanne, 1994. *Allies and Adversaries in International Trade.* Princeton, NJ: Princeton University Press.

Hawksley, H., 2006. China's new Latin American revolution. *Financial Times,* 4 April.

Hill, Christopher R., 2005. *Emergence of China in the Asia–Pacific: economic and security consequences for the US,* Testimony before the Senate Foreign Relations Committee, Subcommittee on East Asian and Pacific Affairs, Washington, DC, June 7. Available online at: http://2001-2009.state.gov/p/eap/rls/rm/2005/47334.htm# (accessed 1 April 2015).

Howard, John, 2004. *Australia's Engagement with Asia.* Address to the Asialink–ANU National Forum, 13 August.

Hu, Richard W., 2008. *Advancing Sino–US Energy Cooperation Amid Oil Price Hikes.* Brookings Northeast Asia Commentary, No. 17, March.

Hu, Richard W. and Zha, Daojiong, 2007. Promoting energy partnership in Beijing and Washington. *The Washington Quarterly,* 30, 105–15.

International Crisis Group, 2008. *China a Thirst for Oil.* Asia Report No. 153, 9 June.

Jaipuriyar, Mriganka, 2010. Update: China cuts Iranian crude oil imports by 30 percent in H1 2010. *Platts,* 26 July.

Kirshner, Jonathan, 2008. The consequences of China's economic rise for Sino–US relations: rivalry, political conflict and (not) war. *In:* Robert S. Ross and Zhu Feng, eds. *China's Ascent: Power, Security, and the Future of International Politics.* Ithaca, NY: Cornell University Press, 238–59.

Klare, Michael T., 2001. *Resource Wars: The New Landscape of Global Conflict.* New York, NY: Owl Books.

Kurlantzick, Joshua, 2007. *Charm Offensive: How China's Soft Power is Transforming the World.* New Haven, CT: Yale University Press.

Litvak, Isaiah A., 2006. China Minmetals Corporation and Noranda Inc. *Harvard Business Review,* 19 January.

Lu, Xiao, Shao, Junyu, and Xu, Nengwu, 2011. On China's short and long-term concern about its interests in African petroleum under hegemony. Paper prepared for the conference on Resource Diplomacy Under Hegemony, Hong Kong University of Science and Technology and University of Macao.

Ma, Xin and Andrews-Speed, Philip, 2006. The Overseas Activities of China's National Oil Companies: Rationale and Outlook. *Minerals and Energy,* 1, 17–30.

Milner, Helen V., 1988. *Resisting Protectionism: Global Industries and the Politics of International Trade.* Princeton, NJ: Princeton University Press.

Ministry of Commerce (Mofcom), 2004. Dui wai touzi guobie chanye daoxiang mulu (An advisory list of sectors in different countries for foreign investment), 21 July. Available online at: www.mofcom.gov.cn/aarticle/bi/200407/20040700252005.html (accessed 1 April 2015).

Mufson, Steven, 2006. As China, US vie for more oil, diplomatic friction may follow. *Washington Post,* 14 April.

National Energy Policy Development Group, 2001. *Reliable, Affordable, and Environmentally Sound Energy for America's Future.* Washington, DC: US Government Printing Office.

Pollack, Jonathan D., 2008. Energy insecurity with Chinese and American characteristics: implications for Sino–American relations. *Journal of Contemporary China,* 17 (55), 229–46.

Secure Energy, 2014. www.secureenergy.org/Oil2025Presser (accessed 1 April 2015).

US Department of Energy, 2011. A progress report by the US Department of Energy. Available online at: http://energy.gov/sites/prod/files/piprod/documents/USChina-CleanEnergy.PDF (accessed 1 April 2015).

US Energy, 2014. US Energy Information Administration. Country Analysis Briefs: Venezuela. February. Available online at: http://www.eia.gov/countries/cab.cfm?fips=VE, June 2014 (accessed 1 April 2015).

van Geuns, Lucia, 2008. *China, Africa and the International Oil Market.* Clingendael International Energy Program. The Hague, Netherlands, 20 May.

Walt, Stephen, 2009. Alliances in a unipolar world. *World Politics,* 61, 86–120.

Wang, Z., 2011. Interview by author with Wang Zhen, Dean, School of Business, China Petroleum University, Beijing.

Wen, Han, 2004. Hu Jintao urges breakthrough in 'Malacca Dilemma', *Hong Kong Wen Wei Po* (Internet Version) in Chinese, January 14. Available online at: http://political-research.blogspot.hk/2004_04_17_archive.html (accessed 21 May 2015).

Wolfers, Arnold, 1962. *Discord and Collaboration: Essays on International Politics.* Baltimore, MD: Johns Hopkins Press.

Wu, Bingbing, 2011. Strategy and politics in the Gulf as seen from China. *In:* Bryce Wakefield and Susan L. Levenstein eds. *China and the Persian Gulf: Implications for the*

United States. Washington, DC: Woodrow Wilson International Center for Scholars, 10–26.

Xiong, Guangkai, 2004. *Zonglun guojin zhanlue xingshi* (Comprehensively Observing the International Strategic Situation), *Guoji wenti yanjiu [International Studies]*, 3, 5.

Xu, Xiaojie, 2013. *Meiguo nengyuan duli yu yingxiang fenxi* (Analysis of US energy independence and its influence), 2012 *nian shijie jingi huangpishu* (2012 Yellow book of the world economy), *Shehui kexue wenxian chubanshe* (Social Science Literature Publishing House) January, pp. 1–18.

Yergin, Daniel, 2011. *The Quest: Energy, Security, and the Remaking of the Modern World.* New York, NY: Penguin Books.

Yergin, Daniel, 2012. US energy is changing the world again. *Financial Times*, 16 November. Available online at: www.ft.com/intl/cms/s/0/b2202a8a-2e57-11e2-8f7a-00144feabdc0.html#axzz3W15URWKoovember 16, 2012 (accessed 1 April 2015).

Yong, Qian, 2002. A Symposium on the International Energy Situation and China's Energy Strategy Held in Beijing. *Guoji wenti yanjiu [International Studies]*, 3.

Zhang, Wenmu, 2006. Sea power and China's strategic choices. *China Security*, Summer, 17–31.

Zhao, Hongtu, 2006. Some thoughts on Sino–US energy cooperation. *Contemporary International Relations*, 1.

2 China's energy needs and energy security

Philip Andrews-Speed

Introduction

In 2010, China overtook the US to become the world's largest consumer of commercial energy (BP, 2014). In just ten years between 2002 and 2012, total commercial energy consumption in China rose by 250 per cent, at an average annual rate of increase of 10 per cent. Such a high rate of increase has presented China's government with severe challenges in maintaining sufficient and reliable supplies of energy to support its rapidly growing economy. This demand had a wide range of international consequences. The country now accounts for nearly one quarter (22 per cent) of total world commercial energy consumption and has become a net importer of oil, gas and coal. This increasing import dependence has exacerbated the government's sense of vulnerability. In contrast, the US over the same period has seen its energy demand fall marginally and is now set to become a net exporter of energy.

The scale of China's energy import requirement, along with the short-term unpredictability of the scale of these imports in the cases of oil and coal, has made China a significant player in energy international trade. This situation has affected not only international prices for energy products, but also the direction of energy flows. The concomitant expansion of international investment and service provision by China's national oil companies and of diplomatic initiatives by the government has drawn China into almost every major oil and gas province around the world.

This chapter will examine China's current and future energy needs and the strategies the government and enterprises are pursuing to enhance the security of energy supply.

China's energy policy challenges

Securing an adequate supply of energy to support a growing economy has been a constant concern of China's government since 1949. Until the early 1990s, this concern focused on the need to exploit domestic sources of energy (Dorian, 1994). The year 1993 was a turning point in the government's attitude to energy security as the country became dependent on imported oil for the first time in its history.

Growing dependence on oil imports since then, and the high levels of international oil prices from 2003, saw oil rise steadily up the agenda of the central government. But the continued ability of the international markets to supply these imports, and the country's ability to pay for them, meant that this increasing vulnerability did not bring security of oil supply to the top of the agenda.

In contrast to this relatively relaxed attitude to international energy security, the government's realization in 2004 that the country faced a major shortfall in *domestic* energy supplies, particularly of electricity, brought energy security right to the top of the agenda. Immediate and radical action was needed to ensure that the economy and people's livelihoods were not seriously damaged by a shortfall in energy supply. Attention switched from the production of energy to its consumption and also to the challenge of reducing waste in all parts of the energy supply chain (Andrews-Speed, 2009).

At the same time as the government was formulating new policies to reduce national energy intensity, it realised that climate change was also an urgent challenge. The Chinese government has long recognised the negative environmental impacts of the country's dependence. Although action has and continues to be taken to constrain these impacts, environmental concerns alone have not been sufficient to push energy up the government agenda. Indeed, even the recently enhanced enthusiasm for addressing climate change builds mainly on the energy efficiency programmes, which are themselves driven by security of supply objectives (Niederberger *et al.*, 2006).

The international components of China's energy policy arise from a combination of energy, industrial, diplomatic, strategic and other considerations. On the one hand, the government appears to have an innate distrust of international energy markets and feels increasingly vulnerable to accidental disruptions to the market, as well as to deliberate action that might be taken by other states such as the US. On the other hand, energy provides a useful vehicle for China to build diplomatic relations across the world.

Recent strategies for domestic energy supply

Since 2003, China's government has been struggling to supply the country's booming economy with sufficient energy. At that time, shortages of electricity and of transport fuels had spread across most of the country and imports of oil were soaring, partly due to a rush to buy diesel generators to produce electricity. Late in 2004, the government launched a largely successful six-year campaign to reduce national energy intensity by 20 per cent between 2005 and 2010 (Andrews-Speed, 2009). Seven years later, the 12th Five-Year Plan for 2011–2015 emphasised the need to restructure the economy, constrain the annual rate of growth of energy demand to just 4.3 per cent and reduce energy intensity by a further 16 per cent (Interfax, 2013).

Along with these measures to constrain the rate of growth of demand for energy, the government and the energy companies continued their efforts to raise the domestic supply of energy, both from conventional sources and from new and renewable sources of energy.

Conventional sources of energy

Coal continues to account for nearly 70 per cent of China's primary commercial energy consumption and the country is pre-eminent in the world's coal industry. In 2012, it accounted for 50 per cent of the world's consumption and 48 per cent of the production of coal (BP, 2014). Coal is likely to remain China's single most important primary source of energy for the foreseeable future, although its use will be progressively restricted to power generation as the industrial, commercial and residential sectors switch to natural gas. With the third largest reserves in the world, China should remain almost self-sufficient in coal for many years to come.

Until 2006, the country could supply its own coal needs and was a minor net exporter of coal to neighbouring countries. Excess production led to a sudden rise of net exports to a high of 90 million tonnes in 2001, equivalent to about 9 per cent of total internationally traded coal (Sagawa and Koizumi, 2008). By 2009, China had become a net importer of just over 100 million tonnes of coal per year. Annual net imports rose steadily to 280 million tonnes in 2012, representing 7.5 per cent of national coal consumption and 20 per cent of world seaborne traded coal (Du, 2013). The country is likely to remain a significant net importer of coal for several years, not least due to the long-standing domestic transportation problems (International Energy Agency, 2009). Mongolia and Australia are the main suppliers of coking coal to China, whilst Indonesia builds its lead over Australia as a supplier of thermal coal to the Pacific market. Further afield, coal mines in the US, South Africa and Colombia have switched their exports away from the depressed economies of the Atlantic region to the more vibrant economies of the Pacific, especially China. The 12th Five-Year Plan sets the objective of reducing to 65 per cent the share of coal in the energy mix by 2015, as well as an absolute cap of 4.0 billion tonnes of annual consumption (Interfax, 2013).

In 2012, China was the world's fourth largest producer of oil, after Saudi Arabia, Russia and the US, and accounted for about 5 per cent of world production. Yet existing proven reserves amount to just 11 years of present-day consumption. Demand has been rising at an annual rate of 7–9 per cent, despite government moves to restrict the use of oil to the transport and petrochemical sectors, whilst domestic production has been growing at an average of just 1–2 per cent per year. New discoveries are being made, but they barely replace ongoing production rather than adding to the remaining reserves. As a consequence, the gap between demand for oil and domestic supply grows each year (Figure 2.1).

China has been a net importer of oil since 1993. Net imports of oil rose four-fold between 2002 and 2012, reaching 6.5 million barrels per day in 2012, equivalent to 67 per cent of the nation's total oil consumption. In order to source these rising imports, China has sought new suppliers and its list of crude oil providers in 2012 exceeded 35 in number[i] (Tian, 2013). First the Middle East and then Africa became progressively more important (Figure 2.2). Since 1999, these two regions have consistently accounted for 75–80 per cent of China's crude oil imports, up from

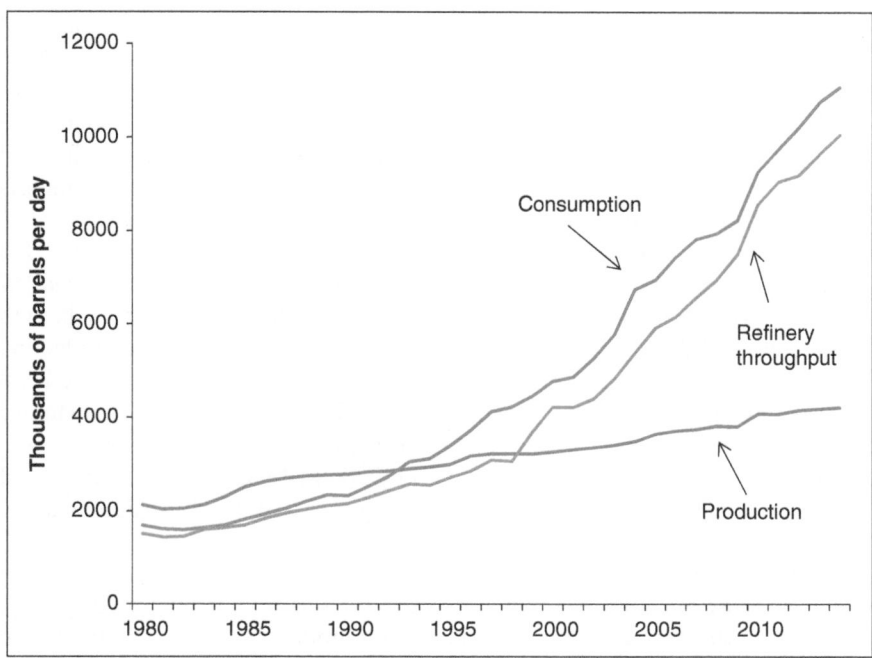

Figure 2.1 Oil production, consumption and refinery throughput in China, 1980–2014, in thousands of barrels per day.

Source: BP (2014) and unpublished Chinese data for 2014.

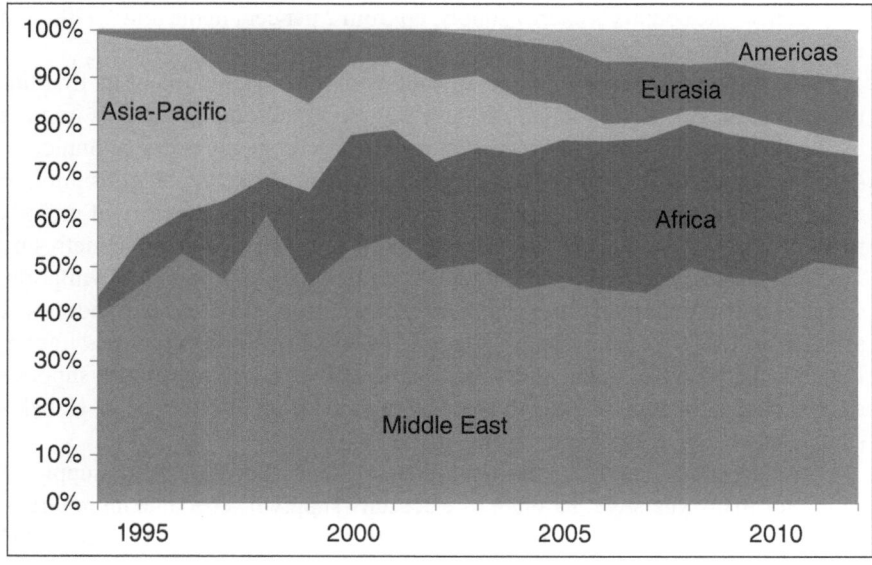

Figure 2.2 Sources of crude oil imports to China, 1994–2012.

Sources: Tian (2000, 2005, 2010, 2013).

50–60 per cent in the early and mid-1990s. The roles of Russia and Kazakhstan as oil suppliers have also grown because import pipelines to China have been commissioned in 2005 and 2010, respectively. In contrast, the proportion of crude oil imports supplied by the Asia–Pacific region has declined. Despite these changes, the sources of supply have remained relatively concentrated, with Saudi Arabia, Angola and Iran accounting for 40–50 per cent of China's crude oil imports since 2007, although Iran's contribution has been affected by sanctions.

Demand for liquid fuels will continue rising to 2020 and beyond, although the rate of increase will depend on domestic economic growth rates, international oil prices and on measures to constrain the use of liquid fuels, for example through the promotion of electric vehicles and public transport. Although the production of crude oil may continue to rise slowly, it is likely to reach a peak before 2020 unless substantial new reserves are discovered. Net imports of oil are likely to rise from 327 million tonnes (6.5 million barrels per day) in 2012 to as much as 450 million tonnes (9 million barrels per day) by 2020 (Reuters, 2013). This would take China's net oil imports to levels equivalent to that of the US or Europe today and to an import dependency of 70–80 per cent.

Since 1997, the government and the state companies have together tried to raise the level of use of natural gas. Three considerations have underpinned this policy: to use domestic sources of primary energy, to introduce a cleaner fuel to replace coal and to diversify the energy supply mix (International Energy Agency, 2002). Annual increases of 15–20 per cent in the domestic production of natural gas have allowed China to raise its domestic production of natural gas from 19 billion cubic metres in 1997 to 107 billion cubic metres in 2012. Production is due to rise to 160–170 billion cubic metres by 2015. Despite this rapid growth, gas provides just 5 per cent of the country's energy supply, up from 2 per cent in the mid-1990s, and demand outstrips domestic supply.

China became an importer of gas in 2006 with the construction of its first liquefied natural gas (LNG) plant in Guangdong Province, and its first gas imports by pipeline arrived from Turkmenistan in 2010. Total gas imports amounted to 41 billion cubic metres in 2012, nearly 30 per cent of the annual consumption of 143 billion cubic metres. Current plans will take the total quantity of imports to 90 billion cubic metres by 2015 and to more than 130 billion cubic metres by 2020 (Li, 2013). Although Australia has been the main source of LNG imports, China is now drawing on other suppliers such as Qatar, Malaysia and Indonesia, as well as UAE, Yemen, Oman, Algeria, Egypt, Equatorial Guinea, Nigeria, Ecuador, Peru and Trinidad and Tobago (BP, 2014). Additional future supplies of imported natural gas are due to come by pipeline from Turkmenistan, Uzbekistan, Kazakhstan, Russia and Myanmar.

Hydroelectricity has been the main source of primary electricity supply in China for many years, yet its share of electricity supply has become more erratic and declined from nearly 20 per cent in 1990 to around 16 per cent since 2004 (Andrews-Speed, 2013). These trends arise from a combination of changing weather patterns and the inability of dam construction to keep pace with rising electricity demand.

New and renewable sources of energy

The need to produce more energy from domestic sources and to reduce the inten-sity of carbon emissions has led the government to encourage investment in a range of new and renewable forms of energy. These have included developing substitutes for oil, exploiting unconventional gas reserves, expanding the role of nuclear power and building capacity for wind and solar power.

In light of the growing requirement for oil imports, the government has sup-ported the development of two sources of alternative supply of liquid fuels for transportation and petrochemicals: biofuels and coal-to-liquids, but enthusiasm for these technologies has waned in recent years as the environmental costs have become increasingly apparent. Instead, attention is being directed to explore the potential for shale oil.

Despite the projections of rising domestic production of natural gas, it is clear that what is called 'conventional natural gas' is only likely to provide a modest proportion of the national energy supply. As a result, the last ten years have seen significant efforts in China to identify and exploit domestic sources of 'unconventional gas', and these are now starting to bear fruit. 'Tight gas' (gas produced from poor quality reservoirs) is already being exploited in the Ordos Basin of northern China and in the Sichuan Basin. Collaboration with foreign companies such as Shell and Total will continue to be crucial for this success (Chen, 2010).

The country's energy companies have been working jointly with foreign com-panies since the early 1990s to develop coal-bed methane resources, which are abundant in some of the major coal basins of northern China. Progress has been constrained by a combination of technical and transportation challenges. Annual production from coal mines and surface wells was 14 billion cubic metres in 2012, and the government is aiming for coal-bed methane production to reach 30 billion cubic metres by 2015.

China may have even larger resources of shale gas than the US. Systematic exploration started in 2009 and the government has set production targets of 6.5 billion cubic metres in 2015 and 60–100 billion cubic metres by 2020. Two licensing rounds have been held that resulted in 20 blocks being awarded to Chinese companies other than the major national oil companies. PetroChina and Sinopec have met with some exploration success on their own territories, but as of December 2013 it is too early to assess the significance of these discoveries. If China can even partially replicate the US 'shale gas revolution' then this new source of gas supply could constrain its rising requirement for imported gas.

The electrical power industry possesses great scope for diversification away from coal towards fuels that yield few or no carbon emissions. In addition to continuing to pursue the construction of large-scale hydroelectric dams, China is aggressively developing nuclear, wind and solar power. In 2012, these forms of energy accounted for just 2 per cent of China's total consumption. Their contribution to the energy mix is set to grow steadily and dampen the demand for coal.

Actors in China's energy policy

This dependence of future energy needs and of import requirements on energy policy decisions requires us to examine the environment and processes of energy policy-making and implementation, starting with an examination of the actors in these processes. Three sets of actors directly or indirectly affect the development of energy policy in China: the government, industry and the military.

Government

Below the level of the State Council, China's energy sector has a disaggregated structure and lacks a strong and well-resourced agency at the central government level. One symptom of these deficiencies was the progressive decline of central government control over the energy sector, which was highlighted by the energy crisis that faced the new government in 2003 (Downs, 2006; Kong, 2006).

Two key priorities for the government at this time were to regain and centralise control over the energy sector and to provide for more coherent policy-making. Three institutions were established in order to achieve these objectives. In March 2003, the Energy Bureau was created within the National Development and Reform Commission (NDRC). Its functions of the Energy Bureau included formulating policy and drawing up plans for sector reform, as well as routine oversight of the country's energy sector. Two years later, in 2005, the government set up an Energy Leading Group within the State Council, supported by a State Energy Office. Their roles were to set strategic directions and to improve policy coordination (Downs, 2006).

In 2008, the Energy Bureau was renamed the National Energy Administration (NEA), and a National Energy Commission (NEC) was created from the pre-existing National Energy Leading Group. The NEC retained the overall roles of coordinating energy policy and setting strategic direction that were previously held by the Leading Group. Meanwhile the NEA took on the functions of the former State Energy Office, the NDRC's Energy Bureau and its Department for Energy Efficiency, and the nuclear power administration of the Commission of Science, Technology, and Industry for National Defence. The NEA's functions were to develop energy strategy, draft plans and policies, make proposals for energy industry reform, oversee the country's oil, natural gas, coal and power industries, manage the strategic oil reserves, formulate policies for renewable energy and energy conservation, and carry out international energy cooperation. Despite this apparent consolidation of energy sector governance, the responsibility for energy pricing has remained with the NDRC's Department of Price Administration. Further, the key actor responsible for driving the national energy intensity reduction campaign was not the NEA but the NDRC's Department of Environmental Protection and Resource Utilisation.

Other government agencies at or close to ministerial level have also continued to play a role in the domestic energy sector. The Ministry of Land and Resources manages resource extraction and exploitation, and the environmental protection of the land. The newly created Ministry for Environmental Protection is responsible for controlling the pollution of air and water, and also for nuclear energy safety. The State Asset Supervision and Administration Commission took the ownership role for government over the large state-owned enterprises. The leading research centres in the field of energy are the State Council's Development Research Centre and the NDRC's Energy Research Institute (Downs, 2004).

Although according to the constitution, China's government is unitary, substantial economic power was delegated to successive levels of government at provincial, city and district levels, especially in the early years of reform (Naughton, 2007). Unlike in a truly federal state, these subordinate levels of government have no rights over natural resources, except for those powers delegated to them by the central government. Indeed, in most aspects of energy policy, the major responsibility of sub-national governments is to adapt national polices, laws and regulations to local conditions and enforce them. All the ministries and most government agencies have equivalents at lower levels of government.

Industry

China's energy sector continues to be dominated by Chinese state-owned enterprises, although they are now commercialised and partially privatised. Three companies dominate the oil industry, each comprising a wholly state-owned holding company and a listed subsidiary (Zhang, 2004). These companies – China National Petroleum Corporation (CNPC)/Petrochina and Sinopec – are very large, vertically integrated corporations, with the former focusing on upstream (oil and gas extraction) and the latter on downstream (refining and distribution). In terms of reserves, production, refining output, revenue and profits, these two corporations rank alongside the largest international oil companies (IOCs) in the world such as Exxon, Shell and BP. The smaller China National Offshore Oil Corporation (CNOOC) is predominantly an offshore exploration and production company, and is equivalent in size to medium-sized IOCs and national oil companies (NOCs) such as BG from the UK and India's ONGC. A large number of smaller companies play a minor role in China's domestic oil refining and distribution sectors.

In the electricity industry, the State Power Corporation has been progressively commercialised and broken up into two grid companies and five large, listed national power generating companies. These account for more than 50 per cent of China's power generation capacity (Andrews-Speed, 2013). In addition, there are a number of smaller power generation companies that are owned by a range of entities at local government level. The coal industry is even more highly fragmented, with a small number of large companies and a very large number of private and locally owned mining companies (World Bank, 2008).

Despite this apparent diversity and the number of partially listed companies, the majority of enterprises operating in China's energy sector are owned mainly by the state, at central or local levels (Andrews-Speed and Cao, 2005). With the exception of the upstream oil and gas industry, foreign investment in China's energy sector has been very limited, despite official encouragement, due largely to policy ambiguities, regulatory weaknesses and the power of the state-owned incumbents (Andrews-Speed, 2004).

The military

The People's Liberation Army (PLA) plays many roles in China's energy sector. As with any large military organisation, the PLA uses much energy and wants to secure its own supplies of energy in case of crisis. As a consequence, it is improving its energy efficiency and has constructed its own emergency oil stocks, separate from those built by the national oil companies (Mulvenon, 2008). The PLA and the People's Armed Police Force are both routinely involved in constructing energy infrastructure, such as hydroelectric projects (State Council, 2013). In the past, the PLA undertook international oil trade through companies such as Zhuhai Zhenrong Corporation. As part of a wider programme of divestment of the PLA's commercial assets, Zhuhai Zhenrong became a private company in 1999 and is now China's major vehicle for importing oil from Iran (Wu, 2012). The nature of the company's current links to the PLA, if any, is unclear. Another energy company with apparent links to the PLA is China Huaxin Energy Company. As well as being involved in domestic energy projects, such as oil storage and inland liquefied nature gas plants, Huaxin Energy also runs the China Energy Fund Committee, a think tank which promotes international dialogue relating to energy (Stokes and Hsiao, 2013).

The PLA's precise role in making energy policy remains opaque (Mulvenon, 2008), for example the Deputy-Chief of the General Staff joined the NEC on its creation in 2010 (Bo, 2010). However, its role as a tool to enhance national energy security has been explicit for several years. As the country's dependence on seaborne imports of oil, gas and coal rises, its reliance on free passage along critical sea lines of communication (SLOCs) will grow concomitantly. The latest defence white paper explicitly mentions the need to protect both SLOCs and overseas investments in energy and resources (State Council, 2013). Today, China relies on the US Navy for much of its sea lane security, but the PLA Navy, unwillingly depending on the US in the long-term, continues to build its own blue-water capacity, focusing principally on defending national interests in Northeast and Southeast Asia (Cole, 2008; Collins *et al.*, 2008; Mulvenon, 2008).

Implications for policy-making and implementation

Progressive decentralisation and pluralisation has resulted in a system of policy-making and implementation that has been characterised as 'fragmented authoritarianism', which balances the forces of fragmentation and unification (Lieberthal and Oksenberg, 1988). These unifying forces include the power of the Communist Party, the role of personal relations and informal institutional networks, and the traditional preference for decision-making by consensus (Lieberthal and Oksenberg, 1988; Shirk, 1993; Xia, 2000). Although formulated more than

20 years ago, the concept remains applicable today. Recent analyses have added insightful nuances to this understanding. The term 'pluralistic elitism' recognises the enhanced pluralisation of the policy-making process, whilst ultimate power is retained by the political elite (Liao, 2006). This tense pairing of forces has also been referred to as 'Consultative Leninism' (Tsang, 2009). A symptom of this retention of power by the elite is that policy proposals that are aligned with key priorities of the top leadership are agreed to relatively rapidly, whilst those proposals that may ignore these key concerns and threaten the interests of other key actors may take years to be enacted (Kong, 2009).

Many of these general features of policy-making can be seen in China's energy sector, for example, a lack of strong leadership over the energy sector; the dispro-portionately high degree of influence over policy-making held by state-owned energy companies; the prevalence of bargaining in policy-making; and marked contrasts between some policies that are launched within a few weeks or months of their inception and others that remain under debate for ten years or more. The active role of the NOCs can be seen in the success with which they have persuaded the central government to support actively their massive investments in overseas oil and gas resources (Andrews-Speed and Dannreuther, 2011). Nevertheless, the central government retains the power to set the overall direction of energy policy and, if necessary, to take steps to force the NOCs to comply.

The implementation of energy policy also suffers from active resistance on the part of local governments to new initiatives. This resistance has two sources. First, local governments, even provincial governments, are rarely formally involved in the formulation of national policy, or only marginally. They may push for national policy changes or be recipients of national policy, but – except if the policy goes through the National People's Congress – representatives from lower levels of government do not formally draft national policies nor do they have the chance to debate them in an open forum. Second, in the case of energy, natural resources and the environment, the interests of local governments are often diametrically opposed to policy initiatives from the central government. The latter seek the effective long-term management of energy, natural resources and the environ-ment, whilst the local governments focus on short-term economic growth. The implementation of national policy is further constrained because the local bureaus of the ministries report to and are paid by the local governments and not by the central government (Economy, 2004).

Poor local implementation is enhanced by the involvement of local officials in the very businesses they should be regulating, by a shortage of trained staff in certain areas, by the low level of penalties for most offences and by rent-seeking and corruption. Further, the immaturity of the legal system protects local state-owned enterprises, local governments and even local private businessmen from prosecution by private parties (Andrews-Speed, 2012).

That being said, all domestic policy initiatives in the field of energy are defi-nitely not doomed to fail or generate undesirable side effects. The reason for this is that the authority of the Chinese Communist Party (CCP) is all pervasive, at all levels of government, in state-owned enterprises and in the private sector.

The CCP is the glue that holds the fragmented system together, along with networks of relationships between individual officials and managers (Lieberthal, 1995; Lieberthal and Oksenberg, 1988). The importance of the CCP and of loyalty to national interests is best illustrated by the career progression of officials both in government and in state-owned enterprises. Successful and loyal officials at lower levels of government may be promoted to higher levels of government, as may presidents of state-owned enterprises (Lieberthal, 1995; National Congress of the Communist Party of China, 2007). This process has seen powerful groups emerge from the Daqing and Shengli oil fields that have dominated the state planning sector in both 1979 and 2012, respectively.

Steps taken by China to secure international supplies of oil and gas

Separate from domestic measures to enhance energy security, since the early 1990s, China's government has taken a number of steps to enhance the nation's international security of supply for oil and gas. The key priority has been to raise the level of imports of oil and to diversify the sources of these imports, as described earlier. These supplies from major resource rich states (RRS) are underpinned by both long-term supply agreements and by wider economic and diplomatic engagement.

A further tactic employed in recent years by China's government to secure oil supplies has been to provide financial loans to the national oil companies of the RRS. Russia has been the main beneficiary. In 2005, CNPC lent Rosneft US$6 billion in return for deliveries of 350 million barrels of oil between 2005 and 2010. Three years later, CNPC provided Rosneft with an additional loan of US$15 billion and provided Transneft with US$10 billion in return for a guarantee that a planned oil pipeline to China would be completed and that shipments would reach 300,000 barrels per day by 2011. Further substantial loans have been made by China to Venezuela, Brazil, Angola and Kazakhstan, and some 80 per cent of Iraq's outstanding debt to China was forgiven (Andrews-Speed and Dannreuther, 2011). China Development Bank has also been extremely active, loaning US$65 billion to NOCs in Russia, Brazil, Venezuela, Turkmenistan and Ecuador in 2009–2010 alone (Downs, 2011).

A second priority for China has been to construct new infrastructure to deliver imports of oil and gas to China. This work has included rail capacity to bring oil from Russia, port capacity to receive marine oil tankers, re-gasification facilities to receive seaborne LNG and its own oil tanker fleet. But the most important and costly measure has been to embark on the construction of an extensive network of oil and gas import pipelines: from Russia in the north, from Central Asia in the west and from Myanmar in the southwest. These pipelines have three objectives: first to bring oil and gas from key neighbouring suppliers to China by a direct route; second to reduce China's dependence on seaborne imports and thus its vulnerability to disruption of shipping anywhere in the world; and finally, in the case of the pipelines through and from Myanmar, to reduce dependence on the sea lanes off South and East Asia, especially the Malacca Straits, through which some 80 per cent of China's oil imports flow (Blanchard, 2010). As of 2013, oil now flows via pipelines from Kazakhstan and Russia, and gas is imported through a pipeline network from the central Asian states of Turkmenistan, Uzbekistan and Kazakhstan, as well as from Myanmar.

The third component of China's international oil and gas strategy has been the investment in overseas oil and gas assets by its national oil companies. After 20 years of expansion of their overseas activities, Chinese oil companies by 2013 had a stake in more than 200 projects in about 50 countries. Since 2008, the aggregate value of new acquisitions by China's NOCs has exceeded US$100 billion. In 2013 alone, they spent US$32 billion (Lee, 2013). This compares to total annual capital investment outlays in recent years by the major IOCs of between US$20 billion and US$40 billion, but little of this capital was deployed for acquisition because the IOCs have been divesting assets in recent years, notably to the Chinese NOCs.

By 2009, sub-Saharan Africa and the Former Soviet Union each accounted for 30 per cent of the Chinese overseas upstream investment, with the Middle East and North Africa amounting to just 25 per cent, and the Americas just 5 per cent (Kong, 2010). The investments made since then have placed the Middle East and North Africa ahead of the other regions and have greatly enhanced the relative importance of the Americas for the Chinese NOCs. By 2012, the overseas equity production for CNPC alone amounted to 42 million tonnes (840,000 barrels per day) of oil and 14 billion cubic meters of gas, 27 per cent and 14 per cent, respectively, of its total output (China National Petroleum Corporation, 2013). In that year, CNPC accounted for 58 per cent of China's overseas oil and gas production of 90 million tonnes of oil equivalent (1.8 million barrels of oil equivalent per day) (Hang, 2013).

Investment by NOCs in overseas oil and gas reserves is not considered to be a normal part of energy security policy in Organisation for Economic Co-operation and Development (OECD) countries, not least because most national oil companies have been privatised and these privately owned companies are less beholden to their governments than are NOCs. But there is also strong scepticism that access to oil and gas reserves and production in remote countries can contribute to national security of supply in the event of an international supply crisis (Zha, 2006). In China's case, not only does the belief persist in some quarters that these investments do enhance national security of supply, but they are also seen to satisfy other national policy objectives, such as promoting national industrial champions, national independence, increasing employment and international diplomacy. Indeed, a close symbiotic relationship exists between these investments and China's increasingly active diplomacy on all continents (Andrews-Speed and Dannreuther, 2011).

Partly in support of its energy diplomacy, China has expanded its naval capabilities. This expansion includes the development of a blue-water navy capable of deployment across the world's oceans and cooperation with governments in South Asia to construct ports that can receive these vessels, as well as oil tankers and cargo vessels, for example in Pakistan, Bangladesh, Myanmar and Sri Lanka (Cole, 2008; Ebel, 2005; Odgaard, 2002; Pehrson, 2006; Storey, 2008).

This multi-pronged approach to securing overseas supplies of oil and gas has given China and its NOCs access to growing supplies of oil and gas and to substantial oil and gas resources, as well as enhancing its diplomatic profile in many parts of the world. China has been accused of tying up the world's oil and gas resources to the detriment of other nations, but its acquisitive strategy is of no real long-term consequence to the rest of the world, given the international nature of oil and gas markets. China will import the oil and gas it needs, and its choice of suppliers should have little global impact in purely economic terms. In contrast, the continuing success of its NOCs in concluding ever-larger and more diverse deals for access to oil and gas reserves does pose a challenge for other oil companies – IOCs and NOCs. Although Chinese NOCs do compete with NOCs from other countries and with IOCs and independent oil companies, they are also collaborating with them. In this way, they are behaving like other oil companies by both competing and collaborating.

International implications

As the world's largest user of energy, China's strategies to secure energy supplies have a truly global impact, on energy markets, on the flow of investment in resource extraction, on international and national energy companies from other countries, on the global environment and on international diplomacy. These impacts will expand as China's energy consumption and energy imports rise further, and as its energy companies expand their activities around the world.

Due to these uncertainties, security of energy supply will remain the top priority in its energy policy. This target is predicated on the desire for continuing economic growth and the need to maintain the CCP's legitimacy. Despite the clarity of this policy priority, and the high degree of path-dependency of China's energy policy, the exact nature of future policies and the success with which they will be implemented remain highly unpredictable. As a consequence, although it may be possible to sketch general patterns of future policies and behaviours of actors in the domestic energy sector and abroad, these patterns will be subject to sudden and unexpected changes which, in turn, may have significant international impacts.

For the foreseeable future, China's imports of oil and gas are certain to keep growing and its imports of coal will probably rise. We should expect to see China enhancing its diplomatic and economic measures to secure these supplies, and its NOCs will play an active part in this strategy, although with their own commercial agendas; however, the number of RRSs that are able to supply China with large quantities of fossil fuels will decline.

As the domestic demand for energy in the world's current oil-exporting nations rises, these countries will become net importers and the number of net exporters of oil will become progressively more limited. By 2020, most of today's major oil exporters will have entered a period of declining oil exports. After 2030, only a small number of countries will be major exporters of oil (Mitchell and Stevens, 2008), mostly nations with large oil reserves and small populations, such as Norway and Kuwait, or countries such as Brazil, which have alternative transport fuels.

The regional markets for natural gas are also becoming more integrated and China is drawing its gas imports from an increasing geographic range of suppliers. However, as with oil, the number of major net exporters of gas is likely to decline between 2020 and 2030.

New discoveries of conventional and unconventional oil and gas are being made. Some of these are large enough to be 'game changers' in a regional sense, for example the shale gas in North America (see Chapter 6, this volume) and the sub-salt oilfields of Brazil (see Chapter 10 by Moreira). The Arctic and the South Atlantic hold the promise of vast new resources, as do oil sands and shale gas, but they all face formidable obstacles relating to technology, cost and environmental protection. These obstacles will delay or even prevent their exploitation and, therefore, the vision of a decline in the number of net exporters of oil and gas remains valid. As competition amongst companies and amongst governments to gain access to these supplies increases, the economic and political significance of these remaining exporters will grow. They are likely to become focal points for tension and conflict between importers and exporters and amongst different importers. China is certain to be deeply involved through both its government and NOCs.

The US is currently the world's largest net importer of oil, but as more supplies become available from Canada, from oil sands and oil shale in the US itself, and possibly from Mexico, the North American continent is likely to become self-sufficient in oil, as is happening with gas. Thus, the US's direct interest in global oil and gas supplies may decline as China's continues to grow. At the same time, many developing and emerging nations will be seeking supplies from international markets. At present, the US is the main player in providing the political and military resources to assure the (relatively) efficient functioning of these markets. In the future, China is likely to demand a role, and other major net importers such as the EU, Japan and India may have to work with China as they have done with the US. However, these and other nations continue to distrust China, not least with respect to energy and climate change.

Note

1 See Table 1.1 in Chapter 1 by Zweig.

References

Andrews-Speed, P., 2004. *Energy Policy and Regulation in the People's Republic of China.* London: Kluwer Law International.

Andrews-Speed, P., 2009. China's ongoing energy efficiency drive: origins, progress and prospects. *Energy Policy,* 37, 1331–44.

Andrews-Speed, P., 2012. *The Governance of Energy in China. Transition to a Low-Carbon Economy.* Basingstoke: Palgrave Macmillan.

Andrews-Speed, P., 2013. Reform postponed. The evolution of China's electricity markets. *In:* F. Sioshansi, ed. *Evolution of Global Electricity Markets. New Paradigms, New Challenges, New Approaches.* Waltham, MA: Elsevier, 531–69.

Andrews-Speed, P. and Cao, Z., 2005. Prospects for privatisation in China's energy sector. *In:* S. Green and G.S. Liu, eds. *Exit the Dragon? Privatization and State Ownership in China.* London: Royal Institute for International Affairs, 196–213.

Andrews-Speed, P. and Dannreuther, R., 2011. *China, Oil and Global Politics.* London: Routledge.

Blanchard, B., 2010. China's risky steps with Myanmar pipelines. *Reuters Beijing,* 3 February 2010. Available online at: http://in.reuters.com/article/2010/02/03/idINIndia-45868120100203 (accessed 8 March 2010).

Bo, Z., 2010. China's new National Energy Commission: policy implications. EAI background Brief No. 504. Singapore: East Asia Institute.

BP, 2014. *BP Statistical Review of World Energy 2014.* London: BP.

Chen, A., 2010. Reforms needed as China plans unconventional gas push. *Reuters Beijing,* 7 May 2010. Available online at: http://in.reuters.com/article/2010/05/07/idINIndia-48309520100507 (accessed 16 June 2010).

China National Petroleum Corporation, 2013. *Annual Report 2012.* Beijing: China National Petroleum Corporation.

Cole, B.D., 2008. The energy factor in Chinese maritime strategy. *In:* G.B. Collins, A. Erickson, L.J. Goldstein and W.S. Murray, eds. *China's Energy Strategy. The Impact on Beijing's Maritime Policies.* Annapolis: Naval Institute Press, 336–51.

Collins, G.B., Erickson, A.S. and Goldstein, L.J., 2008. Chinese navy analysts consider the energy question. *In:* G.B. Collins, A. Erickson, L.J. Goldstein and W.S. Murray, eds. *China's Energy Strategy. The Impact on Beijing's Maritime Policies.* Annapolis: Naval Institute Press, 299–335.

Dorian, J.P., 1994. *Minerals, Energy, and Economic Development in China.* Oxford: Clarendon Press.

Downs, E.S., 2004. The Chinese energy security debate. *The China Quarterly,* 177, 21–41.

Downs, E.S., 2006. *The Energy Security Series: China. The Brookings Foreign Policy Studies.* Washington DC: The Brookings Institution.

Downs, E.S., 2011. *Inside China, Inc: China Development Bank's Cross-Border Energy Deals.* Washington, DC: The Brookings Institution.

Du, J., 2013. China coal imports continue affecting global process: *Platts, China Daily,* 22 May 2013. Available online at: www.chinadaily.com.cn/cndy/2013-05/22/content_16517987.htm (accessed 3 January 2014).

Ebel, R.E., 2005. *China's Energy Future. The Middle Kingdom Seeks its Place in the Sun.* Washington, DC: CSIS Press.

Economy, E., 2004. *The River Runs Black. The Environmental Challenge to China's Future.* Ithaca, NY: Cornell University Press.

Hang, D., 2013. China targets 100 mt share of overseas oil and gas output. *Interfax Natural Gas Daily,* 25 March 2013, 9.

Interfax, 2013. 12th Five-Year Energy Plan Emphasizes Consumption Reduction. *Interfax China Energy Weekly,* 19–25 January 2013, 56.

International Energy Agency, 2002. *Developing China's Natural Gas Market. The Energy Policy Challenges.* Paris: OECD/IEA.

International Energy Agency, 2009. *Cleaner Coal in China.* Paris: OECD/IEA.

Kong, B., 2006. Institutional insecurity. *China Security,* Summer, 64–8.

Kong, B., 2009. China's energy decision-making: becoming more like the United States. *Journal of Contemporary China,* 18, 789–812.

Kong, B., 2010. *China's International Petroleum Policy.* Santa Barbara, CA: Praeger.

Lee, Y. 2013. Higher debt at China's oil firms could slow M&A. *Wall Street Journal,* 6 May. Available online at: www.wsj.com/articles/SB10001424127887323368760457846 6541966919904 (accessed 21 September 2014).

Li, X., 2013. China's gas plan ends 2013 with mixed results. *Interfax Natural Gas Daily,* 19 December 2013, 7–8.

Liao, J.X., 2006. *Chinese Foreign Policy Think Tanks and China's Policy Toward Japan.* Hong Kong: Chinese University Press.

Lieberthal, K.G., 1995. *Governing China. From Revolution through Reform.* New York: W.W. Norton.

Lieberthal, K.G. and Oksenberg, M., 1988. *Policy Making in China. Leaders, Structures and Processes.* Princeton, NJ: Princeton University.

Mitchell, J.V. and Stevens, P., 2008. *Ending Dependence. Hard Choices for Oil Exporting States.* London: Chatham House.

Mulvenon, J., 2008. Dilemmas and imperatives of Beijing's strategic energy dependence: the PLA perspective. *In:* G.B. Collins, A. Erickson, L.J. Goldstein and W.S. Murray, eds. *China's Energy Strategy. The Impact on Beijing's Maritime Policies.* Annapolis: Naval Institute Press, 1–12.

National Congress of the Communist Party of China, 2007. Documents of the 17th National Congress of the Communist Party of China. Beijing, Foreign Languages Press.

Naughton, B., 2007. *The Chinese Economy. Transitions and Growth.* Cambridge, MA: MIT Press.

Niederberger, A., Brunner, C.U. and Zhou, D., 2006. Energy efficiency in China: Impetus for a global climate policy breakthrough? *Woodrow Wilson International Center for Scholars, China Environment Series,* 8, 85–6.

Odgaard, L., 2002. *Maritime Security between China and Southeast Asia.* Aldershot: Ashgate.

Pehrson, C.J., 2006. *String of Pearls: Meeting the Challenge of China's Rising Power Across the Littoral.* Carlisle, PA; Strategic Studies Institute.

Reuters, 2013. China oil imports to overtake US by 2017: Woodmac. 20 August 2013. Available online at: www.reuters.com/article/2013/08/20/us-china-oil-idUSBRE97J0BB20130820 (accessed 3 January 2014).

Sagawa, A. and Koizumi, K., 2008. *Trends of Exports and Imports of Coal by China and its Influence on Asian Markets.* Tokyo: Institute of Energy Economics Japan.

Shirk, S.L., 1993. *The Political Logic of Economic Reform in China.* Berkeley, CA: University of California Press.

State Council, 2013. *The Diversified Employment of China's Armed Forces.* Beijing: Information Office of the State Council.

Stokes, M. and Hsiao, R., 2013. *The People's Liberation Army General Political Department. Political Warfare with Chinese Characteristics.* Arlington, VA: The Project 2049 Institute.

Storey, I., 2008. Securing Southeast Asia's sea lanes: a work in progress. *Asia Policy,* 6, 95–127.

Tian, C., 2000. Analysis of China's oil and gas imports and exports in 1999. *International Petroleum Economics,* 8(3), 5–9 (in Chinese).

Tian, C., 2005. Analysis of China's oil and gas imports and exports in 2004. *International Petroleum Economics,* 13(3), 10–16 (in Chinese).

Tian, C., 2010. Analysis of China's oil and gas imports and exports in 2009. *International Petroleum Economics,* 18(3), 4–13 (in Chinese).

Tian, C., 2013. Analysis of China's oil and gas imports and exports in 2012. *International Petroleum Economics,* 21(3), 44–55 (in Chinese).

Tsang, S., 2009. Consultative Leninism: China's new political framework. *Journal of Contemporary China,* 18, 865–80.

World Bank, 2008. Economically, socially and environmentally sustainable coal mining sector in China. Washington, DC: World Bank.

Wu, K., 2012. *Energy Economy in China. Policy Imperatives, Market Dynamics, and Regional Development.* Singapore: World Scientific Publishing.

Xia, M., 2000. *The Dual Developmental State. Development Strategy and Institutional Arrangements for China's Transition.* Aldershot: Ashgate.

Zha, D., 2006. Energy interdependence. *China Security,* Summer, 2–16.

Zhang, J., 2004. *Catch-up and Competitiveness in China – The Case of Large Firms in the Oil Industry.* London: Routledge Curzon.

3 US energy security strategy and China's energy diplomacy[1]

Mikkal Herberg

Introduction

The United States (US) is the reigning superpower, or 'hegemon', of global energy and oil. This fact mirrors in energy its broader power as the reigning hegemon in strategic, military, economic, and financial terms. The US has for decades been the world's largest oil consumer and the largest oil importer. Oil imports peaked at over 12 million barrels per day in 2005, representing 60 per cent of its 20 million barrels per day annual oil consumption. Noticeably, in mid-2013 China surpassed the US as the world's largest net oil importer as US net imports fell dramatically whilst China's continued their relentless rise (US Energy Information Administration, 2014a). However, the US still consumes almost twice as much oil as China. The US was the third largest oil producer behind Russia and Saudi Arabia, but with the boom in US unconventional oil production since 2007, the US became the world's largest producer of oil in 2013 (US Energy Information Administration, 2013b). The US is home to many of the world's largest, most sophisticated and experienced international oil companies. It is the largest vehicle market in terms of total vehicles and, until recently, the largest in terms of annual sales, only lately surpassed by China. Global oil markets are priced in dollars, reflecting its reserve currency status and the liquidity, stability, and transparency of US financial markets, making US monetary and fiscal policy central to global oil prices. The US is the strategic leviathan wielding military power and diplomatic influence across the key centers of global oil production and export, most importantly the Arabian Gulf, and its navy protects the free flow of energy across all the vital sea-lanes and maritime bottlenecks of global oil transit.

Consequently, the US's energy security strategy is at the center of the geopolitics of global oil and gas markets, but the enormous scale of raw US power in global energy markets and geopolitics is accompanied by a conflicted and often confused domestic energy security debate. This confusion has resulted in a less than coherent approach to energy security, reflecting partly the US political system's on-again/off-again focus on energy security. Former US Energy and Defense Secretary James Schlesinger once aptly observed that when it comes to energy policy America has only two modes: complacency and panic (Luft and Korin, 2013). US public focus and political action ebbs and flows with the state of oil prices, resulting in inconsistent and often feckless policies. When President Obama introduced his energy plan in March 2011 he pointed out that the US keeps going "from shock to trance on the issue of energy security, rushing to propose action when gas prices rise, then hitting the snooze button when they fall again" (Ford, 2011: A16). Most recently, as US unconventional oil and natural gas production has grown, the debate has turned to what an increasingly oil and gas self-sufficient US means for our future energy security needs.

The US energy security debate is crosscut by a confused political dialogue that reduces energy security to a notion of 'energy independence'. A deep streak in US populist politics sees energy security in terms of ending US dependence on imported oil. Until recently, this notion seemed out of the realm of possibility, but with enormous oil production increases from 'tight oil' in the US due to new technological developments and reduced oil demand as a result of the economic recession, the US has reduced oil imports dramatically to just 6.2 million barrels per day in 2013, from a peak of 12.5 million barrels per day in 2005 (US Energy Information Administration, 2013a). Indeed, some forecasts suggest that US oil import dependence could disappear as early as 2020 (Citi Global Perspectives and Solutions, 2012). Although this situation would strengthen US 'energy security', insofar as physical security of supply for the US, it would not end dependence on what happens in global oil and energy markets. Oil prices are set in global markets and the potential for severe supply disruptions causing damaging price shocks that would damage the US economy would continue. Energy security for the US will dictate a continuing vital interest in seeking stability in global oil markets and supplies; however, from Richard Nixon in 1973 to Barack Obama in 2012, every Presidential candidate and politician has run for office on a platform of 'energy independence'. Now, as oil and natural gas self-sufficiency approaches us, Americans are debating what this really means for US energy security and what exactly 'energy independence' means for US energy policy and strategy.

Overview: does the US have an energy security strategy?

The US has not had what can reasonably be described as a coherent long-term energy security strategy; instead it has pursued a range of policies (mostly externally oriented) that address the perception of oil 'scarcity', the dependence on imported oil supplies from unstable places, and also the risk of oil price shocks. This policy has sought to reduce vulnerability to external oil shocks and shape the international oil market and strategic environment to provide the prodigious oil supplies required by an oil-intensive lifestyle. It has existed alongside anemic efforts to reduce oil demand growth. This policy mix began to shift in the later stages of the George W. Bush Administration and the Obama Administration as these administrations confronted economically punishing oil prices since 2003.

US energy security policies have responded to four interrelated concerns. First, is the risk of large oil supply or oil transit disruptions due to political events resulting in severe shocks to the global oil market. The paradigm for this is the oil shocks of 1973–74 and 1979–80 caused by political events in the Middle East that led to widespread supply shortages and extremely high prices (Deutch and Schlesinger, 2006). Second – and related – is the fear that sudden extreme oil price shocks will substantially damage the economy, even if availability remains adequate. For example, the 2006–8 oil price crisis, when prices rose to US$147, even though there was no physical supply shock. Third, US policy has been animated by concerns over high oil prices caused by Organization of the Petroleum Exporting Countries (OPEC) policies to deliberately restrict supply in the face of limited alternative non-OPEC supplies.

A fourth, more recent concern, which will be discussed later, is a relatively inchoate fear that China's new and aggressive efforts to secure ownership and control of overseas oil supplies somehow risks undermining future US and Western access to global oil supplies. This type of zero-sum view of future oil availability reflects fears over oil 'scarcity' driven by earlier supply shocks and reinforced by sharply rising oil prices. This concern, although largely unfounded because the scale of China's oil investments remains modest on a global scale, nevertheless influences how US policymakers perceive and respond to China's efforts to secure its future oil supplies. Furthermore, China's overtly mercantilist, overseas energy strategy runs counter to US approaches to energy security.

Four broad strands of US energy security policy are aimed at strengthening the availability and reliability of oil supplies, whilst keeping oil and energy prices affordable. The central pillar has been advocacy of free, transparent markets and the free flow of investment to drive global supply, demand, and prices. The thrust is markets over mercantilism, which involves heavy state involvement in oil markets elsewhere (Morse and Myers, 2001). This international focus mirrors US domestic energy market arrangements, where oil and gas markets are largely market-driven. Officially, the US government accepts that national control over strategic resources is a legitimate concern of resource-producing countries; however, the US supports the broadest possible market access to reserves for new exploration and development. The US has opposed OPEC's policies over the years to restrain supplies in order to raise prices. It pioneered new and transparent market mechanisms in the 1980s, such as the NYMEX futures market and the 'West Texas Intermediate (WTI)' marker price, which added to the flexibility and efficiency of global oil markets. Along with the later development of the London 'Brent' crude oil price benchmark, this eventually ended the direct administration of world oil prices by Saudi Arabia's 'Arab Light' marker price.

There is, however, an element of hypocrisy in US advocacy of free resource markets. Since the early 1970s, the US has maintained a ban on crude oil exports in the name of its own energy security. When new oil supplies came on line from Alaska in the late 1970s, federal law mandated that Alaskan oil must be shipped to the US lower-48. Moreover, exports of natural gas have been tightly constrained by law and only allowed in modest exceptions to North American Free Trade Agreement (NAFTA) partners, Canada and Mexico and a modest exception of Alaskan liquefied natural gas (LNG) exports to Japan. These constraints now come under new scrutiny in the wake of the unconventional oil and gas boom, which will be discussed later.

A second strand of policy has been promoting multilateral and bilateral cooperation to promote collaboration with other oil consumers to manage the potential impact of major supply disruptions and promote increased investment and transparency. The US led the development of the International Energy Agency (IEA) in the mid-1970s in response to the 1973–74 oil shock. The IEA has developed a vital strategic petroleum reserve system of oil stocks held by the member states that can be used in case of an oil supply disruption. The US helped make energy security an important topic in G-8 and G-20 meetings to broaden the range of multilateral energy cooperation. It also has a wide range of bilateral energy cooperation programs, including a robust bilateral energy dialogue and set of cooperative programs with China.

A third strand has been US energy diplomacy aimed at boosting global oil supplies and developing more diversified oil transit routes, in particular new supplies outside the OPEC countries. More supplies from different transportation routes bolster the flexibility of oil markets and weaken the power of OPEC to influence prices. Since the collapse of the Former Soviet Union, the US government has actively encouraged new oil supplies and new export routes from the Caspian region and Russia. Also, the US government has been deeply involved in the Caspian region promoting new oil investment in Azerbaijan and Kazakhstan and developing major new oil and gas pipeline export routes to the West, including the Caspian Pipeline Consortium (CPC) from Kazakhstan to the Black Sea, the Baku–Ceyhan (BTC) pipeline from Azerbaijan to Turkey's Mediterranean coast, and the new gas export pipeline from Azerbaijan to southern Europe. Finally, the US has strongly supported new investment in oil and gas development in Africa and Latin America as alternative supplies to OPEC.

When the first three market-oriented approaches fail, the fourth or default option becomes utilizing the 'hard power' of a global hegemon. The US has forged a series of strategic alliances and partnerships with key producing countries, combined with a long history of military interventions, to try to stabilize oil production, exports, and transit. The Arabian Gulf has been at the center of most of these efforts and central to this has been the historic strategic alliance with Saudi Arabia (Myers Jaffe, 2003). The US has provided strategic security for the Saudis and, in return, the Saudis have maintained large spare oil production capabilities that it has used to moderate large oil price movements. The Saudis have also been generally a voice for price moderation in OPEC. The US has intervened militarily in the Gulf repeatedly since the 1970s to try to ensure the flow of oil from the region, including helping secure Kuwaiti oil tanker shipments during the Iran–Iraq War of 1980–88, the first Iraq war in 1990 to expel Saddam Hussein from Kuwait, and the second Iraq war in 2003 aimed at removing Saddam Hussein and establishing a regime less threatening to its key oil-exporting neighbors. US-led pressure for sanctions on Iran is aimed fundamentally at limiting Iran's ability to threaten its Arab oil-exporting neighbors, as well as the potential threat to Israel and broader Western and US interests in the region and globally.

There is some misunderstanding of this aspect of US energy diplomacy, which is often portrayed by Chinese and other analysts as US hegemonic domination of the

Gulf region to secure control over world oil supplies for its own use (Liangxiang, 2013). However, the US gets relatively little oil from the Gulf – presently less than 20 per cent of its oil imports, which equates to only about 10 per cent of its total oil needs – and with rising US mid-continent oil production, even that share is declining. Ironically, Europe, Japan, and China depend far more heavily on the Gulf for their oil supplies. In fact, it is Asia that is exposed to the greatest risks to reliable supplies from the Gulf. US policy, on the other hand, is predicated on the importance of Gulf oil flowing to the world market and the potential damaging impact of large supply disruptions on global and, therefore, US economic growth. As the hegemon, the US is seeking to stabilize the global oil market in the interest of supporting global economic growth, which is central to US economic prosperity. The other major oil importers, including China, are therefore 'free riders' on US strategic power.

Others argue that US intervention is about securing US oil company interests. Again, this is more political myth than reality or, at least, it is an artifact of the past. Historically, post-World War Two US strategic and economic power did pave the way for US oil company investments in the Middle East during the major oil investment and production boom of the 1950s and 1960s. The same was true much earlier in Mexico and Venezuela. But since the oil shocks and nationalizations of the 1970s, the US focus on energy security through markets and global multilateral cooperation has led it largely to back away from directly supporting US oil companies – and US oil companies have increasingly sought to distance themselves from the US government as its power has become increasingly fraught and controversial. Moreover, relations between the large oil companies and the US government has become more adversarial. The US government is generally quite reticent to 'go to bat' for US companies abroad, except in the most general sense of protecting them from capricious and arbitrary treatment by host governments. By contrast, home governments of other large international oil companies, including London, Paris, and now Beijing, are far more willing to use diplomacy and economic efforts to support the success of their own national champions. This situation was quite apparent in the recent awarding of large Iraqi oil development contracts. In the two latest rounds of huge, new oil development contracts awarded by Iraq in 2010, the Chinese national oil companies (NOCs) were the largest national group of companies to win large contracts (Arango and Krauss, 2013). US-based oil companies were only modestly successful in this bidding and were not favored by Iraqi authorities in any way.

Another aspect of US strategic 'hard power' put to the service of energy security has been US naval power over the major oil and gas shipping lanes and key transit bottlenecks around the globe. The US Navy commands the Sea-Lanes of Communication (SLOCs) throughout the Middle East and Arabian Gulf, the East and South China Seas, the Indian Ocean, and other key sea-lanes. The free flow of energy through the Straits of Hormuz and through the Malacca Straits is a critical strategic mission of the US Navy, part of the larger goal of the US Navy's global reach to ensure the free flow of trade, goods, and energy essential to global economic growth and US prosperity (Blair and Lieberthal, 2007).

These four strands of energy security policy all aim at shaping the external, international environment and reinforcing the stability and flexibility of the global oil market to support US oil interests. They have evolved over the past six decades as the US became the global strategic and economic superpower in the wake of World War Two, the collapse of European and especially British power in the Middle East, the continued rise in US oil import dependence that began in the late 1950s, and the rising scale and power of US oil companies and their expanding investments globally.

The main weakness in US energy security efforts has been the lack of a serious, consistent policy on the oil demand side. Cheap energy, in some ways, has become the 'third rail' of US politics. Energy taxes have been kept very low and cheap energy is seen as crucial to the US's global economic competitiveness. Until recently, there has been very little real political support for strong measures to reduce energy and oil demand growth. Low energy and oil prices have been supported across the board by the auto industry, oil industry, construction industry, and industry more broadly. Due to low taxes, US gasoline prices remain generally one-half or less than gasoline prices elsewhere in Europe and Japan. As a result, US oil demand has continued to rise, particularly after the 1986 oil price collapse, while demand in Europe and Japan has remained essentially flat for the past 30 years. The only serious demand measure enacted in the mid-1980s was the new vehicle mileage standards that mandated a vehicle fleet average of 27.5 miles per gallon, which was then left unchanged for the next 20 years. Actual average fleet mileage in the US reached that target by 1987 and flattened out at that level until the late 2000s.

Two recent developments have invigorated domestic policy efforts to rein in oil demand growth. First, the run-up in oil prices since 2002 has forced Washington to finally revisit vehicle mileage standards. In 2007, the Bush Administration raised mileage standards to achieve a fleet average of 35 miles per gallon by 2020 (United States Congress, 2007). That target has been boosted dramatically under the Obama Administration to 54.5 miles per gallon for cars and light duty trucks by 2025 (The White House Office of the Press Secretary, 2012). The Obama Administration has also boosted support for development of hybrid and electric vehicle technology that could decisively impact US oil demand over the long-term. Both Administrations have promoted the development of alternative transportation fuels, such as ethanol and more advanced bio-fuels, in order to cut dependence on imported crude oil.

A second set of drivers for these changes is the growing concern over greenhouse gas emissions and climate change. In announcing the new 54.5 miles per gallon standards, the Obama Administration said that, 'The standards also represent historic progress to reduce carbon pollution and address climate change…and will cut greenhouse gas emissions from cars and light trucks in half by 2025' (The White House Office of the Press Secretary, 2012).

The US energy security dilemma, 1973–2010

Domestic political resistance to efforts to slow US oil demand growth has, until recently, left US energy security policy to focus largely on massaging the external environment to supply rising US oil needs reliably. But even with the power of a global hegemon, the results have been mixed. Despite US efforts, global investment in new oil supplies continues to lag demand growth because oil exporters use oil revenues to support social and political spending rather than new investment in oil capacity. Slow-moving, inefficient national oil companies among the big oil exporters undermine the pace of new oil supply development even further. Saudi Arabia is the notable exception to the rule. Political instability in key producing countries continues to undermine oil supplies and leads to frequent and unpredictable production outages that whipsaw global oil prices (US Energy Information Administration, 2014a). Ironically, US interventions in the Arabian Gulf aimed at stabilizing the region have taken large chunks of production capacity off line for long periods. Sanctions on Iran reduce its oil production and export capacity. US energy diplomacy has boosted investment, supply, and new transit routes in the Caspian region and Russia, and the IEA's emergency oil stocks system is a critical insurance policy to manage potential supply disruptions. The US–Saudi alliance has supported the development of ample Saudi spare oil production capacity that has been crucial during disruptions of supplies from Iraq, Kuwait, and Iran, and US naval power has succeeded at preventing any sustained maritime transportation disruption in oil flows through key bottlenecks.

Broadly speaking, however, the US's ability to manage a more stable global oil market has been mixed and frustrating, as the recent inexorable oil price rise of 2011–14 demonstrates. The image of Gulliver comes to mind. This probably helps explain the oscillation of panic and complacency in policy and why the siren call of 'Energy Independence' is so persistent in US politics.

Implications of the US unconventional shale oil and natural bas boom

The US energy security narrative of 'scarcity and vulnerability' is in the midst of a substantial re-evaluation in the wake of the enormous production shifts in the US and North America due to the unconventional shale oil and gas revolution. A new narrative of energy 'abundance' is beginning to take shape as US oil and shale gas production rises and dependence on imported oil plummets.

As a result of the deployment of hydraulic fracturing production technology (fracking), the US is rapidly headed for near self-sufficiency in oil supplies by 2020–25, only 15 or so years after the peak in oil imports at 60 per cent of total oil demand in 2005 (Citi Global Perspectives and Solutions, 2012). After a steady and relentless decline in US oil production for nearly 40 years, production has now risen from a nadir of 6.8 million barrels per day in 2008 to 9.2 million barrels per day in 2013. Oil imports have dropped from a peak of 12.5 million barrels per day in 2005 to just 6.2 million barrels per day in 2013 (US Energy Information Administration, 2013a). When conservative estimates of US oil production growth are combined with rising Canadian oil production later in the decade, the two countries on a combined North American basis seem likely to become a net oil exporter by 2020 or earlier (BP, 2014; Citi Global Perspectives and Solutions, 2012; Myers Jaffe, 2011). Similarly, shale natural gas production has risen by more than one-third since 2005, from 50 billion cubic feet per day to nearly 70 billion cubic feet per day (US Energy Information Administration, 2014b). The US is poised to become a large exporter of natural gas later this decade as new LNG export projects get approval from the Department of Energy, although just eight years ago it was expected that the US would become a very large *importer* of LNG (US Energy Information Administration, 2006).

The transition toward quasi-self-sufficiency in both oil and gas has occurred so rapidly that the implications for US energy security interests have only begun to be debated in Washington policy circles, Congress, and the Obama Administration (Jones *et al.,* 2014; Ladislaw *et al.,* 2014). As US oil import demand declines, *actual requirements* for Middle East crude oil imports are rapidly disappearing (Mitchell, 2010, 2012). A large and increasing share of Middle East oil is now moving to Asia and China, rather than to the US and Europe. Will the US reduce its costly geopolitical commitments to Gulf security taken up over the past 40 years? Other changes that could influence future US strategic commitments to the Gulf include the Arab Spring, tightening defense budget constraints, an increasingly war-weary and inwardly focused US public – what some call 'hegemony fatigue' – and the need to commit more strategic and diplomatic resources toward the rebalance to Asia (Alterman, 2013). Many of these concerns are prevalent among Arab Gulf rulers, particularly the Saudis, who are deeply dependent on the US security guarantee. On the other hand, US commitments to ensuring the reliable flow of Gulf oil are also predicated on their importance to world oil markets and the prosperity of the global economy and the economies of the US's key allies and trading partners. In a deeply globalized, world oil market, prices depend on continued Gulf oil supplies to the world. Many argue, therefore, that the US retains a vital interest in stable oil flows from the Gulf, regardless if any of the barrels actually end up in the US (Cordesman, 2013).

US policymakers have repeatedly emphasized that the US plans to maintain its strong engagement in the Middle East and that the drawdown in US forces in the region in the wake of the Afghanistan and Iraq wars is simply returning to its traditional level of engagement (The White House Office of the Press Secretary, 2013). The US has continuing interests in the region, including defense of its key allies, such as Israel and Saudi Arabia, the fight against terrorism, containing nuclear proliferation in Iran, and traditional defense of energy sea-lanes. Despite the changes wrought by the unconventional oil and gas boom in the US, the US will likely continue to focus strong diplomatic and 'hard power' efforts in the Arab Gulf and the energy sea-lanes to Asia.

US perceptions of China's expanding energy diplomacy and footprint

As China expands its energy investments, economic relations, and energy diplomacy in key energy exporting regions on an enormous scale, it is inevitable that China and the US will be 'bumping up against' one another in a growing number of places. China is already increasingly occupying strategic space that historically has been monopolized by US power. This is increasingly true in the key energy exporting parts of the world, raising a new set of issues in Washington about the implications of China's growing diplomatic involvement in countries and regions where the US for decades has been the dominant outside player. Not surprisingly, Washington's views of the implications of China's expanding energy and diplomatic footprint parallel its views toward China's broader economic and strategic rise. US views vary along the cooperation–competition continuum that mirrors the broader divide in Sino–US strategic relations between engagement and containment. At its center of gravity, US views reflect deep, underlying strategic distrust in the bilateral relationship that colors the prism through which Washington views Beijing's growing global energy reach and influence.

Markets versus mercantilism

Washington's concerns over China's energy expansion reflect two dimensions of global energy and geopolitical dynamics. First, the US views its energy security as closely tied to ensuring open, flexible, and transparent global oil markets, competitive and market-driven oil investment, and market access to reserves and contracts. The result is today's dynamic oil market that, although recently plagued by enormous imbalances and price volatility, nonetheless rapidly readjusts oil flows to reflect fundamental supply and demand conditions. The second was the development of multilateral energy cooperation through the IEA to create collectively managed, emergency oil stocks that provide an insurance policy in the event of severe supply disruptions. These efforts have been aimed at bolstering the stability and flexibility of the global oil markets and ensuring open access to supplies, rather than each country competitively seeking to ensure its own privileged access to supplies through political and bilateral deals.

The US and Western approach contrasts sharply with Beijing's natural impulse to seek privileged access, physical control of 'equity' oil, and bilateral political, economic, and financial ties to secure oil supplies. China's 'go out' strategy is predicated on a strong collaboration between the state and China's NOCs, which the US views as deeply mercantilist and statist and an approach that could undermine open and competitive oil investment flows (Andrews-Speed and Dannreuther, 2011; Downs, 2010; Kong, 2010). Among conservative and populist politicians and in conservative think tanks in Washington, which are suspicious of China's strategic intent, China's energy and resource mercantilism appears to be a direct state-sponsored challenge by Beijing to the energy security of other major oil consumers (Lohr, 2005; US–China Economic and Security Review Commission, 2013). In fact, across much of the political spectrum, this strategy is seen as a mercantilist attempt to gain control of oil supplies ahead of other states. As Downs (2007: 52) has written, "Even the Bush administration has joined the chorus, taking the Chinese government to task for 'attempting to follow a mercantilism borrowed from a discredited era through its efforts to 'somehow lock up' energy supplies around the world.'" In 2005, US Deputy Secretary of State Robert Zoellick accused China of seeking to 'lock up' energy supplies and contributing to a 'cauldron of anxiety' in the US and other countries over its efforts (Kessler, 2005).

Closer analysis suggests the Beijing–NOC collaboration is less strategic than it often appears and many investments are much more market-driven than state-driven (Kong, 2010); however, from the outside the process appears opaque and strategic in intent (Herberg and Lieberthal, 2006). This is not to say that Beijing does not have direct control of its NOCs when it chooses to exercise it. Senior executives of the major NOCs are appointed directly by the Chinese Communist Party (CCP) and they are expected to tow the political line of the leadership. Promotions from the NOCs to more important political roles depend on loyalty to CCP leaders. Major overseas projects and investments must be approved by the government and State Council, which feed the perception of 'China Energy Inc.' at work. More recently, the heavy involvement by the China Development Bank (CDB) in bringing to bear China's huge foreign assets by financing US$77 billion worth of large, long-term oil-backed loans with Russia, Kazakhstan, Brazil, Venezuela, and Ecuador reinforces the perception in Washington that Beijing's moves are undermining the competitive access to future oil supplies (Downs, 2011; Jiang and Sinton, 2011). This view is quite widespread even among senior officials, although a number of studies suggest it is an exaggeration of the impact of China's energy investments. For example, a 2006 US Energy Department study of the implications of China's overseas energy acquisition concluded it was not a threat to US energy security interests and could actually be beneficial because they will enlarge the total global oil supply (US Department of Energy, 2006). A 2011 IEA study reached the same conclusion that, "The Chinese NOCs substantial investments and pursuit of service contracts and loans to resource-rich countries have contributed and will continue to contribute to global upstream investment and global oil supply" (Jiang and Sinton, 2011: 8).

The jaundiced prism through which both official and unofficial Washington views China's approach to energy investment and competition explains the strong Congressional reaction to China National Offshore Oil Corporation's (CNOOC) attempt in 2005 to acquire Unocal in competition with Chevron. It reflected a strong perception in Washington, rightly or wrongly, that China's NOCs are instruments of Beijing's broader strategic agenda and potential long-term challenge to US power and interests. Recent signs, however, suggest that these fears are easing somewhat. In late 2012, China's CNOOC bid US$15 billion to acquire Canadian oil company Nexen, which also owned a major offshore oil field and multiple deep-water leases in the US Gulf of Mexico. The acquisition was subject to review by the Committee on Foreign Investment in the US (CFIUS). Despite modest opposition from some members of Congress, the acquisition was approved (Rampton, 2012); however, CNOOC was required to relinquish operating control of its offshore assets to quell national security concerns (Penty and Forden, 2013). In 2010, CNOOC also secured two joint venture investments in shale natural gas properties in the US in partnership with Chesapeake Company, without any serious reaction from Congress or CFIUS review (Winning and Yang, 2011). A number of similar deals by Chinese NOCs have since followed (Daily and Driver, 2012). Such investments involve taking minority shares in domestic natural gas production properties, and not the outright acquisition of a large US oil company, and therefore they may not represent a fair test of possible change in attitudes in Washington.

China's energy diplomacy and US power

Beijing's energy security drive is accelerating its emergence as a regional and global power. With expanding investments, oil import and LNG supply deals, and active pipeline diplomacy, China will inevitably become a key diplomatic and economic player in virtually every major oil and gas-exporting region of the world. Beijing will gradually acquire a significant capability to influence developments in these strategic energy regions (Friedberg, 2006; Herberg, 2011). For the US, this ability represents new challenges in its relations with China. Will Beijing's future diplomacy support US and Western energy, economic, and security policies in these key energy-exporting regions, or will Beijing challenge or seek to reshape diplomacy in these regions? In other words, in the energy sphere is China likely to be a 'status quo' or 'revisionist' power (Johnston, 2003)?

The first decade of energy emergence has been challenging for Beijing as it tries to reconcile its pursuit of energy resources with the new diplomatic pressures in many of the countries and regions where it must go to secure those supplies. In order to understand what this may mean for US energy and strategic interests, it seems sensible to analyze China's investment and strategic impact by 'triangularizing' among three different types of countries according to their relationship with the US. Although China's NOCs are active in over 40 countries, the US has focused largely on their involvement in a small number of 'pariah' states – Iran, Sudan, and Myanmar – as well as in Africa, where China's involvement intersects directly with Western strategic, human rights, and governance interests. Beijing has come under withering criticism for its reluctance to support fully Western efforts to isolate these governments or to press for reforms. For the most part, Beijing has reacted defensively, citing its traditional policy of 'non-intervention in the internal affairs of other states'. Despite the fact that non-intervention is a strongly held long-term policy position for China, this explanation for inaction exposes Beijing to the charge that it is a rationalization for blindly pursuing its energy interests at the expense of human rights and nuclear non-proliferation, as well as governance, economic, and financial reforms in Africa (Kleine-Ahlbrandt and Small, 2008). Recently, Beijing has begun to respond to these pressures with new initiatives to support Western reform efforts in Sudan and UN pressure on Iran to open up its nuclear program to inspections.

As suggested by the terms used in this book and the 'triangularization' analytical construct, there is a general discomfort in Washington with China's growing involvement in key energy-producing countries. Such concerns are based on the US's strategic unease with China's growing power capabilities and a lack of confidence in China's long-term strategic intentions. But the energy dimension of such concerns varies distinctly depending on the case. Among the pariah states of

Iran, Sudan, and Myanmar, Washington sees China's energy interests affecting its relations with these key countries and often undermining US interests and goals.

China's energy relations with Iran have probably been the most controversial. China has large oil supply contracts with Iran and potentially large oil and gas investment opportunities, and there is strong suspicion in Washington that this explains an important part of China's limited support for increasing sanctions on Iran (Harold and Nader, 2012). Chinese companies have also supplied significant amounts of oil products to Iran at a time when the US is trying to tighten availability of oil products to put new pressure on the regime (Downs and Maloney, 2011). However, Beijing also seems to recognize that too close a relationship with Iran risks seriously damaging Sino–US relations (Garver, Chapter 11, this volume). In 2010, the US agreed to forego direct sanctions on Chinese energy companies for doing business with Iran because Beijing apparently pressured its own NOCs not to follow through with potential large energy investments (Aizhu and Buckley, 2011). With the more recent imposition of tough oil export sanctions on Iran by the US and Europe, Beijing appears to have reduced its imports of Iranian oil, at least sufficiently to be granted a series of six-month waivers from the US State Department (Hook, 2012; Lelyveld, 2013). Reflecting its nuanced balancing act, Beijing insists it does not support the sanctions and it imports the oil that it needs regardless. Japan, South Korea, Turkey, and India have also reduced their imports and received similar treatment. But the key point here is that China's major role as an importer of Iranian oil and potential large energy investor gives Beijing the capacity to frustrate key US strategic goals if it chooses to do so.

China's growing energy relationship with a key US ally in the Gulf – Saudi Arabia – demonstrates the complex diplomatic interplay of China's energy interests and Sino–US relations when it comes to allies. China now imports roughly one million barrels per day from Saudi Arabia and the Saudis have made major refinery investments in the Chinese market. In some months, China imports more oil from the Saudis than does the US. Beijing therefore faces a delicate balancing act, wanting to keep good relations with both Iran and Saudi Arabia, while not severely damaging Sino–US relations (Blumenthal, 2005a; Gulf News, 2014). One example of this balancing act comes from the impact of Iran sanctions on China's oil supplies. As fears mounted in Beijing in 2013 that cutting Iran's exports would drive global prices higher and undercut its crude supplies, the US worked to convince the Saudis to raise oil production to reassure other buyers, including Beijing, that it would make up for lost Iranian barrels.

Other cases of China's energy involvement with US allies, such as Canada or Australia, are generally viewed as a normal investment process with few serious security implications for the US. Nevertheless, energy and security does get linked for some in Washington (Blumenthal, 2005b; Hou, 2008). For example, there has been some concern that China's growing energy and mineral contracts and new investments in Australia have become so important to its economy that Australians may rethink some aspects of their security cooperation with the US, for example in the case of a Taiwan confrontation. When Australian Foreign

Minister Alexander Downer suggested in Beijing in August 2004 that Australia might not automatically come to the aid of the US in the case of a Taiwan conflict, he set off a strong reaction in Washington (Thomas, Chapter 5, this volume). The US State Department demanded an explanation and retorted that Australia's treaty obligation under the ANZUS Pact (Australia, New Zealand, United States) to assist the US in a military defense of Taiwan was clear (McDonald, 2004; Yaxley, 2004). Prime Minister John Howard was forced to clarify Downer's remarks.

Similarly, China's NOCs have been investing substantially in western Canada's heavy oil industry. Sinopec is a partner in a proposed oil pipeline to Canada's west coast where oil would flow to China and Asian markets. According to the Premier of Alberta, in bilateral discussions, US policymakers have expressed concern that oil exports to China could reduce the availability of supplies to the US (Gillies, 2011; Chapter 6, this volume). This zero-sum attitude is also evident in the politics surrounding the US decision over the proposed Keystone XL oil pipeline, which could carry heavy oil from Alberta to the US Gulf coast. Although US environmentalists are battling against Canada's heavy oil growth, proponents of the pipeline argue that it is an important secure supply source. Other supporters argue that, if the US turns down the southbound pipeline proposal the oil will go instead to China, thereby enhancing its energy security whilst weakening that of the US (Goldenberg, 2010).

As for China's energy involvement in neutral cases – countries that are neither opponents of the US nor traditional allies, such as Kazakhstan, Nigeria, or Angola – the discomfort in Washington revolves mainly around whether China is taking such a large and important energy role that it could alter that supplier's commitment to allow competitive access to its oil and gas, including competition for investment opportunities. This could result in changes to the direction of crude and gas flows to suit political, rather than market logic. In some cases, these countries may alter their foreign policy to suit Chinese interests. For the most part, however, such developments have not yet occurred significantly, but it is certainly something that US policymakers should watch. The US has also focused on the impact of China's oil and resource investments on countries that seriously violate human rights and democratic values. For example, in a 2012 speech in Senegal, Hillary Rodham Clinton, then Secretary of State, implicitly warned Africa about China. "The continent," she argued, needs "a model of sustainable partnership that adds value, rather than extracts it, adding that unlike other countries," America will stand up for democracy and universal human rights even when it might be easier to look the other way and keep the resources flowing (French, 2014: A14).

Another real concern for some Washington policymakers is that China's state-led, mercantilist approach to investments and support for its NOCs risks altering the oil industry's competitive landscape and undermining competitive opportunities for US oil companies. Kazakhstan suggests that there are complex issues potentially at play in US–China energy relations. US policymakers will state officially that the oil and gas resources in the region are large enough to meet the energy needs of both east and west. The US has been content to watch China's growing energy supply relationships with Kazakhstan and Turkmenistan because these ties and pipelines have undermined Russia's strong control of Central Asian oil and gas transport routes (Peyrouse, Chapter 9, this volume). US efforts over the past two decades to get Caspian oil and gas flowing west have been substantial; however, China's increasing efforts to pull oil and gas to the east via newly built pipelines – and others still to come – could undermine the viability of some US-favored pipeline projects. A key test over the next five years will be the disposition of crude oil from Kazakhstan's giant Kashagan field in the Caspian and how much of that oil will go east to China.

The notion of triangularization and the interaction of Chinese and US energy interests in resource-rich countries remains a very uncertain territory because little objective research has been carried out to date. This project, therefore, is vital to put some 'meat on the bones' on the rhetoric we have heard over the past decade that has been largely un-informed by real research. The various country papers should add immensely to our understanding of the complex interactions among the three sides of the triangle.

Note

1 I would like to thank David Noel, Graduate School of International Relations and Pacific Studies, UCSD, for research assistance on this chapter.

References

Aizhu, C. and Buckley, C., 2011. Exclusive: China curbs Iran energy work. *Reuters*, 2 September.

Alterman, J., 2013. What should the Middle East expect from the United States and its allies? Global Forecast 2014: US Security Policy at a Crossroads. Washington, DC: Center for Strategic and International Studies, November.

Andrews-Speed, P. and Dannreuther, R., 2011. *China, Oil and Global Politics*. London: Routledge.

Arango, T. and Krauss, C., 2013. China is reaping biggest benefits of Iraq oil boom. *New York Times,* 2 June.

Blair, D. and Lieberthal, K., 2007. Smooth sailing: the world's shipping lanes are safe. *Foreign Affairs,* May–June, 7–13.

Blumenthal, D., 2005a. Providing arms: China and the Middle East. The Middle East Quarterly, Spring: 11–19.

Blumenthal, D., 2005b. Strengthening the US–Australian alliance: Prospects and pitfalls. AEI Outlook Series, April–May.

BP, 2014. BP Energy Outlook 2035: Focus on North America. Available online at: www.bp.com/content/dam/bp/pdf/Energy-economics/Energy-Outlook/North_America_Energy_Outlook_2035.pdf (accessed 14 May 2014).

Citi Global Perspectives and Solutions, Citi Commodity Research, 2012. Energy 2020: North America, the New Middle East? Available online at: www.citivelocity.com/citigps/ReportSeries.action?recordId=6 (accessed 17 May 2014).

Cordesman, A., 2013. American strategy and US 'energy independence'. Washington, DC: Center for Strategic and International Studies, 21 October.

Daily, M. and Driver, A., 2012. Sinopec, total invest $4.5 billion in US shale. *Globe and Mail,* 3 January.

Deutch, J. and Schlesinger, J., 2006. National security consequences of US oil dependency. Council on Foreign Relations, Independent Task Force Report, No. 58.

Downs, E., 2007. China's quest for overseas oil. *Far Eastern Economic Review,* September, 52–6.

Downs, E., 2010. Who's afraid of China's Oil companies? In: C. Pascual and J. Elkind, eds. *Energy Security: Economics, Politics, Strategy, and Implications.* Washington, DC: Brookings Institution Press, 73–102.

Downs, E., 2011. Inside China, Inc: China Development Bank's cross-border energy deals. The Brookings Institution, 21 March.

Downs, E. and Maloney, S., 2011. Getting China to sanction Iran: the Chinese–Iranian oil connection. *Foreign Affairs,* March–April, 15–21.

Ford Jr., H., 2011. Washington vs. energy security. *The Wall Street Journal*, 11 May.

French, H., 2014. Into Africa: China's wild rush. *New York Times,* 16 May.

Friedberg, A., 2006. Going out: China's pursuit of natural resources and implications for the PRC's grand strategy. The National Bureau of Asian Research, NBR Analysis, September.

Gillies, R., 2011. China eyes Canada oil, US's energy nest egg. *Boston Globe,* 26 June.

Goldenberg, S., 2010. Canada looks to China to exploit oil sands rejected by the US. *The Guardian,* 14 February.

Gulf News, 2014. China wants to play bigger role in the Middle East: Foreign Minister outlines 'new policies' where China wants to increase involvement. *Gulf News,* 9 January.

Harold, S. and Nader, A., 2012. China and Iran: economic, political, and military relations. RAND Center for Middle East Public Policy, Occasional Paper.

Herberg, M., 2011. China's energy rise and the future of US–China relations. *New America Foundation,* 21 June.

Herberg, M. and Lieberthal, K., 2006. China's search for energy security: implications for US policy. The National Bureau of Asian Research, NBR Analysis, April.

Hook, L., 2012. US exempts China from Iran oil sanctions. *Financial Times,* 29 June.

Hou, M., 2008. *The Role of the US in the Post-Cold War China–Australia Relations.* Washington, DC: Georgetown University Center for Australian and New Zealand Studies.

Jiang, J. and Sinton, J., 2011. Overseas investments by Chinese national oil companies: assessing the drivers and impacts. *International Energy Agency,* 10, 48pp.

Johnston, A., 2003. Is China a status quo power? *International Security,* 27(4), 5–56.

Jones, B., O'Brien, E. and Steven, D., 2014. Fueling a new order: Geopolitical and security consequences of energy. The Brookings Institution, 15 April.

Kessler, G., 2005. US says China must address its intentions. *The Washington Post,* 22 September.

Kleine-Ahlbrandt, S. and Small, A., 2008. China's new dictatorship diplomacy: is Beijing parting with pariah's? *Foreign Affairs,* 87(1), 38–56.

Kong, B., 2010. China's petroleum diplomacy. In: B. Kong, ed. *China's International Petroleum Policy.* Santa Barbara, CA: Praeger, 229pp.

Ladislaw, S., Leed, M. and Walton, M., 2014. New energy, new geopolitics: balancing stability and leverage. Washington, DC: Center for Strategic and International Studies, April.

Lelyveld, M., 2013. China faces challenge dodging US sanctions over Iranian oil imports. *Radio Free Asia,* 2 September.

Liangxiang, J., 2013. Fallacies about the US military presence in the Middle East. *China. org.cn,* 21 October. Available online at: www.china.org.cn/opinion/2013-10/21/ content_30352616.htm (accessed 16 January 2014).

Lohr, S., 2005. Unocal bid denounced at hearing. *The New York Times,* 14 July.

Luft, G. and Korin, A., 2013. The myth of U.S. energy dependence. *Foreign Affairs,* October.

McDonald, H., 2004. Downer flags China shift. *The Age,* 19 August.

Mitchell, J., 2010. *More for Asia: Rebalancing World Oil and Gas.* London: Chatham House, Royal Institute of International Affairs.

Mitchell, J., 2012. Asia's new role in global energy security. Oil and Gas for Asia: Geopolitical Implications of Asia's Rising Demand. The National Bureau of Asian Research, Special Report No. 41.

Morse, E. and Myers Jaffe, A., 2001. Strategic energy policy: Challenges for the 21st century. James A. Baker III Institute for Public Policy of Rice University and The Council on Foreign Relations, Independent Task Force Report No. 33.

Myers Jaffe, A., 2003. The United States and the Middle East: Policies and dilemmas. James A. Baker III Institute for Public Policy of Rice University.

Myers Jaffe, A., 2011. The Americas, not the Middle East, will be the world capital of energy. Foreign Policy, Special Report, 15 August.

Penty, R. and Forden, S., 2013. CNOOC said to cede control of Nexen's US Gulf assets. *Bloomberg,* 1 March.

Rampton, R., 2012. US lawmaker asks for conditions on CNOOC–Nexen deal. *Reuters,* 30 July.

The White House Office of the Press Secretary, 2012. Obama Administration Finalizes Historic 54.5 MPG Fuel Efficiency Standards. 28 August. Available online at: www. whitehouse.gov/the-press-office/2012/08/28/obama-administration-finalizes-historic-545-mpg-fuel-efficiency-standard (accessed 12 April 2014).

The White House Office of the Press Secretary, 2013. Remarks by Tom Donilon, National Security Advisor to the President at the launch of Columbia University's Center on Global Energy Policy, 24 April.

United States Congress, 2007. One Hundred and Tenth Congress of the United States of America, First Session. 4 January. Available online at: www.gpo.gov/fdsys/pkg/BILLS-110hr6enr/pdf/BILLS-110hr6enr.pdf (accessed 15 November 2013).

US Department of Energy, 2006. Energy Policy Act 2005: Section 1837: National Security Review of International Energy Requirements, US Department of Energy.

US Energy Information Administration, 2006. Annual Energy Outlook 2006. Available online at: www.eia.gov/oiaf/archive/aeo06/index.html (accessed 17 April 2014).

US Energy Information Administration, 2013a. US Net Imports of Crude Oil and Petroleum Products. Available online at: www.eia.gov/dnav/pet/hist/LeafHandler.ashx? n=pet&s=mttntus2&f=a (accessed 12 April 2014).

US Energy Information Administration, 2013b. US Expected to be Largest Producer of Petroleum and Natural Gas Hydrocarbons in 2013. Available online at: www.eia.gov/todayinenergy/detail.cfm?id=13251 (accessed 13 November 2013).

US Energy Information Administration, 2014a. Short Term Energy Outlook. February. Available online at: www.eia.gov/forecasts/steo/archives/Feb14.pdf (accessed 20 February 2014).

US Energy Information Administration, 2014b. China is Now the World's Largest Net Importer of Petroleum and Other Liquid Fuels. Available online at: www.eia.gov/today-inenergy/detail.cfm?id=15531 (accessed 30 June 2014).

US–China Economic and Security Review Commission, 2013. *2013 Annual Report to Congress* 20 November. Washington, DC: US Government Printing Office.

Winning, D. and Yang, J., 2011. CNOOC, Chesapeake agree on deal. *The Wall Street Journal,* 31 January.

Yaxley, L., 2004. Diplomatic wrangling over Taiwan. *The World Today, ABC,* 20 August.

Part II

US and its allies

4 Sweet and sour

Sino–Saudi crude collaboration and US-crippled hegemony

Yitzhak Shichor

Introduction

In late 2009, for the first time in its history, Saudi Arabia exported more crude oil to China than to the US. At over 843,000 barrels per day, the Saudi share in China's 2009 oil imports (nearly 4.1 million barrels per day) had reached over one-fifth, making Saudi Arabia China's leading oil supplier (*Financial Times*, 21 February 2010). Saudi crude oil exports to the US in November 2009 were the lowest in 21 years (*Emirates Business*, 18 February 2010). This trend continued and in 2013 China has overtaken the US as the leading crude oil importer in the world (Blas, 2013; *Arab News*, 2013a). In May 2010, Saudi Arabia's state-run Aramco (Arabian–American Oil Company), held its board of directors meeting for the first time in China. Symbolically, if not practically, this was another indication of the US's *perceived* decline and China's *perceived* upsurge. As a Saudi editorial put it, "America's fall from grace opened the door to a resurgent China to make its presence felt on the international stage in a way it had never done before" (*Saudi Gazette*, 2010b).

Many regard the pursuit of energy, especially oil, as the leading incentive underlying Chinese foreign policy (Jaffe and Lewis, 2002: 115), primarily in the Middle East, yet, while there is no doubt that China's energy needs are growing, relations with oil producing and exporting countries had begun long before Beijing became aware of its oil shortage. Indeed, China's increased international presence has been motivated by economic considerations, but also by strategic ones that were only later supplemented by energy needs. These considerations are likely to continue determining China's international behaviour in the future. Oil diplomacy is *not* an independent variable of China's foreign policy but is dependent on relations with other countries, not only with oil suppliers but also primarily with the US (Shichor, 2014). As Yergin (2006) commented, "Energy security does not reside in a realm of its own, but is part of the larger pattern of relations among nations." This is also true of US–Saudi relations, which had been initially founded not so much on oil but primarily on strategic Cold War interests (Bronson, 2006).

This chapter explores the role of oil and the interaction of oil and politics in the triangular relations between China, Saudi Arabia and the US, including the role (if any) of US–Saudi time-honoured 'special relationship' on Sino–Saudi oil collaboration and the impact (if any) of Sino–Saudi oil collaboration on Sino–US (and US–Saudi) relations. These triangular ties have undergone a number of stages in the last four decades, witnessing China's transformation from a marginal, nearly outcast, nation in the international system to a global power – at least in economic terms, if not (yet) in political and military ones. Regarding this transformation as positive when it began in the late 1960s and early 1970s, the US China policy, however, has become more and more negative since the late 1980s and early 1990s, turning the two powers from virtual partners into actual rivals and competitors. Sino–Saudi oil collaboration should be interpreted in this perspective, although initially oil was not an issue in Sino–Saudi–US triangular relations. To what extent does the US accept the expanding Sino–Saudi energy (and other) relations and why? Are the Saudi motivations in turning to China economic or also political and strategic and to what extent do they affect relations with the US? The first part of the chapter discusses the growing Sino–Saudi oil partnership and the second part deals with the trilateral perspectives related to the implications of this partnership. Potential future obstacles to these triangular relations are discussed in the conclusion.

China and Saudi Arabia: the oil connection

Although China was a net oil exporter until the mid-1990s, its oil imports had already begun to grow in the 1980s, picking up from the early 1990s. Most of this oil came from Southeast Asia with some from south Arabia (Yemen and Oman), but not from Saudi Arabia, which was an insignificant oil supplier to China until the mid-1990s. The Chinese were aware of Saudi Arabian oil prominence (Beijing Daxue, 1985), but they initially downplayed the Middle East as an oil supplier for a number of reasons. First, China adopted the principle of diversifying its oil resources so as not to become dependent on one country or region, given its bitter experience with the Soviet Union. This policy led to the creation of its own 'oil bases' abroad by acquiring oilfield concessions. Second, because of the endemic wars and unrest and the related restrictions on oil exports, China was concerned that Middle Eastern 'instability' would disrupt oil supplies and could be used as a weapon by Middle Eastern producers, Saudi Arabia included. Third, despite the relative honeymoon in Sino–US relations in the 1980s, the Chinese were still suspicious about the perceived US control of Saudi Arabia and its potential of blocking oil supplies to China. Fourth, the Chinese capacity of refining the sour Saudi crude oil was limited so that increased import of Saudi oil required huge investments in new refining technologies whilst less expensive oil supply alternatives still existed. Finally, Beijing's oil import strategy may have been determined by a long-term outlook to use whatever short-term and diminishing oil resources were available overseas, before it turned to Saudi crude oil and knowing that this will be available for a long time.

By the mid-1990s, both countries had indicated their interest in increasing the volume of oil transactions (Mei Dan, 2013: 210). When Ali bin-Ibrahim al-Nua'ymi, former Saudi minister of oil and mineral resources, visited China in December 1995 he revealed that China had begun importing Saudi oil two years earlier and added: "Saudi Arabia is ready and capable of providing China with all its oil needs" (Saudi Press Agency, 20 December 1995). This pledge was reiterated by the late King Abdullah when he visited China in January 2006 (the first stop in his Asian tour and the first visit by a Saudi head of state since diplomatic relations had been established in 1990) and again in November 2009 by al-Nu'aymi: "Let me be as explicit as possible: China can rely on Saudi Arabia to provide it with the oil it will need to continue its projected growth for the coming decades" (Al-Naimi, 2009: 5). Indeed, after 1998 Saudi oil exports to China picked up quickly, but its offer to supply all of China's oil needs was ignored – for the reasons mentioned earlier. Still, the small Saudi share in China's oil import (ranking sixth in 1999) grew gradually, reaching fourth in 2000, second in 2001 and first in 2002. This increase in Saudi crude oil import forced China to expand and modify its refineries to adapt to the high-sulphur Saudi crude (Mei Dan, 2013: 211–14).

Given the potential of Saudi oil supply, Beijing decided to upgrade a number of refineries, all designed to process Saudi crude. US companies contributed to these efforts. The first upgrade was the Fujian Integrated Refining and Ethylene Joint Venture Project, which went into operation in November 2009 in Quanzhou, Fujian Province. With investments of over US$4.9 billion, the complex is owned by the Fujian Petrochemical Company, itself a joint venture between Sinopec–Fujian Province (50 per cent), Saudi Aramco Sino Company (25 per cent) and ExxonMobil China Petroleum and Petrochemical Company (25 per cent). As China's first refining and petrochemicals venture with foreign participation, it is fully integrated with the Fujian Fuels and Marketing Joint Venture (SenMei), which operates around 750 downstream retail service stations and 18 product distribution terminals in Fujian and is owned by Sinopec (55 per cent), ExxonMobil and Aramco (22.5 per cent each) (*Saudi Gazette*, 12 November 2009; *China Daily*, 23 September 2010). Tripling the existing refinery capacity to 240,000 barrels per day (about one-quarter of China's current Saudi crude imports), the complex produces transportation fuel and other refined products.

In May 2010, a one-million-ton refining and ethylene complex plant was launched in Tianjin. This was a joint venture between Sinopec and Saudi Basic Industry Corporation (SABIC), called Sinopec–SABIC Tianjin Petrochemical Company (SSTPC), where both held a 50 per cent stake. The complex is the largest ethylene production base in China with investments of nearly US$5 billion. In April 2012, SSTPC laid the foundation for a polycarbonate production complex with a 260,000-tons annual capacity (*Saudi Gazette*, 12 July 2009; *Arab News*, 4 April 2012). Aramco has also been interested in investing in the Qingdao refinery (*Arab News*, 25 May 2007), and in March 2011 Aramco and China National Petroleum Corporation (CNPC) signed a Memorandum of Understanding (MOU) for the construction of a 200,000-barrels-per-day refinery in Yunnan Province. Aramco will provide crude oil that the refinery will process into low-sulphur petrol and diesel. The new refinery will be fed by a US$2 billion 1,100-km pipeline from Myanmar, shortcutting the Malacca Strait, and although the pipeline became operational in October 2013 (*Myanmar Times*, 27 October 2013), the tender notice for the refinery was only published on 12 December

2013 following public opposition that was overruled (Phillips, 2013; Saudi Arabia, 2011).

Directly related to imported Saudi oil and investments, this dramatic expansion of China's oil refining infrastructure represents mutual interests (Alexander's Gas and Oil Connections, 2009): the Chinese prepare for increased future imports of sour Saudi oil, and the Saudis want to commit China to long-term Saudi oil exports at the expense of other potential suppliers. D. Kirsch, director of market intelligence at PFC Energy in Washington, said:

> No single producer is really going to challenge the Saudi position in China in the long run. [...] China is a key market and Saudi [Arabia] doesn't want to lose market share there. It doesn't want to lose out to Russia and Iran. And that's part of why they will continue to push for long-term refinery deals in China.
>
> (Analysis, 2010)

Khalid al-Falih, Aramco chief executive, put it in more comprehensive terms: "We don't consider ourselves simply sellers of oil to China, but rather strategic partners whose many relationships in that important country are founded on mutual respect, independence and mutual benefit" (*The Wall Street Journal*, 2011b).

China's reliance on foreign oil companies not only for investments, technologies, and know-how but also for the supply of oil, apparently contradicts its fundamental drive to *reduce* its dependence on foreign suppliers and to secure oil by overseas production-sharing agreements; however, the East Asian economic crisis of 1997 and the decline of international oil prices undermined much of China's overseas ventures. Although oil supply from Chinese-controlled oilfields in countries such as Kazakhstan and Sudan may be secure, it offers smaller quantities that occasionally decline. Also, it appears that oil suppliers like Kazakhstan and Russia are reluctant to substitute their already institutionalized and prospering oil (and gas) relations with Europe for the China market. They cannot supply both (*China SignPost*, 21 February 2011). However, Russia's economic failures and its tense relations with Europe forced it to turn to China. Consequently, China's imports of Russian oil skyrocketed by 36 per cent in 2014 (Cunningham, 2015), although Saudi Arabia remained China's leading oil supplier, nearly 54 per cent more than Russia. It is substantially cheaper and simpler for the Chinese to buy Saudi crude, because its potential output is stable and production costs are low (and likely to remain so), than to invest in unpredictable oilfields, some of which are already exhausted and are located in unstable countries where exploration, construction, production and transportation costs are likely to be prohibitive. Some Chinese experts admit that 'oil bases' and pipelines may offer greater *security* but are less sound *economically* (Chen, 2002: 49).

Indeed, as the Chinese are laying the infrastructure to refine greater amounts of sour Saudi crude, they appear somewhat less concerned about diversifying their oil import sources and more reconciled to the Saudi option as China's leading oil supplier. In the long run, Beijing has no choice. Given its future oil needs, there is no supplier more reliable than Saudi Arabia – not only because of the potential of

Saudi oil but also and primarily because of its relative domestic stability and the Saudi capacity for balancing the global market.

> Riyadh is the only holder of significant spare oil output capacity, the main recourse to deal with any surprise major outage in global supply. [...] Saudi Arabia has enough spare capacity to replace the next largest exporter and is willing to adjust production up and down to meet the needs of the market.
>
> (Analysis, 2010)

China seeks stability. However, despite the 2011 upheavals in Libya disrupting its oil supply to China (in 2010 nearly 150,000 barrels per day, or about 3 per cent of China's oil imports), Unipec (Sinopec trading arm) still declined the Saudi offer to replace Libya's oil exports. Sinopec decided to rely on its own stockpiles – at least for the time being, "We are still studying it. After all, Libyan crudes are different from Saudi crudes in terms of pricing and quality. Currently we have no shortage" (*Reuters,* 28 February 2011).

Since the beginning of the 2000s, China systematically increased its crude oil import from Saudi Arabia. If by the late 1990s the Saudi share in China's oil import had been around 7–8 per cent, ten years later it increased to 20 per cent on an annual basis and in some months even more: 27 per cent (September 2008 and February 2009), 24 per cent (July 2009) and 23 per cent (November–December 2009). Crude oil import from Saudi Arabia in 2013 was over ten times more than in 2000 (Table 4.1), somewhat declining in 2014.

To somehow offset Saudi oil imports, Beijing has increased exports and investment, although there is still a growing deficit in China's trade with the Saudis – about one-quarter exports to three-quarters imports. Trade with China is growing fast. In 2010, Sino–Saudi trade was anticipated to grow to US$60 billion in 2015 (*China Daily,* 23 September 2010); in fact it reached over US$64 billion already in 2011 and over US$73 billion in 2012 (Table 4.2) and went slightly down in 2013 (US$72.2 billion) and in 2014 (US$71.3 billion). China is now Saudi Arabia's number one source of imports. As in other cases where they depend on commodities and energy import (for example, Sudan and Iran), the Chinese try to

Table 4.1 China's import of Saudi crude oil, 1996–2013 (rounded to thousand barrels per day and per cent share of China's total oil import)

Year	1996	1997	1998	1999	2000	2001	2002	2003	2004
Quantity	5	10	36	50	115	175	229	317	346
Share	1.0	1.4	6.6	6.8	8.2	14.6	16.4	17.3	14.0

Year	2005	2006	2007	2008	2009	2010	2011	2012	2013	2014
Quantity	445	479	529	730	843	896	1,005	1,009	1,170	997
Share	17.5	16.4	16.1	20.3	20.5	18.9	19.8	19.2	21.0	16.1

Source: Adapted from Almanac of China's Foreign Economic Relations and Trade, various years; *Reuters*, 2014a, 2015b.

Table 4.2 China's trade with Saudi Arabia, 2000–2012 (in billion US dollars)

Year	2000	2001	2002	2003	2004	2005	2006
Imports	1.95	2.72	3.43	5.17	7.52	12.23	15.08
Exports	1.14	1.35	1.67	2.15	2.77	3.82	5.06
Total	3.09	4.07	5.10	7.32	10.29	16.05	20.14
Deficit	0.81	1.37	1.76	3.02	4.75	8.41	10.02

Year	2007	2008	2009	2010	2011	2012	2013[1]
Imports	17.56	31.02	23.57	32.81	49.46	54.86	54.00
Exports	7.81	10.82	8.98	10.37	14.85	18.45	22.00
Total	25.37	41.84	32.55	43.18	64.32	73.31	76.00
Deficit	9.75	20.20	14.55	22.44	34.61	36.41	32.00

Source: Adapted from *China Statistical Yearbooks*, various years.

Note:
1 Preliminary.

create counter-dependencies by investing in these economies and exploiting their membership in international organizations to provide political services. Saudi Arabia is no exception.

China is a latecomer in the Saudi investment market. By early 2011, the value of licenses granted by the Saudi Arabian General Investment Authority to Chinese companies had reached US$8.5 billion, but their actual investment did not exceed US$400 million (Al-Maimouni, 2011). By late 2014, about 158 Chinese firms were working in Saudi Arabia in projects whose value was estimated at US$18 billion, covering construction, telecom, infrastructure and petrochemicals (Fattah, 2006; *People's Daily*, 30 November 2007; *Arab News*, 2013b), the most significant being the Yanbu refinery. The deal for the newly built Red Sea Refining Company (later renamed YASREF – Yanbu Aramco Sinopec Refinery) was signed on 16 March 2011 between Sinopec (37.5 per cent) and Aramco (62.5 per cent) after the withdrawal of the US Houston-based oil firm ConocoPhilips a year earlier. Trial runs started in September 2014, reaching full capacity (400,000 barrels per day) in February 2015 (*Gulf Daily News*, 17 January 2015), the Yanbu refinery is Sinopec's first investment abroad in oil *processing*, in addition to its investments abroad in oil *production*.

China has also been involved in gas exploration through the Sino–Saudi Gas Company, a joint venture established in 2004 between Sinopec (80 per cent) and Aramco (20 per cent). After seven drills and costs that exceeded the original projection of US$300 million, they did not find any flow of commercial value (*The Wall Street Journal*, 26 March 2008; *AMEinfo*, 18 June 2009). In November 2007, China Aluminium Corporation (Chinalco) bought a 40 per cent stake in the Binladin [Bin Laden] Metal Aluminium project in Jasan for US$1.2 billion (*Bloomberg*, 4 October 2007). The Mecca Metro, built by China Railway Construction Corporation at a cost of US$1.8 billion, was inaugurated in November 2010.

By late 2012, nearly 30,000 Chinese nationals were stationed in Saudi Arabia. In 2006, Saudi Arabia ranked tenth on China's foreign economic relations list (at US$1.04 billion), and in 2012 it was fifth (at US$4.62 billion), an increase of nearly 4.5 times in four years (National Bureau of Statistics of China, 2013: 253), reflecting China's efforts to gain a foothold in the country and to tie it to China in a long-term view (Al-Tamimi, 2014: 144–65).

Although Saudi Arabia is perceived as a rich country, its per-capita gross domestic product (GDP) is relatively small. The Saudis also have smaller oil reserves than they claim. Based on Saudi sources, US classified cables released by WikiLeaks indicate that Riyadh will be able to slightly – and slowly – raise oil production only until about 2025. This peak oil output may then remain steady for 15 years, followed by an inevitable decline, meaning that there will be less oil available for export. Moreover, given Saudi Arabia's economic and demographic growth over the years, energy demand has increased, resulting in more oil being diverted for domestic consumption (WikiLeaks, 2011). And the falling demand for its oil in the US and Europe, forced Saudi Arabia to find new markets in the east, notably in China.

The two countries have therefore become interdependent, based on common interests. By selling its oil, Riyadh does not do any favours for Beijing – possibly it is the other way around (Han, 2008; Zhang and Zhou, 2007). Whereas this increase in Sino–Saudi economic relations is directly related to oil, this policy may have been determined not only by economic considerations (theoretically China could still get its oil from other sources, see later) but also by political ones. It is doubtful that China's relations with Saudi Arabia reflect only oil interests. Much like in Mao's time, China worries about the role of the US in the world. In the past, Beijing tried to challenge the US by revolutionary means; now Beijing tries to challenge it by economic means, and much more effectively.

China and Saudi Oil: trilateral perspectives

Since the mid-1990s, Washington has been criticizing Beijing often for supporting and sustaining 'pariah' governments such as Sudan and Iran. These governments are top oil exporters to China and have been abusing human rights, underwriting global and regional terrorism, promoting Islamic radicalism, and occasionally persecuting religious and other minorities – and their own people. Although Saudi Arabia has been involved in *all* these evils one way or another, Washington has been careful to overlook them and, therefore, could not criticize Beijing for expanding oil relations with Riyadh. Also, although criticizing China for its aggressive acquisition of oil assets around the world by using government funds to overpay for these assets to outbid US and other competitors (Shichor, 2014: 63–4; Baltimore, 2006), Washington admits that in the case of Saudi Arabia, China does act in full accordance with international norms. The US does not appear to be concerned about the implications of China's intrusion into Saudi Arabia and the Persian Gulf. In fact, Washington may have facilitated China's intrusion when it encouraged Sino–Saudi collaboration against the Mujahidin in Afghanistan (Ali, 2005:

175–6) and turned a blind eye to the Sino–Saudi 1988 missile deal, of which it was undoubtedly aware (Shichor, 1989, 1991). In fact, recent official and unofficial publications on US–Saudi relations do not really mention China (for example, Prados, 2006; WikiLeaks, 2009; Blanchard, 2011).

Put differently, promoting economic *and* political relations with Saudi Arabia, despite or because of its close association with the US, may paradoxically shield China from US condemnation. Washington considers the Sino–Saudi oil connection not only legitimate but also expedient, and despite its non-intervention policy and rivalry with Washington, Beijing has been instrumental in promoting US policy in the Middle East (particularly in Iraq and Iran), which affected Saudi Arabia. The US can't block China's access into the Saudi oil market because, reluctantly and implicitly but still effectively, China allowed the US a free hand in its 1991 and 2003 offensives against Saddam Hussein. Forsaking its non-use of force stand and overlooking Washington's sidestepping the UN, China's policy indirectly helped protect and consolidate Saudi Arabia.

Saddam Hussein invaded Kuwait barely twelve days after diplomatic relations had been established between China and Saudi Arabia. To guard its assets in Kuwait and especially in Saudi Arabia, Washington suddenly needed Beijing to facilitate the military offensive against the Iraqi aggression under UN Security Council auspices, if not by voting for it at least by abstention. Usually rejecting the use of force in international politics particularly under UN auspices, Beijing could not support the US-orchestrated offensive, but at the same time it could not use its veto power. After all, China condemned Iraq's invasion as an act of aggression directed at its partners, Kuwait and now Saudi Arabia. Beijing, which had already voted for milder sanctions against Iraq, did not have a choice. Undoubtedly aware of Beijing's predicament, the US still preferred to conduct negotiations with the Chinese to convince them not to block the forthcoming offensive, which they had no intention of doing. Washington therefore agreed to pay (by lifting some of the post-Tiananmen sanctions against China imposed just a few weeks earlier) for the Chinese abstention it could have received for free (Solomon cited in Tucker, 2001: 453; Shichor, 2005: 191–228; Mann, 2000: 248–9). In retrospect, this deal amounted to a de facto recognition of the People's Republic of China as equal to the US – perhaps not (yet) as an actual power but already as a virtual one. US hegemony was further crippled, but still nothing to do with oil. This, however, changed in the case of Iran.

China's agreement to the US policy of imposing sanctions on Iran was likely related to Saudi oil. Reportedly, Washington tried to enlist Saudi support in encouraging Beijing to avoid blocking sanctions against Iran. Despite Riyadh's official denials, it appears that both Secretary of State Hilary Clinton and Secretary of Defence Robert Gates, while visiting Saudi Arabia in February and March 2010, asked King Abdullah to offer long-term oil supply guarantees to China in return for gaining its support for further sanctions against Iran (Zambelis, 2010: 5). The issue must also have been raised in Sino–Saudi discussions. Prince Saud al-Faisal, the Saudi Foreign Minister, sounded reassured when he said, "China is perfectly aware of the scope of its responsibilities and its obligations, including the position it holds on the international stage and as a permanent member of the [UN] Security Council" (*Saudi Gazette*, 2010a).

Indeed, during Chinese Foreign Minister Yang Jiechi's visit to Saudi Arabia in January 2010, the Saudis 'pressed him hard' to adopt a more active and cooperative role against Iran's nuclear threat, which they considered 'a critical security issue' for Riyadh. Deputy Foreign Minister Prince Torki al-Saud al-Kabir said that they understood China's concern that changing its Iran stand could affect the flow of Iranian oil to China and added that Saudi Arabia "is willing to take actions to address those concerns [i.e. sign long-term contracts], but must have Chinese cooperation in stopping Iran's development of nuclear weapons as a quid pro quo" (WikiLeaks, 2010). In other words, Riyadh offered Beijing all the oil it needs, but only in exchange for tangible Chinese actions to restrain Iran. There are no free lunches. Oil is being used by Saudi Arabia (probably with US blessings) to engage Beijing against Iran, forcing the Chinese to do what they hate most: take sides. This policy was moderately successful. China's crude import from Iran in 2013 was nearly 23 per cent lower than in 2011 before the sanctions had begun. Iran, who occasionally had been China's second crude oil supplier, was not amongst the five leading suppliers in 2013 and 2014. In January 2015 China's crude oil imports from Iran fell a further 17 per cent (*Reuters*, 2014b, 2015a).

Allowing China access to Saudi (and other) oil resources is not only an expedient US policy; to begin with, for all its 'special relationship' with Saudi Arabia, there are limits to what the US can do. Given the Saudi crucial dependence on US arms, Washington could have applied pressure on Riyadh not to sell oil – a highly strategic commodity – to China, but it has not. The Saudi's turn to Asia, and especially China, is based on long-term economic considerations partly related to the decline in the US import of Saudi oil. Moreover, the US has in the past tried to restrict Chinese access to oil by supporting Japan's oil interests in Russian East Siberia or to the Caspian and Gulf of Mexico oil. Washington must have realised (perhaps based on the oil embargo it had imposed on Japan that later led to the attack on Pearl Harbour) the negative implications of its policy, and also the principles of free trade and the fact that China today is much more powerful economically than Japan was and substantially more integrated in the US economy. Protecting the sea lanes that carry around 88 per cent of China's oil shipments, the US Navy could also intercept them – certainly in case of a conflict. This is, however, a virtual and potential threat that is highly unlikely because of the regional and international situation and the Sino–US economic interdependence.

Some believe that Sino–US oil competition in Saudi Arabia is equally unlikely. One reason is the so-called 'shale revolution' that will not only eliminate US oil imports from the Persian Gulf but will also lead to the US retreat from the Middle East. This shift is still questionable. The implications of the shale oil revolution are somewhat overstated. It is expected to contribute 15 per cent and up to one-third of US total liquid output by 2035 (British Petroleum, 2014: 30–4). By that time, the US population will have increased by 40–60 million, leading to an increase in oil demand. It is also anticipated that the US oil import will drop by 75 per cent, but an increase in US oil output does not necessarily imply it will import less oil because the US will need Persian Gulf oil for processed oil products that it exports (Nocera, 2013; *Oil & Energy Insider*, 2014). US oil companies, which occasionally pursue policies that do not conform to official government policy, are likely to stay in the Middle East. EIA data point out that in 2014 the US still imported an average of 43 per cent of its oil consumption, of which an average of over 25 per cent still came from the Persian Gulf. Persian Gulf oil is much cheaper than shale oil – and always will be. Put differently, the US is unlikely to leave the Middle East, at least for some time to come – and not only because of oil. No other power, certainly not China, can force it out. In fact, some believe the US has already pivoted back to the Middle East (Rachman, 2013).

But, while the US does not seem to be terribly concerned about Sino–Saudi oil (and other) relations, the Chinese are aware of the long and friendly US–Saudi historical relations. Beijing still suspects and realizes that Washington's hold on Riyadh, while perhaps eroded, is nevertheless strong enough primarily because of the US arms deliveries that China cannot match, now or in the future. The 1988 missile deal was an aberration whose military value was dubious, and since then Chinese arms sales to Saudi Arabia, which were mainly artillery, were tiny (valued at US$66 million in 2008–2009) compared to those of other suppliers, notably the US (SIPRI, 2015). Reports that China sold Saudi Arabia DF-21 solid-fuel, medium-range ballistic missiles "with CIA approval" (Keck, 2014; Stein, 2014) remain unconfirmed, as are reports that Saudi Arabia is considering buying the JF-17 Thunder fighter, a Sino–Pakistani co-production (*The Nation*, 25 January 2014). The Chinese, however, are still concerned that

> the US will probably treat the Sino–Saudi alliance as a threat to its interests in the Middle East. In addition to increasing competition with China for Saudi oil, the US also worries that the Sino–Saudi oil ties may lead to more active Chinese involvement in Middle Eastern issues.
>
> (Zhang and Zhou, 2007: 38; author's translation)

Although the prospects of greater Chinese activism in the Middle East are rather small, Riyadh (and perhaps Washington as well) wish that Beijing would become more involved in the region – on their side of course.

However, China will not turn against Iran. Also, China cannot and does not want to replace the US as Saudi Arabia's political-strategic guardian, let alone as the regional hegemon. Given the long-term US–Saudi relations, Beijing cannot and, more importantly, would not step in to fill the void if and when created by the so-called 'America's withdrawal from the Middle East'. China is a powerful and growing economic power that does not want to undertake regional or global responsibilities that might undermine its relations with other nations, least of all with the US. Indirectly, by using economic means Beijing is gradually and primarily eroding US traditional spheres of influence, first and foremost in East Asia but also in West Asia. This does not mean, however, that China would welcome a US withdrawal from the Middle East (Friedman, 2014).

Because Sino–US competition for oil resources is likely to intensify, China and the US will have to regulate their respective demands for crude oil in the Persian Gulf region and in Saudi Arabia. Shale oil (in the US and perhaps in China as well) may improve the situation somewhat – and for a while – but oil scarcity will increase (Qasem, 2010a). Oil substitutes and renewable resources will not replace oil for long, and China's demand for oil, mainly for transportation, will grow further. China and the US are already the leading oil consumers, and their GDPs and populations, especially China's, are also bound to grow. With oil supplying around 18 per cent of its energy consumption, China is relatively less dependent on oil than the US at 38 per cent; however, as China's economy grows, the *amounts* of oil (still at around 18 per cent of total energy) will grow enormously until 2035 (British Petroleum, 2014: 24, 26, 38). As China's economy continues to grow, even at a slower but still respectable pace, and given the anticipated stagnation or a two per cent decline of domestic oil output, the share of import in China's oil consumption will accelerate (at least to 76 per cent by 2035, if not before). With oil reserves in many countries diminishing, most of the future global oil supply potential will concentrate in the Persian Gulf. Although "Saudi Arabia alone will account for almost half of all Gulf production" (US Government, 2008: 41–2), its domestic consumption is growing fast, from 610,000 barrels per day in 1980 to 25 million in 2013 – nearly five times – (Saudi Arabia, 2015; US EIA, 2014), and its crude oil reserves may have been overstated. Less oil means tougher and more expensive competition.

Conclusion

Looking ahead, Beijing should be concerned less about US competition for Saudi oil and more about Saudi Arabia's political and economic future. A critical issue is leadership succession. Born in 1924 and coming to the throne in 2005, King Abdullah died on January 23, 2015. His successor, King Salman, will be 80 years old in December 2015 and is in poor health. Most of the future successors are also old and ill and the younger generation is inexperienced and basically pro-American. A change in government, a democratic revolution or radical Islamic subversion could undermine Beijing's long-term partnership with Saudi Arabia (Bi and Wang, 2010). To ensure the continuity of its good relations with Beijing, Riyadh sent Crown Prince Salman Bin Abdulaziz al-Saud to Beijing in mid-March 2014.

Suspicious about the recent unrest in the Middle East, Beijing points out that the so-called 'Arab Spring' has led less to democracy and more to Arab–Islamic radicalism. Although Saudi Arabia was not (yet?) part of the Arab Spring, it has been or may still be affected (Xu, 2008: 5–7; Chen and Xu, 2008: 95–6). Beijing is also well aware of the personal and ideological links between Saudi Arabia and al-Qaeda. Having frustrated Saudi attempts in the 1990s to interfere in China's Islamic affairs, the Chinese are vigilant about Islam in general and about what they call *Wahhabiyya* in particular. Interpreted as Islamic radicalism and fundamentalism, *Wahhabiyya* is the dominant form of Islam in Saudi Arabia. Although Chinese academics differentiate between Saudi Arabia and Islamic radicalism and terrorism in most publications (Lu, 2004; Qian, 2002), they have been nervously watching the penetration of Islamic radicalism (*wahabizhuyi*) into Xinjiang via Central Asia. Chinese scholars (and policymakers) instinctively associate terrorism with Islam (Ma, 2004). Following September 11, China denied visas to Saudi (and other Middle East) nationals, but the restrictions were gradually lifted (Teitelbaum, 2000; Trofimov, 2007). Similarly, Beijing must be aware of terrorist (primarily al-Qaeda) threats to Saudi energy infrastructure and installations that could disrupt future oil supply. Beijing can do little about these threats, but Washington can and does (Cordesman, 2009; Scheuer *et al.*, 2006; *The Wall Street Journal*, 2011a; WikiLeaks, 2008), creating another incentive for Sino–US cooperation in Saudi Arabia in the future.

Although the Chinese cannot offer Saudi Arabia military protection or even active diplomatic and political support, as does the US, they have two advantages over the US and other rivals and competitors in the region. One is what is known as the Chinese Model. Hardly applicable to most, if not all, developing countries, the Model is still attractive because it demonstrates that brisk economic growth is not incompatible with political authoritarianism. Strangely enough, this aspect of Saudi rule appears acceptable to the US, which has consistently failed to criticize ongoing Saudi abuse of human rights, personal freedoms and democratic principles. Nor does China insist on any political conditions, let alone democratisation, as a basis for bilateral relations other than a diplomatic break with Taiwan. The other Chinese advantage relates to its Middle East record: "China's friendship comes risk-free. Unlike the US, which has long been perceived as pro-Israel and anti-Arab, often leaving the Saudi family at odds with its own society and with other Arab governments, China's record in the region is untarnished" (Qasem, 2010b). Some Saudis view China as a counterbalance to the US that has abandoned and betrayed its friends in the Middle East, such as President Mubarak of Egypt and the Syrian people butchered by Assad (Friedman, 2014).

At the same time, Saudi Arabia should be concerned about its future oil sales to China because of Iraq's emergence as a major oil supplier. China is already a dominant player in Iraq's oil sector. Iraqi sweet crude is not only lighter than Saudi crude but also cheaper to produce. Indeed, Chinese oil imports from Saudi Arabia in 2014 declined to 997,000 barrels per day (from 1.17 million barrels per day in 2013), but oil imports from Iraq, that had been expected to jump dramatically in 2014, reached only 566,387 barrels per day in 2014, making Iraq China's fifth oil supplier (9.3 per cent, up from 8.0 per cent in 2013). Still, in January 2015 Iraq became China's second oil supplier, following Saudi Arabia (Bloomberg, 2015). Between 2005 and 2013, the Chinese invested in Iraq's oil sector twice as much as they invested in Saudi oil (US$14 billion compared to US$6.8 billion), or 97 per cent of their total investments in Iraq (compared to 39 per cent in Saudi Arabia). Chinese investments in the oil sector of at least ten other countries, including the US (US$15.6 billion) were higher, occasionally considerably higher, than in Saudi Arabia (*China Global Investment Tracker*, 2014; Scissors, 2013). Yet the security crisis in Iraq, and doubts about the quality of its crude oil have led to a decline in Chinese imports.

Riyadh must also be concerned about China's non-intervention principle. Entitled "Power and Prejudice", a Saudi editorial harshly criticized Beijing's principle that no outsider has a right to interfere in China's internal affairs:

> Whenever there is adverse comment about human rights abuses in the country or about Tibet or the subjugation of the Uyghurs in Xinjiang Province, Beijing rushes to pontificate about sovereignty, internal affairs and foreign interference...Beijing should realize that its perpetual complaints about interference only damage its own image. They make it look both arrogant and immature. No other state acts so pompously in such circumstances...China is now an economic superpower, and increasingly a political and military [one] as well. Criticism goes with the territory. It cannot be a major player on the international stage and expect immunity from censure.
>
> (*Arab News*, 7 April 2011)

In sum, China's increasing oil acquisitions from Saudi Arabia do not necessarily contradict US interests. Indirectly and paradoxically, Beijing contributes not only to Saudi Arabia's defence, but also to sustaining the US military industrial complex, or even to the US protection of sea lanes. In 2013, the cost of China's oil imports from Saudi Arabia was over US$42.3 billion, easily subsidizing Saudi Arabia's entire arms acquisitions from 2008 to 2012 (SIPRI, 2015) supporting US defence production. At the same time, these triangular relations still reflect trilateral suspicions, as well as understandings. Beijing remains suspicious about Washington, but expects it to go on protecting Saudi Arabia. China is also upset about Riyadh, which continues to block further investments in the oil sector but can do little about it. Washington remains suspicious about Beijing's motivations in the Middle East and the Persian Gulf but respects its right to acquire crude oil. And Riyadh remains suspicious about the US behaviour in the region, turning its back on Mubarak; lifting the sanctions imposed on Iran to stop its nuclear programme for meaningless words; and opting for a Chinese-style non-intervention policy in the Syrian civil war (Shalom and Guzansky, 2014; *Reuters*, 22 October 2013; *New York Times*, 23 October 2013). Is Saudi Arabia next in line? There is, however, little the Saudis can do about it, not to say that ultimately the US is more important for them than China. In the triangular oil relations, the US, China and Saudi Arabia are stuck in three-way interdependencies that involve both positive and negative aspects.

References

Al-Maimouni, M. (2011) Govt support needed to facilitate Chinese investment in Kingdom. *Saudi Gazette*, 24 January.

Al-Naimi, A.I. (2009) Energy for a new Asian century. Overseas Exchange Centre, Peking University, 13 November: 5.

Al-Tamimi, N.M. (2014) *China–Saudi Arabia Relations, 1990–2012: Marriage of Convenience or Strategic Alliance?* London: Routledge.

Alexander's Gas and Oil Connections (2009) China's refinery growth to outpace oil demand. Available online at: www.gasandoil.com/goc/company/cns90452. htm (accessed 9 April 2015).

Ali, S.M. (2005) *US–China Cold War Collaboration, 1971–1989*. London: Routledge.

AMEinfo (2009) *Sino–Saudi gas drills final exploratory well*. Available online at: www. gasandoil.com/goc/company/cns90452.htm (accessed 9 April 2015).

Analysis (2010) Saudi Arabia hosts Chu but woos China. *Reuters News*, 19 February. Available online at: www.reuters.com/article/2010/02/19/saudi-usa-asia-idAFLDE61701Q20100219 (accessed 21 March 2015).

Arab News (2007) Saudi Aramco, Sinopec set to seal Qingdao refinery deal. *Arab News*, 25 May.

Arab News (2011) Translation in MEMRI (The Middle East Media Research Institute). *Arab News Special Dispatch* 3771, 19 April 2011.

Arab News (2012) Sinopec, SABIC form joint venture in Tianjin. *Arab News*, 4 April.

Arab News (2013a) China set to become world's biggest net oil importer. *Arab News*, 13 August.

Arab News (2013b) Saudi–China trade surged to SR273.7 billion in 2012. *Arab News*, 21 November.

Baltimore, C. (2006) US–China talks progress. *Reuters*, 19 December.

Beijing Daxue (1985) Beijing daxue yafei yanjiusuo, xiya yanjiusho [Beijing University, Afro-Asian Research Institute, West Asia Office] (1985) *Shiyou wangguo, shate alabo* [Saudi Arabia: the Oil Kingdom]. Beijing: Beijing daxue chubanshe.

Bi, J.K. and Wang, Y.F. (2010) Shate Alabo wangwei jicheng wenti chulun [Discussion of Saudi Arabia's Royal Succession Problem]. *Xiya Feizhou* [West Asia and Africa] 2: 31–7.

Blanchard, C.M. (2011) *Saudi Arabia: Background and US Relations*. Washington, DC: Congressional Research Service.

Blas, J. (2013) China Becomes World's Top Oil Importer. *Financial Times*, 4 March.

Bloomberg (2007) MMC, Chalco plan $3 billion Saudi aluminum project. *Bloomberg News*, 4 October.

Bloomberg (2015) Iraq takes second spot among China oil sellers as Russia cedes. *Bloomberg News*, 27 February.

British Petroleum (2014) *BP Energy Outlook 2035*, January. Available online at: Energy_ Outlook_2035_booklet.pdf (accessed 9 April 2015).

Bronson, R. (2006) *Thicker than oil: America's uneasy partnership with Saudi Arabia*. New York, NY: Oxford University Press.

Chen, M. (2002) China's dependence on Middle East oil. In: *The Middle East and East Asia: Political Transition, Economic Reform and Energy Security*. Kyoto: Doshisha University, 49.

Chen, L.M. and Xu, X.D. (2008) Shate shixian minzhu zhengzhi de kenengxing yu jianjin xing [The progress and possibility of Saudi Arabia to achieve political democracy]. *Heilongjiang jiaoyu xueyuan xuebao* [Heilongjiang College of Education Journal], 27 (3): 95–6.

China Daily (2010) A year of milestones in Saudi Aramco's relations with China. *China Daily*, 23 September.

China Global Investment Tracker (2014) American Enterprise and Heritage Foundation. Available at: www.heritage.org/research/projects/china-global-investment-tracker-interactive-map (accessed 9 April 2015).

China SignPost (2011) Twilight in the tundra: Russian and Kazakh oil production cannot keep up with China's rising demand. *China SignPost*, 21 February.

Cordesman, A.H. (2009) *Saudi Arabia: National Security in a Troubled Region*. New York, NY: Praeger.

Cunningham, N. (2015) Russia and China growing energy relationship, 28 January. Available online at: http://oilprice.com/Energy/Energy-General/Russia-and-Chinas-Growing-Energy-Relationship.html (accessed 9 April 2015).

Emirates Business (2010) Saudi oil exports to US lowest in 21 years. *Emirates Business*, 18 February.

Fattah, H.M. (2006) Avoiding political talk, Saudis and Chinese build trade. *New York Times*, 23 April.

Financial Times (2010) China taps more Saudi crude than US. *Financial Times*, 21 February.

Friedman, B. (2014) Saudi alternatives to US power in the Middle East. *Tel Aviv Notes*, 8 (2): 1–7.

Gulf Daily News (2015) New Aramco-Sinopec refinery set to hit full capacity. *Gulf Daily News*, 17 January. Available at: www.gulf-daily-news.com/NewsDetails.aspx?-storyid=394083 (accessed 9 April 2015).

Han, J.W. (2008) Xiandaihua de beihua: dui shate ziyuan yifu xing jingji de fansi [The Paradox of Modernization: Reflections on the Resource-Dependence Model of the Saudi Economy]. *Lishi Jiaoxue* [History Teaching], 16: 72–7.

Jaffe, A.M. and Lewis, S.W. (2002) Beijing's oil diplomacy. *Survival*, 44 (1): 115.

Keck, Z. (2014) China secretly Sold Saudi Arabia DF-21 Missiles with CIA Approval. *The Diplomat*, 31 January.

Lu, Y.X. (2004) Qiantan wahabi jiao yu xiandai shate wangguo zhengquan de hudong [Interaction between Wahhabi Religion and the Political System of Saudi Arabia]. *Shaanxi jiaoyu xueyuan xuebao* [Journal of Shaanxi Institute of Education], 20 (2): 47–50.

Ma, T. (2004) Yisilan shiye lide kongbuzhuyi [Terrorism in an Islamic Perspective]. *Journal of the Second Northwest University for Nationalities*, 3: 96–9.

Mann, J. (2000) *About face: a history of America's curious relationship with China from Nixon to Clinton*. New York, NY: Vintage.

Mei Dan, M. (2013) Muddling Through with Chinese Characteristics: Beijing's Energy Policy and Diplomacy in West Asia and North Africa. Unpublished doctoral dissertation, Science Po, Paris.

Myanmar Times (2013) Controversial pipeline now fully operational. *Myanmar Times*, 27 October.

National Bureau of Statistics of China (2013) *China Statistical Yearbook 2013*. Beijing: National Bureau of Statistics of China.

New York Times (2013) Criticism of United States' mideast policy increasingly comes from allies. *New York Times*, 23 October.

Nocera, J. (2013) Fracking's Achilles' Heel. *New York Times*, November 18.

Oil & Energy Insider (2014) US oil exports; what goes unnoticed. January 2014.

People's Daily (2007) Chinese chemical giant tenders winning bid for Saudi phosphorous project. *People's Daily*, 30 November.

Phillips, T. (2013) Chinese protesters take to the streets in Kunming over plants plans. *The Telegraph*, 17 May.

Prados, A.B. (2006) *Saudi Arabia: current issues and US relations*. Washington, DC: Congressional Research Service.

Qasem, I. (2010a) *Resource scarcity in the 21st century: conflict or cooperation*. The Hague Centre for Strategic Studies.

Qasem, I. (2010b) *A Sino–Saudi axis on the horizon?* The Hague Centre for Strategic Studies. Available at: http://hcss.nl/news/a-sino-saudi-axis-on-the-horizon/7/.

Qian, X.W. (2002) Shate de wahabizhuyi [Saudi Arabia's Wahhabiyya]. *Alabo shijie* [Journal of Arab World], 3: 36–40.

Rachman, G. (2013) The year the US pivoted back to the Middle East. *Financial Times*, 23 December.

Reuters (2011) China's Unipec yet to take up Saudi offer for more oil. Reuters, 28 February. Available online at: www.reuters.com/article/2011/02/28/china-crude-libya-idAFTOE 71R05Z20110228 (accessed 21 May 2015).

Reuters (2013) Saudi Arabia warns of shift away from US over Syria, Iran. *Reuters,* 22 October.

Reuters (2014a) Saudi to keep 2014 China crude contract volumes steady. *Reuters,* 14 January.

Reuters (2014b) China's Iranian crude imports drop 2.2 pct in 2013. *Reuters,* 21 January.

Reuters (2015a) China's Iran crude oil imports down 17 percent on year in January. *Reuters,* 27 February.

Reuters (2015b) Almanac of China's Foreign Economic Relations and Trade (Beijing: Zhongguo zecheng jingji chubanshe, various years); 'Record high Chinese imports of Russian oil in 2014'. Available online at: http://rt.com/business/225687-saudi-oil-china-russia/ (accessed 21 March 2015).

Saudi Arabia (2011) *The Economist: Intelligence Unit*, 6 April.

Saudi Arabia (2015) Saudi Arabia crude oil consumption by year. Available online at: www.indexmundi.com/energy.aspx?country=sa&product=oil&graph=cosumption (accessed 9 April 2015).

Saudi Gazette (2009) Aramco, ExxonMobil, Sinopec JV petroleum plant fully operational. *Saudi Gazette*, 12 November.

Saudi Gazette (2010a) China 'knows its duties' in Iran nuke tussle: Saud. *Saudi Gazette*, 17 March.

Saudi Gazette (2010b) Editorial, Sino–Arab cooperation. *Saudi Gazette*, 15 May.

Saudi Press Agency (SPA) (1995) 20 December 1995, in FBIS-CHI, 20 December 1995: 10–11.

Scheuer, M., Ulph, S. and Daly, J.C.K. (2006) *Saudi Arabian Oil Facilities: The Achilles Heel of the Western Economy*. Washington, DC: The Jamestown Foundation.

Scissors, D. (2013) China's global investment rises: the U.S. should focus on competition. Baxkrounder #2757 (The Heritage Foundation, January 9). Available online at: www. heritage.org/resaerach/reports/2013/01/chinas-global-investment-rises-the-us-should-focus-on- competition (accessed 9 April 2015).

Shalom, Z. and Guzansky, Y. (2014) US–Saudi relations: on the verge of a crisis? *INSS Insight* No. 504. Tel Aviv: The Institute for National Security Studies.

Shichor, Y. (1989) *East wind over Arabia: origins and implications of the Sino–Saudi missile deal*. Berkeley, CA: Center for Chinese Studies, Institute of East Asian Studies, University of California.

Shichor, Y. (1991) *A multiple hit: China's missiles sale to Saudi Arabia*. Kaohsiung: Sun Yat-sen Centre for Policy Studies, National Sun Yat-sen University.

Shichor, Y. (2005) Decision-making in triplicate: China and the three Iraqi wars. In: A. Scobell and L. Wortzel eds. *Chinese National Security Decision-Making under Stress*. Carlisle, PA: Strategic Studies Institute, 191–228.

Shichor, Y. (2014) Regulation or strangulation: US–China–India competition over crude oil. In: Z.Q. Zhu [chief ed.] and J. Li J [volume ed.] *Globalization, Development, and*

Security in Asia. Vol. 3: Political Economy of Energy. Hackensack, NJ: World Scientific Publication Company, 41–68.

SIPRI (Stockholm International Peace Research Institute) (2015) Arms transfers database. Available at: http://armstrade.sipri.org/armstrade/html/export_values.php (accessed 9 April 2015).

Stein, J. (2014) The CIA was Saudi Arabia's personal shopper. *Newsweek*, 29 January.

Teitelbaum, J. (2000) *Holier than thou: Saudi Arabia's Islamic opposition*. Washington, DC: Washington Institute for Near East Policy.

The Nation (Pakistan) (2014) Saudi Arabia may buy JF-17 thunder jets. *The Nation (Pakistan)*, 25 January.

The Wall Street Journal (2008) Saudi Desert's Gas Mirage? *The Wall Street Journal*, 26 March.

The Wall Street Journal (2011a) China's Mideast Headache. *The Wall Street Journal*, 14 March.

The Wall Street Journal (2011b) Aramco agrees to supply oil to Chinese refinery. *The Wall Street Journal*, 20 March. Available online at: www. Wsj.com/articles/SB1000142405 27487044339045762120436695351456 (accessed 20 May 2015).

Trofimov, Y. (2007) *The siege of Mecca: the forgotten uprising in Islam's holiest shrine and the birth of al Qaeda*. New York, NY: Doubleday.

Tucker, N.B. ed. (2001) *China confidential: American diplomats and Sino–American relations 1945–1996*. New York, NY: Columbia University Press.

US EIA (2014) Energy Information Administration, Country Analysis Brief: Saudi Arabia. 10 September, page 3. Available online at: www.eia.gov/countries/analysisbriefs/Saudi_Arabia/saudi_arabia.pdf (accessed 9 April 2015).

US Government (2008) *Global Trends 2025: A Transformed World*. Washington, DC: US Government.

WikiLeaks (2008) SAG [Saudi Arabia Government] agrees to USG [US Government] steps to protect oil facilities. US Embassy Cable 08Riyadh1619 (Secret). 29 October.

WikiLeaks (2009) US Embassy in Riyadh Cable 09Riyadh1557 ('For official use only') 23 November.

WikiLeaks (2010) US Embassy in Riyadh cables 10Riyadh118, 26 January, and 10Riyadh123, 27 January, both 'Confidential'.

WikiLeaks (2011) US cables from the embassy in Riyadh (2007–2009). In: *The Guardian*, 8 February.

Xu, B.X. (2008) Minzhu hua langchao xiade shate alabo [Saudi Arabia under the Wave of Democratization]. *Kunming daxue xubao* [Journal of Kunming University], 19 (3): 5–7.

Yergin, D. (2006) What does 'energy security' really mean? *Wall Street Journal*, 11 July.

Zambelis, C. (2010) Shifting sands in the Gulf: the Iran calculus in China–Saudi Arabia relations. *China Brief* (The Jamestown Foundation), 10 (10): 5.

Zhang, L.Z. and Zhou, Y.H. (2007) Shixi Zhongguo yu shate alabo de sheyou hezuo [Analysing China-Saudi Arabia Oil Cooperation]. *Alabo shijie yanjiu* [Arab World Studies], 5: 34–9.

5 Resourcing Sino–Australian relations[1]

Nicholas Thomas

Introduction

Between 1980 and 2010, China has undergone a profound transformation. From an economy whose gross domestic product (GDP) was only worth US$147.3 billion in 1978, it grew to be worth US$1.414 trillion by 2003 (the start of the current resources-led boom) (Wang, 2004), and was valued at US$8.28 trillion by the end of 2012. Alongside this economic transformation, social modernisation has seen a rapidly expanding middle class and a significant decrease in the number living in poverty. As China's socio-economic development has accelerated, two intertwined processes have emerged. Down one track, the Chinese government has had to meet increasing domestic demand for goods and services. Down the other track, of capital and resources, with Chinese companies increasingly investing in markets around the world, even as international companies seek opportunities in China.

But this rapid rise has not been without controversy. Concerns have been raised as to what China's rise will mean for the future of the US-led international order. As Mearsheimer (2010: 2) stated, 'Australians should be worried about China's rise, because it is likely to lead to an intense security competition between China and the US, with considerable potential for war.... To put it bluntly: China cannot rise peacefully'.

One key focus of concern is China's involvement in the energy and resources sector. China needs more energy and other resources than it can produce domestically. Without these supplies, the capacity of the Chinese economy will shrink and the viability of the ruling Chinese Communist Party (CCP) will be called into question. The US (and other countries) also need these resources, but, as Zweig points out, everywhere that China goes it finds the US. In some cases, such as Australia, the US was there first, but Chinese investment has displaced the US as the major trading partner and largest investor in the resources sector.

The purpose of this chapter is to apply Zweig's model to the trilateral case of the US–Australia–China. Due to the resource sector, the Sino–Australian axis is becoming Australia's key foreign relationship, even as it seeks to accord primacy to the normative and strategic alliance with the US. The challenge of balancing these 'equal but different' relations is an on-going foreign policy test that is further complicated by the rebalancing of power and authority between the US and China.

To fulfil these goals, the chapter explores this triangular relationship, focusing particularly on Chinese investment in the Australian resources sector and how this deepening economic relationship is affecting US–Australia ties.

In doing so, this chapter argues that, as a major US ally in the Pacific region with a large resources sector important to China's development, Australia is caught between its strategic imperatives and its economic necessities. Neither the US nor China can meet both priorities. The US can guarantee Australia's strategic security but, in this period, is unable to deliver greater economic prosperity. Conversely, trade relations with China have underwritten an economic boom in Australia, but China remains ill equipped to become a strategic guarantor.

To understand this trilateral relationship, I will first assess Australian–China bilateral ties, and then review contemporary Australian–US relations and how these are affected by enhanced Sino–Australian linkages. This chapter concludes with an analysis of the utility of the triangular model to this case, as well as the current and likely future state of the relationship from an Australian perspective.

Australia and China (and the US)

The China–Australia relationship will dominate our world stance for the next half-century; it will define the nature and outcome of the great quadrilateral of Australia's most important ties abroad – relations with Indonesia, Japan, the West Pacific region and the US.

(Whitlam, 2002: 323–36)

Relations between China and Australia stretch back to the 1800s, when Chinese labourers came to work the goldfields in New South Wales and Victoria. In some cases, these workers arrived directly from China, while in other instances they came from Chinese communities in Southeast Asia. In the modern period, Australia has seen the relationship grow from one of reluctant aversion to a reliance on Chinese economic support for continued national wellbeing. The shift between these positions has not been linear, with economic and trade interests competing with political and strategic concerns. In the closing years of the 2010s, economic concerns melded into the national strategic discourse, creating a policy contradiction as the country tried to balance its economic needs (with China) against its politico-strategic interests (with the US). This section first reviews the significance of these economic and commercial ties before moving on to explore the bilateral political and strategic relationship. The analysis is framed against a backdrop of US–Australian relations, which are brought to the fore in the subsequent section.

Economic and commercial relations

No country other than China has a greater reason to look back with gratification and satisfaction over those 60 years of the remarkable development of China than Australia. We have been an extraordinary beneficiary of China's economic growth.

(Bob Hawke, former PM, 2009)

In the four decades (1972–2012) since diplomatic recognition, Sino–Australia trade ties have been transformed from a bilateral relationship, where the economic power once resided with Australia, to a relationship where China's economic pull has delinked Australia's economy from its traditional US/European base. Economic ties with China allowed Australia to weather the global financial crisis of 2008–10, and provided successive federal governments the capacity to pursue counter-cyclical growth policies. This delinking has deeply affected Australia's relationship with the US, splitting the country's economic security from its traditional security. Given the former's role in guaranteeing national security at the domestic level, this has also created a perception of a reorientation away from the US to China (see later).

In 1973, bilateral trade between Australia and China was only A$113 million but grew to A$122.2 billion by 2012 (*Xinhua*, 2013). Helped by a mining boom that started in 2003, China's share of Australia's two-way trade in goods and services increased rapidly, from just 5.1 per cent in 1999–2000 to 28.1 per cent in 2012–13 (Department of Foreign Affairs and Trade (DFAT), 2013a). This trade is largely complementary, with Australia exporting mainly primary resources and importing developed products from China. If not for the resource component of Australia's trade with China, levels of bilateral trade would not be so significant today and China's surplus would be far greater. Between 1999 and 2009, the content of the bilateral trade has shifted significantly, with Australia's natural resources comprising the vast majority of Australian exports, and light industrial goods making up the majority of China's exports. By 2012, the three largest component trades were iron ore, coal and gold. Trade in these three components accounted for 37.2 per cent of all exports. In contrast, the three largest imports in 2012 from China were computers, telecommunications equipment and furniture-related items (DFAT, 2013b). Looking forward, Australia's reserves of uranium are also going to become increasingly valuable, as are its holdings of natural gas and rare earths.

The increase in the trade in goods has had a positive impact on the Australian domestic economy. The combination of lower prices and cheaper parts in such industries as the automotive sector directly reduces household expenditure and assists local manufacturing to remain competitive by reducing unit costs (Allen Consulting Group, 2009). China's demand for Australian resources has also contributed to a reduction of overall levels of unemployment, even as the domestic focus on meeting China's energy needs causes distortions in the local economy – especially the manufacturing and housing sectors.

Bilateral investment

Two-way investments play an important role in the relationship. Even though trade is important, it is a short-term activity. Investment implies a longer-term commitment to a partner country, which must be based on greater trust and economic certainty. Although the early investment relationship was dominated by Australian outbound flows, China's rise has brought a wave of new investment. Chinese overseas direct investment (ODI) into Australia now significantly exceeds that of Australian ODI into China. As a DFAT (2014) briefing noted, while Australian investment into China stood at A$29,576 million in 2013, the foreign direct investment (FDI) component was only A$6,350 million. In contrast, overall Chinese investment into Australia stood at A$31,899 million, but Chinese inbound FDI was valued at A$20,832 million. Investment is, however, the one sector of the economic relationship that China does not dominate, with the US, the UK, Canada and India all larger sources of ODI (Foreign Investment Review Board, 2012: xv).

Despite this, Chinese investors in the resources sector have a major presence (including mineral exploration, development and processing), largely through mergers and acquisitions. Chinese companies have also established new companies with Australian investors to explore for deposits in Australia as well as overseas; however, efforts by Chinese state-owned enterprises (SOE) to buy Australian companies with large resources holdings, such as Chinalco's desired purchase of Rio Tinto, often run into government resistance, encapsulating many of the contemporary hopes and fears of Sino–Australian ties. Australia's public and private sectors saw Chinalco's attempt to buy a controlling stake in Rio Tinto as an effort to ensure lower prices for Chinese buyers. Even after Chinalco opted for a lower stake in the company and additional restrictions on its presence on Rio Tinto's internal committees (Chambers, 2013), concerns still abounded that its involvement would allow the company – and by extension the Chinese government – to exert a disproportionate influence upon Rio Tinto (Callick, 2009; Pratley, 2009). Given the importance of high-value commodities trade to the Australian economic model, this bid was rejected for running counter to the country's national interest and a possible (hostile) flexing of China's economic strength in the bilateral relationship (Coorey, 2009), but Chinese officials attributed the negative outcome to anti-Chinese sentiment stemming from 'political deliberation' and 'Cold War thinking' (Guo and Liu, 2009; Sainsbury and Chambers, 2009).

The Chinese perception that Australia is seeking to limit Chinese access to its resources has grown since 2009 when Treasurer Swan revised the threshold for mandatory review by the Foreign Investment Review Board (FIRB). The revised guidelines allowed for foreign investments under A$219 million (annually indexed) to be exempted from mandatory review, unless the investor is an SOE or a sovereign wealth fund (SWF). Similar changes to the Foreign Acquisitions and Takeovers Act (FATA) had earlier provided greater scope for screening foreign investments based on the 'potential voting power' of the investors. Although the FATA did not formally come into effect until early 2010, it was made retroactive to early 2009 when first announced by the Treasurer. But this new definition increased FIRB's latitude to block investments by foreign SOEs or SWFs based on perceptions of Australian national interest. Although no specific country, SOE or SWF was identified by the government as a trigger for these revisions, they were perceived as being targeted against Chinese interests.

This perception was confirmed in a WikiLeaks cable from September 2009 that quotes Patrick Colmer, the Head of the Foreign Investment Division of Treasury and a member of the FIRB, briefing US diplomats that the new rules 'were intended to pose new disincentives for larger scale Chinese investments' (Martin, 2011). However, a number of potentially contentious investments by

Chinese SOEs have been conditionally approved – rather than blocked – since the FATA amendments came into force,[2] suggesting that Australia recognises that Chinese trade and investment is necessary for its economic prosperity. Moreover, Australia is encouraging other resource-hungry countries, such as Japan and India, to invest locally whilst trying to assert a space for economic independence. Within this diversification strategy – and despite signing a Free Trade Agreement (FTA) with the US in 2005 – there is no mention of deepening Australia–US trade and investment, signalling that for the time being the US cannot vie for dominance with China in Australia's economy.

The government's perception that Australia needs to exercise caution in accepting Chinese investments, even as it benefits economically from the resulting development, is one shared by the Australian people. In a 2012 Lowy Institute poll, 70 per cent of respondents agreed with the statement that Chinese investment in the Australian resources sector was the major reason why Australia avoided the global financial crisis (Hanson, 2012). Yet, a majority of respondents also agreed that increases on current Chinese investment levels and intentions were not desirable, with 54 per cent agreeing that 'China is seeking to buy Australian mining and agricultural companies and these need to be kept in Australian hands', and 51 per cent agreeing that 'China has so much money to invest it could end up buying and controlling a lot of Australian companies' (Hanson, 2012). These percentages are higher than the baseline of 46 per cent of respondents who felt that the Australian government was allowing too much foreign investment, indicating that the general population perceives Chinese investors more negatively than most other investors.

The levels of social concern regarding Chinese investment in the Australian agricultural sector spiked in mid-2011, with the revelation that Chinese mining company, Shenhua, had been buying up prime farming land in New South Wales (NSW) to open a major open-cut coal mine called Watermark. This project raised significant local, state and federal concerns as to the appropriateness of the Chinese company's actions in turning productive farmland into a mining site. Reports noted that more than 80 per cent of submissions to the NSW government opposed the proposal (Herbert, 2013), even as the project manager suggested that the company's proposal was 'bulletproof' (Denniss, 2014). Shenhua's purchase of the land also raised concerns as to the effectiveness of the FIRB review mechanisms, particularly the national interest consideration of agricultural land being converted into a mining operation (Massola, 2011). The public awareness of the purchases by Shenhua led to a new Senate inquiry into foreign ownership of agricultural land, and a ruling from the FIRB that the land must be sold if it is not used for mining (Bita and Ryan, 2011).

As Chinese investment in the Australian resources sector continues to diversify into soft resources – such as land or water – this type of social opposition can be expected (ABC News, 2012). Indeed, these types of governmental and social concerns about the impact of Chinese trade and investment temper the desirability of closer economic ties between the two countries. This desire to control investment flows and their local impact led, in part, to Australia agreeing in 2005 to open negotiations with China for a Free Trade Agreement.

China–Australia Free Trade Agreement

With trade between Australia and China increasing rapidly – and given the additional benefits to both sides – in 2003 the two countries established a joint study group between the Chinese Ministry of Commerce and the Department of Foreign Affairs and Trade. In its 2005 report, the Study Group concluded that there were sufficient advantages for a China–Australia Free Trade Agreement (CAFTA) to be pursued. The 2005 report stated that 'an FTA could boost Australia's and China's real GDP in the order of US$18 billion (A$24.4 billion) and US$64 billion (RMB529.7 billion), respectively, over the period 2006–2015' (DFAT, 2005: 4); however, the report also observed that 'there are significant impediments to trade and investment between Australia and China,' (DFAT, 2005: 4) signalling that, although desirable, realising the CAFTA was never going to be easy.

At that same time, China had been lobbying Australia to define it as a 'market economy' to help it avoid World Trade Organization (WTO) anti-dumping penalties, while Australia had been seeking deeper access into the Chinese market and a mechanism to maximise the rising trade between the two states. Following a meeting between PM Howard and Chinese Premier Wen Jiabao in April 2005, a Memorandum of Understanding (MoU) was signed to launch negotiations.

Critically, the MoU recognised China's position as a 'market economy', placing Australia at odds with the position of the US (Calvert, 2004). The DFAT Secretary, Dr Ashton Calvert, claimed that Australia's recognition of China as a market economy did not mean that it was a market economy, but was intended to simply place the two countries on an equal footing for negotiating the FTA. Although this may have been Australia's intent (and may equally have been a clarification for the benefit of the US), it was not China's interpretation (Calvert, 2004).

The US has consistently refused to recognise China as a market economy. As Feldman (2010) observed, 'the US now knows it is holding something that China values highly, and yet is not worth very much, an enviable negotiating position'. The decision to award market economy status to China ahead of a bilateral FTA generated a high degree of opposition from the US. US Trade Representative Bob Zoellick directly lobbied Trade Minister Vaile on the issue and – highlighting the seriousness of the divide – US officials directly approached many top Australian companies operating in the US (Garnaut, 2004).

In considering the progress of the CAFTA negotiations to date against the backdrop of the overall trade relationship, it is clear that although there are strong technical grounds for an FTA to be concluded, the negotiations are hostage to the political strength of the relationship. In October 2013, the new Australian PM, Tony Abbott, tried to restart the CAFTA stating that Australia would take 'whatever we can get'. Despite this public relaxation of Australia's negotiation strategy, however, CAFTA has not progressed. Indeed, between the prime minister's announcement and the first half of 2014, further issues arose – specifically regarding Australian views as to the suitability of inbound Chinese investment. First, was the announcement that the new Coalition government would not reverse the ban on Huawei participating in the National Broadband Network due to national security concerns. As one Chinese diplomat stated, 'The decision on Huawei will no doubt have an impact on the free trade agreement with Australia' (Ireland, 2013).

Second, Australia sent diplomatic signals that, despite the prime minister's announcement to the contrary, it was not inclined to relax the FIRB screening thresholds for private and state-owned Chinese investments. Although the Chinese negotiators were seeking parity between the thresholds required for Chinese and US investments, the Australian government opposed such a move. In early 2014, PM Abbott led a large trade mission to Northeast Asia, with the intention of speeding up the FTA negotiations with Japan, South Korea and China. Although agreements were quickly reached with both Japan and South Korea, finalising an accord with China was more difficult. This was expected. As Callick (2010) noted earlier, given the political problems with the CAFTA and that China has already secured 'market economy' status from Australia, there is no great impetus for China to secure such a complex and domestically problematic FTA. Looking beyond mid-2014, any final-ized version of the CAFTA is only likely to bring Australia into a closer economi-cally strategic partnership with China to the detriment of other possibilities.

Even though Australia and China benefit from closer economic interests, the relationship is not comfortable. Undercutting the commercially led and the politi-cally driven aspects of these economic ties is the lack of complete confidence they have towards each other. Australia remains wary of getting too close to China, while China is concerned that its access to vital Australian resources is far from guaranteed. Over the last decade, the Australian public have also become more sensitised to the socio-economic impact of the relationship on local communities. This new variable to the relationship cannot be entirely predicted or controlled by either government, and so given these economic complexities, what are the polit-ical and strategic interests that bind the two countries together?

Political and strategic ties

> Australian foreign policy since 1972 has undergone an alteration in style and direction probably unprecedented in the experience of any sovereign state which had not been subjected to domestic revolution…changes had taken place in the area of foreign policy which had seemingly shifted Australia's alignment from that of one of the most conspicuously Western-aligned nations to one of the least.
>
> (Barclay, 1975: 1)

The first 25 years of the Sino–Australian relationship saw a rapid evolution in the bilateral acceptance by Australia's major political parties of the importance of China for Australia's future prosperity. All parties quickly jettisoned Cold War rhetoric to embrace China as a newly found old friend. Since 1972, this support has remained unaltered, despite the shock of Tiananmen and other issues relat-ing to human rights abuses and the opaque nature of the Chinese justice system. Labor PM Hawke (1983–91) went so far as to declare Australia to be China's '*lao pengyou*' [old friend]. Indeed, although the Hawke-led government was always sensitive to Australian–US relations, PM Hawke was not adverse to excluding the

US and including China in Australian foreign policy initiatives. For example, the Asia–Pacific Economic Cooperation (APEC) group was originally proposed by Australia without the US as a member. That said, towards the end of this period, Hawke's successor, Labor PM Keating (1991–96), believed that Australia needed to ensure that it could resist 'the pull of gravity from China' by diversifying its political and economic interests (Sheridan, 1997: 137).

This idea of diversification continued under Liberal PM Howard (1996–2007), who sought to balance China against Australia's more traditional partners, the US and the UK. Striking this balance proved difficult for the Howard and subsequent Rudd/Gillard administrations, which had to respond to a more complex policy environment necessitating balancing its national interests against that of China and the US.

In the early days, the Howard government tried to reassert the primacy of the relationship with the US. In March 1996, therefore, when China conducted missile tests in the Taiwan Straits to influence the presidential elections in Taiwan, leading the US to send two carrier groups close to Taiwan, the new Liberal government supported the US, causing a minor chill in the Sino–Australian ties.

The Howard government, however, quickly clarified the importance of bilateral ties with Beijing through a prime ministerial visit to China in March 1997 and the release of a foreign policy White Paper in April 1997, entitled *In The National Interest* (DFAT, 1997). This White Paper equated relations with China to ties with the US, Japan and Indonesia as Australia's four most important bilateral ties. Linking China's economic growth with its strategic importance, the document noted that 'China's economic growth, with attendant confidence and enhanced influence, will be the most important strategic development of the next fifteen years' (Sheridan, 1997: 27). But balancing China with these other relationships, particularly the US, was regarded as problematic. As the Defence White Paper (Department of Defence, 2000: 18) baldly stated, 'US–China relations may be a significant source of tension in the region in coming years. This could be important to Australia's security'.

The need for Australia to balance the two different but competing relationships can be seen in the visits to Australia by President Bush and Premier Hu Jintao in 2003. Both leaders were given the rare opportunity to address parliament: a repeated honour for US presidents but the first time a Chinese leader had been invited to do so; however, the substance of the two visits' speeches varied significantly. President Bush focused on global security and called on Australia to support freedom and democracy in Asia, the rule of law and peace in the Taiwan Straits, signalling that the US expected Australia's support in the event of an attack on Taiwan. In contrast, Premier Hu's address revolved around the increasing economic and commercial ties between the two countries. One report observed that the 'invitation to Bush to address Parliament was steeped in a shared democratic tradition and alliance history. The invitation to Hu was pure political pragmatism and commercial opportunism' (Kitney, 2003).

Such balancing was complicated. The difficulties in maintaining this balance was seen earlier in 1999 during a visit to Canberra by Richard Armitage (a foreign policy advisor to Governor Bush's presidential campaign) who sent the message that 'if Washington found itself in conflict with China over Taiwan it would expect Australia's support. If it didn't get that support, that would mean the end of the US–Australia alliance' (Sheridan, 2000). The result was a strategic reinforcement of Australia's commitment to the ANZUS (Australia, New Zealand, US) alliance, suggesting that the US believed that the Australian-end of the relationship had suffered a strategic drift because Australia now viewed economic security as equal to (if not more important than) national security as comprising wider alliance concerns. For its part, China warned Australia of 'very serious consequences' if it sided with the US in a conflict over Taiwan (Agence France Presse [AFP], 1999) and advocated 'tighter regional relations' to 'defuse international hegemonism by the US and cut into its influence' (Yan, 2000). Caught between the two, Australia recommitted to ANZUS and the 'One China' principle, hoping that a conflict would not break out any time soon (Tow and Hay, 2001).

Five years later, Foreign Minister Downer reignited the issue with a declaration that Australia was not bound to support the US in any conflict with China over Taiwan, despite the ANZUS agreement. As Foreign Minister Downer stated in 2004,

> Well, the ANZUS Treaty is a treaty which of course is symbolic of the Australian alliance relationship with the US, but arguing that the ANZUS Treaty is invoked in the event of one of our two countries, Australia or the US, being attacked, or in conflict over a third territory does not automatically invoke the ANZUS Treaty.
>
> (McDowall, 2009: 34)

The foreign minister went on to categorise Australia and China as developing a 'strategic' relationship akin to that between Australia and more traditionally allied states, thereby moving ties from economic and trade connections to deepen political and strategic relations.

Although a flurry of communications from the US led Australia to reaffirm the primacy of the US–Australian alliance, China's economic significance for Australia reconceptualised Sino–Australian ties and influenced US–Australian ties. Throughout the remainder of the Howard tenure, Australia split its foreign policy – emphasising the historical and contemporary relevance of the strategic relationship with the US but giving (at least) equal weight to economic security, which was increasingly influenced by China.

The election of Labor PM Rudd in 2007 initially led to a renewed embrace of China. At Peking University in 2008, the prime minister declared that Australia was China's '*zhengyou*' or 'true friend', who could – and would – speak frankly to China for the benefit of both states (*The Australian*, 2008). An early indication of these deeper political and strategic ties was the Rudd government's decision in 2008 to withdraw from the planned Quadrilateral Dialogue with the US, Japan and India, which China saw as a form of encirclement, citing Chinese opposition to the plan as one reason for its actions. Foreign Minister Smith also returned to the contentious theme of Australia upgrading its economic relationship into a strategic relationship, stating that we have an emerging relationship based for a long period of time on our early recognition of China as one nation, on the economic complimentarily between our two nations and today we see the relationship going to another level, a strategic dialogue.

(Smith, 2008)

On this occasion, however, this categorisation of the bilateral relationship did not generate the same backlash from the US, as had Foreign Minister Downer's statement four years previously.

Soon after, Australia began publicly to recognise China's threat potential to Australia. The Defence Update 2007 (the last issued by the Howard government) stated that the pace and scope of China's military modernisation could create 'misunderstandings and instability in the region' (Department of Defence, 2007: 19). The 2009 Australian Defence White Paper (the first by the new Labor government) further argued that China holds a military expansionist posture which 'appears potentially to be beyond the scope of what would be required for a conflict over Taiwan' (Department of Defence, 2009: 34).

Chinese officials labelled the 2009 White Paper as 'stupid' and 'dangerous'. Retired Rear Admiral Yang Yi suggested that 'Australia has spawned a new variation of the "China threat thesis" that could be emulated by Japan, South Korea and South-East Asian countries and encourage those nations to accelerate their own re-armament programs' (Garnaut and Nicholson, 2009). A Chinese academic, Shi Yinhong, also voiced opposition to the White Paper's contents, stating that

Kevin Rudd was supposed to be the Chinese-speaking Prime Minister who would provide a bridge between China and America. But now it looks like he wants to act on behalf of America against China. This is going to be hard to explain to the Chinese people.

(Garnaut and Nicholson, 2009)

The second political problem of 2009 involved the detention in China of Rio Tinto executive Stern Hu on the charge of 'stealing state secrets for foreign countries', (Shamim and Lin, 2009), which was understood as information over the [then] ongoing iron ore negotiations. PM Rudd faced a difficult choice. Having labelled Australia as China's *zhengyou*, he expected that Stern Hu would be returned to Australia. So rather than personally intervene, he warned China that 'the world will be watching' (Dodd, 2010). This approach was roundly condemned across Australia by the media, the opposition and some members of the prime minister's Labor party (Sheridan, 2010). The Chinese government was equally quick to denounce Australian criticism of Hu's arrest as 'interference in China's judicial sovereignty', with one official suggesting that the economic impact of Hu's commercial spying was equivalent to 10 per cent of Australia's GDP (Jiang, 2009).

However, even though the White Paper and the fallout from Hu's arrest lingered as negatives for Australia–China ties, the assumption of PM Gillard in June

2010 – to a certain extent – reset the bilateral relationship by taking Rudd, a major political irritant, out of the equation; however, PM Gillard quickly reaffirmed the primacy of the Australian–US relationship during her March 2011 visit to Washington, declaring that 'You have a friend in Australia. And you have an ally. And we know what that means. In both our countries, true friends stick together. …in both our countries, real mates talk straight' (*Sydney Morning Herald*, 2011). In contrast, PM Gillard's visit to China in April 2011 was more low-key with a focus on the domestic impact of the trip rather than on advancing bilateral ties. Although human rights problems and defence transparency were on the agenda, few substantive or controversial matters were raised, with the exception of more defence cooperation and joint exercises (AFP, 2011). As one senior Chinese scholar wrote, 'Australia's China policy is still evolving, but Gillard's diplomacy seems to be moving in a conservative direction' (Han, 2011).

This conservative tilt in Australia's trilateral relationship was reinforced by President Obama's visit to Australia in November 2011. President Obama and his senior officials hardened political rhetoric against China in the weeks leading up to the visit and announced a strategic rebalancing of the US to East Asia.

For Australia–US relations, the regional reorientation of the US was welcomed. The two countries agreed to a rotating deployment of 2,500 marines in the Northern Territory, a revision of the ANZUS Treaty to include a clause on cyber-attacks and a stronger push from the US on the Trans-Pacific Partnership (TPP). Although the first two outcomes generated the most headlines, the TPP announcement held the greatest potential to reorient the triangular relationship since the Whitlam government opened diplomatic relations with China in 1972 because a successful realisation of the grouping necessitated a stronger US involvement that would tie Australia's economic and strategic policies to one dominant partner.

In response to local concerns about Chinese opposition to these developments, PM Gillard said, 'I think it is well and truly possible for us in this growing region of the world to have an ally in the US and to have deep friendships in our region, including with China' (ABC News, 2011). This distinction between allies and friendships confirmed the US-orientation direction of the Gillard government observed earlier. It was also supported by the Australian people. A Lowy Institute poll taken in 2012 confirmed that an overwhelming majority of Australians (74 per cent) saw the US as their most important security partner over the next decade, compared with 10 per cent for China (Hanson, 2012). Interestingly, when asked to consider their first and second most important security partner, China was ranked third behind the US and Great Britain (Hanson, 2012). At the same time, 95 per cent of respondents to the poll agreed with the statement that China is already or will become the leading power in Asia, yet 52 per cent of these respondents were either very uncomfortable (15 per cent) or uncomfortable (37 per cent) with this development. This level of unease has been largely unchanged since the first Lowy polls in 2009, which may suggest that although the economic relationship with China is important for Australians, the dominant perception is that Australia needs to maintain its alliance with the US.

The Chinese overreaction to the Obama visit highlighted China's sensitivity to closer Australia–US ties. The Ministry of Foreign Affairs (MFA) felt that expanding military alliances was not in the interest of the region (Calmes, 2011), while the less diplomatic *Global Times* suggested that 'China cannot remain detached if Australia undermines its security' (*Global Times,* 2011). Interviews by the author in Beijing the following month with Chinese think tanks and academics showed that the real concern was the creation of a TPP with 40 per cent of global GDP, which excluded China.

The return of the federal Coalition government under PM Abbott in 2013 has not signalled an end to these tensions. If anything, the expanding military presence of China into areas considered vital to Australia's national security (such as the waters between Christmas Island and Indonesia) compel a firmer strategic commitment to the US. China's declaration of an Air Defence Identification Zone (ADIZ) in Northeast Asia and the comments by People's Liberation Army (PLA) officers of a need for a similar zone in the South China Sea have been directly rebuffed by Australian ministers and officials, generating more bilateral political tensions. Although China's actions can be seen as the natural expansion of an economically developing power, the critical issue is one of trust. Australia generally trusts the US; it does not trust China. This view is supported by Australian public opinion. As the 2010 BBC World Service Poll concluded with respect to China, 'while Australians leaned positive in 2009 (47% to 37%) they now lean negative (36% positive, 43% negative)' (BBC World Service, 2010). By 2013, these figures worsened to 36 per cent positive but 55 per cent negative (BBC World Service, 2013). One should not conclude that Australian–US relations do not face difficulties nor that, in certain circumstances, greater focus on China might present a favourable policy alternative to Australian governments. Nevertheless, how has Australia's relationship with the US developed against the backdrop of China?

Australia and the US (and China)

Australia's alliance structures have always turned outside the region. Before the Second World War, the UK was the main ally: it was Australia's largest market, its main source of defence and intelligence supplies, and its biggest supplier of immigrants. That robust exchange began to change after the war, with the rise of the US as the world's dominant power. After the UK ended its forward deployment of bases in Southeast Asia in the early 1970s, Australia had to choose between an independent defence policy or one based within an alliance structure. Even though successive defence policy statements throughout the 1980s and 1990s spoke about the need for independence, overarching this ideal was the reality of Australia's position as a US-ally in the Western Pacific.

The turn of the century saw a strengthening of the US–Australia strategic alliance. The terrorist attacks on 9/11 triggered the mutual defence clause in the ANZUS Treaty, the first time this had been invoked since the Treaty's ratification in 1951, and allowed 'relations with the US [dominate] Australian foreign affairs'

(Bell, 2007: 23). Australia was granted unprecedented access to US intelligence, but regional governments also viewed Australia in very much a subordinate relationship to the US, characterised in the regional media as a 'deputy sheriff'. This period also saw a strengthening of the economic relationship with the signing of the Australia–US Free Trade Agreement (AUSFTA) in 2005.

Although the AUSFTA improved the volume and value of Australia's trade relationship, US domestic interests prevented it from reaching its full potential. As the *New York Times* noted,

> the agreement sends a chilling message to the rest of the world. Even when dealing with an allied nation with similar living standards, the administration, under pressure from the Congress, has opted to continue coddling the sugar lobby, rather than dropping the most indefensible form of protectionism.
>
> (*New York Times*, 14 February 2004)

Indeed, between 2004 and 2009, Australia's trade deficit with the US grew from US$6.4 billion to US$11.6 billion (Quiggin, 2010), and by the end of 2011 it was in excess of US$23 billion (DFAT, 2012).

These developments were framed against a vibrant US policy debate over whether China was a strategic partner or a competitive threat to US regional and global interests. At the core of this debate was the key question as to how the US should engage China, and it was on this question that the Australian–US relationship evinced the starkest differences.

As noted earlier, since 1972 Australian policymakers have pursued a largely consistent policy of engagement with China. US policy, in contrast, has been 'heavily influenced by domestic interests and [an] often vicious and partisan debate' (Sutter, 2002: 353). Not surprisingly, different US opinions of China have led to policy disagreements with Australia. The starkest differences have come over strategic clarity – principally, Taiwan and Australia's position in the event of a cross-Straits conflict. As observed earlier, from the mid-1990s to the mid-2000s in particular, the US evinced a high degree of sensitivity whenever Australian policymakers strayed from the traditional role of a strong supportive ally, with the view in Washington being that the alliance was no longer as strong as it once was and needed 'reinvigorating' (ABC Lateline, 2001).

Tensions tended to be from the US to Australia and were framed within the context of alliance duties in a competitive world order. Although they rarely became public, these tensions were often characterised by a personal visit by a high-ranking US official and a subsequent reversal of Australian policy. A reverse of this typical situation was seen in March 2006 when – in response to a call by US National Security Advisor Rice for Australia and Japan to form a common position with the US against China's growing military spending and development – Foreign Minister Downer declared that Australia would not support any policy to contain China. This led to a backdown by the US in support of a policy favouring the greater socialisation of China in international institutions instead (Nicholson *et al.,* 2006).

These policy differences have also caused a reassessment of the bilateral US–Australia relationship in light of the presence of China '*a trois*'. One view is that the US will have to accept the new strategic reality of the importance of China to Australia (and other key US allies), but with the caveat that, in the event of a conflict or sanctions regime imposed on China, Australia would support the US (O'Hanlon and Fullilove 2009: 4). Critically, this view is predicated on Australia avoiding economic dependency with China, which, as the preceding analysis suggest, is not necessarily the case. A stronger interpretation of this view was put forward by Secretary Clinton who, in response to a question about whether or not Australia should reassess its alliance with the US in light of its China ties, stated that,

> our relationship is essential to both of us. That doesn't mean we won't have relationships with others, but it does mean that this will remain the core partnership…And it is important to recognise that just because you increase your trade with China or your diplomatic exchanges with China, China has a long way to go in demonstrating its interest in being – and its ability to become – a responsible stakeholder.
>
> (Hartcher, 2010)

A more equal view is that Australia can act as an interlocutor between the US and China. As Rosencrance (2006: 367) argued, 'Washington and Beijing will continue to have important common interests. This suggests that Australia should strive to support tripartite institutions that bring the three countries together and not be left with the now outdated bipolar pattern of the past'. In this formulation, Australia provides the 'cement' between the US and China. He (2011: 18) articulates a supporting view, arguing that Australia's accommodation of both powers militates against conflict in the triad, encouraging 'nonzero-sum questions, while blurring or undermining friend-enemy assumptions in traditional alliance politics'. There is insufficient evidence, however, to suggest that Australia is significant enough to either major power for this role to be accepted. As White (2013) observed, 'Australia's interest in such a deal is enormous, but Australia can do little to broker this kind of deal – certainly it has no role as an intermediary between America and China'.

This optimistic view can be contrasted against a more demanding position that suggests it is in Australia's own interests to strengthen the US strategic alliance because, ultimately, that is the only 'guarantor that Australians very much will enjoy their lifestyle' (Hartcher and Banham, 2005). As former US Deputy Secretary of State Armitage described,

> If I were Australian and I was sitting here and I was beginning to feel the tectonic plates move a bit by the ascension of China, the ascension of India, then I think I would opt on the side of 'maybe I'll just keep this security alliance a while longer'.
>
> (Hartcher and Banham, 2005)

Responding to this spectrum of views, which change over time and among administrations, is a unique challenge for Australia as it seeks to manage its end of the triangular relationship. It is also a challenge for the US because it takes into account the changing political and social orientations of its long-standing ally. As the 2013 BBC World Service Poll revealed, although the balance of Australian sentiment towards the US is positive it is not overwhelmingly so. Between 2012 and 2013, 'Australian views shifted from leaning positive in 2012 (50% vs 38%) to being divided this year (46% vs 42%)' (BBC World Service, 2013). Keeping both partners happy is clearly a key policy goal for successive Australian governments, but if the US and China become more confrontational then Australia may not have the luxury of choice.

Conclusion

What emerges from this analysis is a complex triangular relationship that has all three actors drawing on multiple levels of engagement to maximise their position – bilaterally, trilaterally, as well as in the wider multilateral environment. For Australia, the attempt to balance China and the US has led to a bifurcated foreign policy. On the one hand, Australia relies heavily upon the ongoing growth of China's economy and is increasingly allowing direct Chinese investment, albeit with heightened concerns over its own resource sovereignty. On the other hand, Australia is unwilling to relinquish its strategic and cultural ties with the US.

China's foreign policy towards Australia is multifaceted. China's critical need is to gain resources for the infrastructure commitments required for realising its modernisation. Australia's depth and diversity of resources, as well as its geographically contiguous location and the relative safety of the shipping lines, makes it a logical choice for Chinese economic and commercial attention. Yet this same attention gives it the capacity to pressure Australia's alliance with the US, potentially shifting the trilateral balance in China's favour.

Since the early 1970s, Australian political parties at the federal and state levels have accepted China as a power of growing economic and commercial significance. Economic and political support for this view has been underpinned by increasing support within Australian society. As China takes more and more Australian exports, more and more Australians come to rely on China for their livelihood. Yet, perversely, as the economic reliance on China increases, there are signs that a social acceptance threshold may be being reached, presenting a new challenge for the bilateral ties.

Most fundamentally, however, the US's re-engagement as an Asian power has been framed as a response to China's rise, and the potential threat this challenger may hold for the US. Even though the US's concerns have created a policy space within which Australia can strengthen its strategic relationship with its traditional ally, as well as collaborate with the US in the development of new economic partnerships (such as the TPP) that are US-led, strengthened strategic ties with the US have not come without cost to its relationship with China. In mid-2012, Chinese strategic officials and scholars warned that Australia was standing with its legs 'in two different boats', a perilous metaphor to describe the risks for Australia of 'deepening military ties with the US while retaining its economic dependency with China' (Garnaut, 2012). Even as Australia tries to balance its relations with each state, both push back against Australia's relationship with the other. The fact remains that the restrengthening of the US–Australia relations is as much a function of US domestic politics as continued economic ties with China are similarly a function of Australia's ability to meet China's domestic needs. Towards both countries, Australian domestic sentiment remains uncertain, unable to provide a clear public foundation for the further direction of the triangular relationship. Yet one point is clear, Australia remains the subordinate power; constricted in its ability to choose its own future, but being increasingly challenged by both powers to do so.

Notes

1 A different version of this case study first appeared in: Thomas, N., 2015. The economics of power transitions: Australia between China and the US. *Journal of Contemporary China,* 24 (95).
2 These include Hunan Valin's bid to acquire 17.6 per cent of Fortescue, the revised Minmetals bid for OzMinerals, and Anshan Iron and Steel Group Corporation's bid for 36.3 per cent of Gindalbie.

References

ABC Lateline, 2001. ANZUS alliance a sacred trust: Armitage. Transcript, 17 August.
ABC News, 2011. Obama visit to focus on military ties. *ABC News,* 17 November.
ABC News, 2012. Chinese company looks to invest in WA land. *ABC News,* 18 January.
Agence France Presse (AFP), 1999. China warns Australia not to side with US over Taiwan. *Agence France Presse,* 5 November.
Agence France Presse (AFP), 2011. Australia to boost military ties with China. *Agence France Presse,* 28 April.
Allen Consulting Group, 2009. The benefits to Australian households of trade with China. A report prepared for the Australia China Business Council. January. Available online at: www.allenconsult.com.au/publications/view.php?id=333 (accessed 15 August 2013).
Barclay, G., 1975. Problems in Australian Foreign Policy, July–December 1974. *Australian Journal of Politics and History,* 21 (1): 1–10.
BBC World Service, 2010. BBC World Service Poll. Global views of US improve while other countries decline. 18 April. Available online at: www.worldpublicopinion.org/pipa/pipa/pdf/apr10/BBCViews_Apr10_rpt.pdf (accessed 5 May 2014).
BBC World Service, 2013. BBC World Service Poll. Views of China and India slide while UK's ratings climb: global poll. 22 May. Available online at: www.worldpublicopinion.org/pipa/2013%20Country%20Rating%20Poll.pdf (accessed 5 May 2014).
Bell, R., 2007. Extreme allies: Australia and the USA. *In:* J. Cotton and J. Ravenhill eds. *Trading on Alliance Security.* Sydney: Oxford University Press, 23–52.
Bita, N. and Ryan, S., 2011. Shenhua hints at 'xenophobia'. *The Australian,* 7 July.
Callick, R., 2009. China and Rio in a torrid affair – Chinalco challenge. *The Australian,* 13 May.

Callick, R., 2010. Trade deal sails on a slow boat to China. *The Australian,* 25 May.

Calmes, J., 2011. A US marine base for Australia irritates China. *New York Times,* 16 November.

Calvert, A., 2004. Closing speech at the Australia China FTA Conference: Future directions. Sydney: August. Available online at: www.dfat.gov.au/news/speeches/Pages/closing-speech-at-the-australia-china-fta-conference-future-directions.aspx (accessed 12 December 2012).

Chambers, M., 2013. Chinalco trims Rio bond deal. *The Australian,* 22 May.

Coorey, P., 2009. Chinalco question not just about investment. *Sydney Morning Herald,* 18 May.

Denniss, R., 2014. Big risk for Liverpool Plains residents. *The Northern Daily Leader,* 25 January.

Department of Defence, 2000. Defence 2000: our future defence force. Canberra: Commonwealth of Australia.

Department of Defence, 2007. Australia's national security: a defence update 2007. Canberra: Commonwealth of Australia.

Department of Defence, 2009. Defending Australia in the Asia–Pacific century: force 2030. Canberra: Commonwealth of Australia.

Department of Foreign Affairs and Trade (DFAT), 1997. *In the National Interest: Australia's Foreign and Trade Policy White Paper.* Canberra: National Capital Printing.

Department of Foreign Affairs and Trade (DFAT), 2005. Australia–China Free Trade Agreement: joint feasibility study. Canberra: Department of Foreign Affairs and Trade. Available online at: www.dfat.gov.au/trade/agreements/chafta/Documents/feasibility_full.pdf (accessed 13 June 2012).

Department of Foreign Affairs and Trade (DFAT), 2012. Trade at a glance 2012. Available online at: www.dfat.gov.au/about-us/publications/Documents/trade-at-a-glance-2012.pdf (accessed 12 March 2013).

Department of Foreign Affairs and Trade (DFAT), 2013a. Australia's trade in goods and services 2012–13. Department of Foreign Affairs and Trade. Available online at: http://dfat.gov.au/about-us/publications/trade-investment/australias-trade-in-goods-and-services/Pages/australias-trade-in-goods-and-services-2012-13.aspx (accessed 4 February 2014).

Department of Foreign Affairs and Trade (DFAT), 2013b. Composition of Australia trade 2012. June. Canberra: Department of Foreign Affairs and Trade: Market Information and Research Section.

Department of Foreign Affairs and Trade (DFAT), 2014. People's Republic of China country brief. Available online at: www.dfat.gov.au/trade/resources/Documents/chin.pdf (accessed 5 July 2014).

Dodd, M., 2010. Kevin Rudd warns China on Stern Hu trial secrecy. *The Australian,* 19 March.

Feldman, E., 2010. China's status as a non-market economy. China–US Trade Law: Baker-Hostetler. 21 September. Available online at: www.chinaustradelawblog.com/2010/09/articles/trade-disputes/wto/chinas-status-as-a-non-market-economy (accessed 19 October 2013).

Foreign Investment Review Board, 2012. Annual Report 2010–11. Canberra: CanPrint Communications.

Garnaut, J., 2004. In trade, Australia looks to China. *Sydney Morning Herald,* 9 August.

Garnaut, J., 2012. China warns on US–Australia ties. *WA Today,* 7 June. Available online at: www.watoday.com.au/national/china-warns-on-usaustralian-ties-20120606-1zwp0.html (accessed 5 October 2013).

Garnaut, J. and B. Nicholson, 2009. Defence plan ruffles China; Exclusive – Beijing 'confused' over Rudd military strategy. *Sydney Morning Herald.* 1 May. Available at: www.smh.com.au/national/defence-plan-ruffles-the-chinese-20090430-aoy2.html (accessed 10 May 2014).

Global Times, 2011. Australia could be caught in Sino–US crossfire. *Global Times,* 16 November.

Guo, X., and G. Liu, 2009. Whom did Chinese firms "offend" in going global? *Xinhua Asia–Pacific Service,* 9 June.

Han, F., 2011. Gillard's first Asian tour. *The China Daily,* 27 April.

Hanson, F., 2012. Lowy Institute Poll 2012: Public Opinion and Foreign Policy, 5 June. Available online at: www.lowyinstitute.org/publications/lowy-institute-poll-2012-public-opinion-and-foreign-policy (accessed 8 May 2014).

Hartcher, P., 2010. Back America over China, Clinton urges. *The Age,* 9 November.

Hartcher, P. and Banham, C., 2005. Don't leave the field to China, US warned. *Sydney Morning Herald,* 19 August.

Hawke, B., 2009. Looking back on China's relations with Australia. East Asian Forum. 27 September. Available online at: www.eastasiaforum.org/2009/09/27/looking-back-on-chinas-relations-with-australia (accessed 3 August 2011).

He, B., 2011. Politics of Accommodation of the Rise of China: the case of Australia. *Journal of Contemporary China,* DOI:10.1080/10670564.2012.627666.

Herbert, L., 2013. Shenhua surprised by lack of EIS responses. *ABC Rural,* 29 May.

Ireland, J., 2013. Tony Abbott rules out change to Huawei ban. *The Sydney Morning Herald,* 1 November. Available online at: www.smh.com.au/federal-politics/political-news/tony-abbott-rules-out-change-to-huawei-ban-20131101-2wn0y.html#ixzz2vGBPkNNs (accessed 5 November 2013).

Jiang, R., 2009. "蒋汝勤：力拓案件折射出什么" [What the Rio Tinto case reflects], *Sina.com,* 8 August. Available online at: finance.sina.com.cn/review/yjfx/20090808/19546591085.shtml (accessed 7 August 2011).

Kitney, G., 2003. No red face in the house of soft welcome. *Sydney Morning Herald,* 25 October. Available at: www.smh.com.au/articles/2003/10/24/1066974316772.html?-from=storyrhs (accessed 10 August 2012).

Martin, P., 2011. Swan denies China targeted. *The Age,* 4 March. Available online at www.theage.com.au/national/swan-denies-china-targeted-20110303-1bggr.html#ixzz1zh-b2lylr (accessed 10 September 2011).

Massola, J., 2011. Chinese miner's land buyout passed national interest test, says Bill Shorten. *The Australian,* 29 June.

McDowall, R., 2009. Howard's long march: the strategic depiction of China in Howard Government Policy, 1996–2006. Canberra: ANU E-press, March. Available online at: epress.anu.edu.au/sdsc/hlm/mobile_devices/ar01.html (accessed 18 February 2010).

Mearsheimer, J., 2010. The gathering storm: China's challenge to US power in Asia. Fourth Annual Michael Hintze Lecture in International Security, 4 August. Sydney: University of Sydney.

New York Times, 2004. A triumph for big sugar. *New York Times,* 14 February.

Nicholson, B., Guerrara, O. and Forbes, M., 2006. Embrace China, Downer tells US. *The Age,* 16 March.

O'Hanlon, M. and Fullilove, M., 2009. Barack Obama, Kevin Rudd and the Alliance: American and Australian perspectives. Lowy Institute Perspectives. Sydney: August.

Pratley, N., 2009. Financial viewpoint: Rio can't disguise the stench of a rotten Chinese deal. *The Guardian,* 22 May.

Quiggin, J., 2010. Lessons from the Australia–US Free Trade Agreement. *Inside Story,* 22 November. Available online at: http://insidestory.org.au/lessons-from-the-australia-us-free-trade-agreement (accessed 15 December 2010).

Rosencrance, R., 2006. Australia, China and the US. *Australian Journal of International Affairs,* 60 (3): 364–8.

Sainsbury, M. and Chambers, M. 2009. Beijing fires up at anti-China 'prejudice'. *The Australian,* 11 June.

Shamim A. and Lin, L. 2009. Australia says no charges Laid against Rio Tinto's Hu (Update2). *Bloomberg,* 20 July.

Sheridan, G., 1997. *Tigers: leaders of the new Asia–Pacific.* Sydney: Allen & Unwin Press.

Sheridan, G., 2000. What if bluff and bluster turn to biff? *The Australian,* 10 March.

Sheridan, G., 2010. Rudd's approach to China and Stern Hu, a lesson in cowardice. *The Australian,* 20 March.

Smith, S., 2008. Joint Press Conference with Chinese Foreign Minister, 5 February. Available online at: www.foreignminister.gov.au/transcripts/2008/080205_jpc.html (accessed 23 January 2009).

Sutter, R., 2002. Thirty years of Australia–China relations: an American perspective. *Australian Journal of International Affairs,* 56 (3): 347–60.

Sydney Morning Herald, 2011. Julia Gillard's Speech to Congress. *Sydney Morning Herald,* 10 March.

The Australian, 2008. Beijing University speech by Australian Prime Minister Kevin Rudd. *The Australian,* 9 April.

Tow, W. and Hay, L., 2001. Australia, the US and a 'China growing strong': managing conflict avoidance. *Australian Journal of International Affairs,* 55 (1): 37–54.

Wang, H., 2004. Official: China's GDP growing 9.1% in 2003. *China Daily,* 20 January.

White, H., 2013. Australia's choice: will the land down under pick the US or China? *Foreign Affairs,* 4 September. Available online at: www.foreignaffairs.com/articles/139902/hugh-white/australias-choice (accessed 11 December 2013).

Whitlam, G., 2002. Sino–Australian diplomatic relations 1972–2002. *Australian Journal of International Affairs,* 56 (3): 323–36.

Xinhua, 2013. Australia eyes free trade deal with China in 2014. *Xinhua,* 4 November.

Yan, X., 2000. Best friends next door. *China Daily,* 7 March.

6 The true north – strong and full of energy

China's resource diplomacy and Canada–US relations

Jiang Wenran, David Zweig and Kang Siqin

As China became the world's second largest economy, its demand for energy as of also propelled it to become the world's biggest comprehensive energy consumer. Accompanying this process has been a sharp upward trend in Chinese foreign direct investment (FDI) focused on energy and other resources. China's quest for energy security in recent years has resulted in the active participation of Chinese national oil companies (NOCs) in energy markets around the world. Canada, with its extremely rich endowment of energy and other resources, has recently become a major target for Chinese NOCs seeking to purchase tangible assets and pursue joint venture deals.

Positioned as the third largest oil reserve country after Saudi Arabia and Venezuela, (US Energy Information Administration [EIA], 2014), Canada's energy sector, especially its unconventional oil sands, has attracted substantial domestic and foreign investment. Although US energy companies are by far the largest share of the more than US$130 billion that has been invested in Canada's oil sands, other international oil companies (IOCs) and investors from numerous countries have also been active in Canada's energy sector. Since late 2009, Beijing has added Canada to its FDI priority list, with US$15 billion worth of Chinese capital pouring into the energy-rich province of Alberta in the year 2010 alone (Liepert, 2011).

China's energy needs in a global context

This study seeks to evaluate the extent to which China's rise and its expanding energy ties with countries around the world has an impact on US ties with that third, energy-rich country, in this case, Canada. According to Zweig's 'triangular' framework, to the extent that China's efforts to purchase oil and energy companies in Canada, a very close ally of the US, is seen by Americans to challenge their energy security, one would anticipate some type of US response to these sales. In this case, Chinese state-owned enterprises (SOEs) are indeed entering a key neighbouring market through FDI, buying oil that is seen as an important source of the US's energy security. Moreover, Canadians are a somewhat conservative breed, with many intelligence officers and common citizens deeply worried about infiltration by companies that are believed to be working at the behest of the

Chinese Communist Party (CCP). The arrival en masse of Chinese investment, particularly the purchasing of several major Canadian companies, should therefore trigger 'pushback' from the Canadian public and hostility from the Canadian political opposition – the left leaning New Democratic Party (NDP) – because economic engagement with China in the energy sector is a useful target for making political gains. All of these factors complicate Chinese efforts to buy Canadian energy and energy companies. We could therefore anticipate, at a minimum, that tensions arise in US–Canadian strategic ties, Sino–Canadian energy ties and perhaps in Sino–US ties.

In fact, such tensions have emerged. First, in 2004, citizens, shareholders and politicians working in Canada's democratic polity pushed back when China's Minmetals Corp entered exclusive talks to buy Canadian mining giant Noranda Inc. in a deal valued at the time at over US$5 billion. The deal floundered due to opposition in Ottawa (Cummins and MacDonald, 2012). Thereafter, the Conservative Party toughened up the Investment Canada Act, adding a component that highlighted risks to national security. Second, in 2012 when China National Offshore Oil Corporation (CNOOC) tried to buy the Canadian energy firm Nexen, similar concerns threatened this sale. Ultimately the Canadian government approved the deal, but not without domestic opposition, a great deal of posturing by the US Congress, and Canadian government handwringing.

Sino–Canadian energy ties

Canada and China defined energy as one of their key bilateral priorities by signing a joint accord on Canada–China energy cooperation during PM Paul Martin's visit to Beijing in 2005. Investments in Alberta's oil sands by two large Chinese energy companies, China Petroleum & Chemical Corporation (Sinopec) and CNOOC, followed immediately. Soon after, China National Petroleum Corporation (CNPC), China's largest energy firm, signed a CAD$2 billion Memorandum of Understanding (MOU) with Enbridge for potential cooperation on the Northern Gateway Pipeline project from Alberta to Canada's west coast.

Although China's investment, merger and acquisition activities in energy sectors around the world intensified after 2005, Chinese energy companies did not invest in large projects in Canada from early 2006 to mid-2009. Not only did they face domestic opposition in Canada, but investment was also stifled by poor Sino–Canadian bilateral relations as the Harper government placed human rights ahead of business ties, causing Chinese SOEs to abjure from investments in Canada's energy and resource sectors. PM Harper did not pursue a visit to China during his first three years in office, resulting in the suspension of bilateral summit diplomacy (Jiang, 2009).

Since fall 2009, after summit diplomacy resumed between the two countries, China has shown renewed interest in the Canadian energy and resource sectors. These fast-paced investment activities in Canada occurred against the backdrop of a sustained high demand for energy, which was forecasted for China, and sluggish prospects for economic recovery in the US. Large investments have materialized and all the large Chinese energy companies are actively seeking potential investment targets, especially in Alberta. In 2012, the two countries penned an *Economic Complementarities Agreement*, which highlighted energy as a key sector where Canada and China shared common interests, albeit the former as producer and the latter as consumer and investor (Economic Partnership Working Group, 2012).

Canadian energy companies have also shifted from overreliance on the US market to the emerging Asian economies, due partly to the financial and economic crises in 2008 and partly to the realization that diversification will serve their long-term interests. Intensive protests from environmental groups in the US against the import of Canadian oil sands products, particularly against the controversial Keystone XL crude oil pipeline from Alberta to the Gulf of Mexico, have made China a valuable partner for Canadian firms seeking to sell their oil.

From China's perspective, Alberta is one of few places in the world where oil companies can invest; most of world's oil reserves are controlled by national governments or their energy arms – the NOCs. In fact, only 22 percent of the total world reserves are accessible to private sector investment, of which 52 percent is in Alberta's oil sands. A country, such as China, that buys equity oil as a key component of its energy strategy, must therefore seriously import Canadian oil.

Since the end of 2009, CNPC, Sinopec, and CNOOC have all made substantial investments in Canadian energy with a particular focus on Alberta's oil sands. The China Investment Corporation (CIC), China's US$300 billion sovereign wealth fund, opened its first office in Canada in 2009 and chose Canada for its only energy sector equity investment.

Before the US$15.5 billion Nexen deal, the largest Chinese investment in the Canadian energy sector was Sinopec's US$4.65 billion takeover of ConocoPhillips' shares in Syncrude Canada Ltd in 2006. The Syncrude Group is Canada's largest oil sands production consortium with most of its production exported to the US market. The Sinopec-Syncrude deal was followed in 2009 by the successful purchase of 60 percent of Athabasca Oil Sands Corporation's (AOSC) MacKay and Dover oil sands projects by PetroChina (a CNPC subsidiary) worth US$1.9 billion. In 2012, Athabasca exercised its right to sell the other 40 percent of the company to PetroChina for CAD$680 million (Dobby, 2012). China has also invested actively in Canada's mining sector since 2009 – notably the US$1.7 billion equity investment in July 2009 by CIC in Teck Resources, a Vancouver-based company with energy and mining assets in North America. The CIC also invested US$1.25 billion in Penn West Energy in July 2011. CNOOC, the third largest Chinese NOC, also acquired the struggling oil-sands producer Opti Canada Inc in July 2011, buying a 35 percent stake in the joint Nexen–Opti oil sands project in Long Lake, Alberta.

A number of factors contribute to renewed interest in Canadian investments. Chinese energy and resource needs have been driving China's foreign investment in these areas over recent years. China's better-than-expected recovery from the worldwide recession in early 2008 also fueled demand for energy and resources. Second, in early 2009, the Canadian government reversed its China policy, culminating in visits by PM Harper in December 2009 and January 2012. During his 2012 visit, he announced in Guangzhou that Canada's energy sources were open for investment from China (Harper, 2012).

> Canada is not just a great trading nation; we are an emerging energy super-power. It has abundant supplies of virtually every form of energy, *and you know, we want to sell our energy to people who want to buy our energy, it's that simple*....And it is increasingly clear that Canada's commercial interests are best served through diversification of our energy markets.
>
> (Harper, 2012)

Third, weak North American stock markets after the 2008 economic crisis presented buying opportunities for cash-rich Chinese firms and selling pressures for cash-strapped Canadian companies (Jiang, 2012). For example, Sinopec's US$2.2-billion purchase of Daylight Energy, the first 100 percent takeover of a North American energy firm by a Chinese oil company, involved a buying price that was more than double Daylight's closing price prior to the announcement. Fourth, high global oil prices made long-term extraction of oil sands appear sustainable and profitable. Fifth, turmoil in North Africa and the Middle East after 2010 taught Chinese investors that putting their fortune into resource-rich but unstable states, or into fast deals with dictators, entails higher costs and greater risks. In a conference in Beijing in 2011, Zhang Guobao, who had just retired as the head of China's National Energy Administration, but was still chairman of the National Energy Security Advisory Committee to then Premier Wen Jiabao, made it clear in his speech that countries, such as Canada and Australia, both resource-rich and democratic, should top the Chinese FDI list due to their stable investment environments.

Sixth, China's energy firms now boldly conduct merger and acquisition as well as joint venture activities around the world, and Chinese energy merger and acquisition and joint venture teams visit Calgary frequently. In 2005, CNPC purchased its first stake in Kazakhstan from Canadian firm PetroKazakhstan for US$4.18 billion, while two months later, it partnered in a US$576 million investment in PetroCanada in Syria. Between February 2006 and December 2009 (the period when Chinese firms abjured from buying in Canada), CNPC, China International Trade and Investment Corporation (CITIC), Sinopec, and CIC, all major Chinese players, bought Canadian held companies in Ecuador, Kazakhstan, Chad, Syria, Lebanon, and Mongolia for a total of US$6.38 billion (Jiang, 2012). Seventh, all Chinese NOCs seek the management skills and technical know-how of extracting heavy oil and shale that the Canadian firms possess. With well over US$24 billion invested in Venezuela's heavy oil exploration (see Chapter 13, this volume), and having a domestic shale reserve that is larger than both the US and Canada combined, Chinese energy companies will benefit technologically from investments in the Canadian energy sector. Finally, Canadians saw the benefits of diversifying into East Asia after the US State Department delayed the approval of the Keystone XL pipeline, with Chinese companies investing in Enbridge's US$100-million Gateway pipeline regulatory approval fund.

'Pushback' against Chinese foreign investment in Canada

Canada is known for its open economy, and merger and acquisition activities involving international entities go on all the time, but Canadians are particularly sensitive to Chinese investment. When China's Minmetals Group was negotiating to take over the largest Canadian metal firm, Noranda, in 2004 (an effort that eventually failed), there was barely any Chinese investment in Canada. Alarm bells rang quickly and loudly on the danger of Chinese SOEs controlling a major Canadian corporation. By 2014, the growth of Chinese investment in Canada was dramatic, which added to the concerns of Canadians about a takeover by China of its natural resources. Whereas Chinese investment in Canada had been almost zero in 2005, the stock of investment had reached CAD$33 billion by the end of 2013 after the Nexen deal (Ivison, 2013).

The Conference Board of Canada, looking at Canada's regime for FDI for China, stated quite boldly that,

> For Canada, China is as politically sensitive as it gets. The political concern starts with the nature of the Chinese regime, a non-democratic, Communist dictatorship for over seven decades. There is an inherent distrust of the regime among many Canadians and this is manifested in its investment relationship with Canada.
>
> (Grant, 2012: 18)

Drawing on a survey by HarrisDécima in February 2012, the Conference Board found that,

> Just 1 in 10 (10 percent) respondents thought Chinese companies taking a majority controlling interest and/or taking over an existing Canadian-owned operation is a good thing. Canadians generally become warmer toward Chinese investment the less it is concerned with control of existing operations (especially Canadian operations) and the more it is about new investments. Yet even in the most positive case of greenfield investment, barely half of respondents think Chinese investment is a good or very good thing.
>
> (Grant, 2012: 18)

In fact, almost 80 percent thought that a Chinese company taking majority interest in a Canadian company was a 'bad' (50%) or 'very bad' (30%) thing, while less than 40% thought that a Chinese company taking minority interest in a Canadian company was 'very good' or 'good' (Grant, 2012: 18).

The best example of pushback to date is the failed effort by China Minmetals to buy Noranda, one of Canada's oldest and largest mining companies, which specialized in zinc, aluminum, copper, and gold (*New York Times,* 2005). While the Minmetals takeover was good for Noranda and its shareholders, the deal had broader political implications. China's interest triggered protests by both the political Right and Left (Zhang and Chen, 2011). The Right's key concern was that the deal was with a 'Communist nation', raising national security concerns (Litvak, 2006). Previous Chinese purchases of Canadian companies were usually of small, privately held operations (*New York Times,* 2004); however, this giant SOE aroused people's concern. On the Left, the NDP believed that the Noranda deal would be the start of a massive Chinese capital invasion, so they called for the establishment of a new subcommittee under the House of Commons Industry Committee, which includes members from all parties in parliament 'to conduct a broader review of foreign investment in Canada, increasing globalization and its effects on Canadians' (Costen, 2004).

The political opposition echoed two other points: environmental protection and human rights. Given the labor and human rights violations common in China, Canadians worried about the possible impact of these two components on the proposed takeover (Zhang and Chen, 2011). David Kilgour, a Member of Parliament, cited human rights as a reason to block a deal with a Chinese company (*Bloomberg,* 2004). Talks eventually failed due to opposition from Canadian labor unions and the federal government's introduction of Bill C-59, an amendment to the Investment Canada Act, which enabled the Canadian Government to screen stringently investment on national security concerns. In late November 2004, Noranda therefore ended the exclusive talks when they could not agree on the price.

A 2011 report by the Norton Rose Company highlighted the impact of these new guidelines on state policy towards foreign investment:

> In 2007, the government added guidelines pertaining specifically to SOEs, targeted presumably at China and others as well, that enable the government of Canada to assess 'the governance and commercial orientation of SOEs' as part of the net benefit calculation. This had been prompted by successive acquisitions of major Canadian mining assets – Alcan, Inco and Falconbridge – and concerns that the country was being 'hollowed out' by foreign takeovers. China MinMetals' interest in acquiring Noranda also played a role in the development of both the national security amendments and the SOE guidelines. What is certain is that the combination of 'net benefit' and 'national security' give ample scope for political considerations to outweigh statistical analyses.
>
> (Burney and Ackhurst, 2011)

As Chinese investment in the Canadian energy sector grew with more than US$16 billion in 2009–2011 alone, so did concerns over whether Chinese money was beneficial or harmful to Canada. As the poll by HarrisDécima cited earlier shows, most Canadians feel uncomfortable with Chinese foreign direct investment (Grant, 2012: 18). This view is understandable. The Canadian economy is mostly integrated with that of the US. The two countries share a common border, language, cultural similarities, and the largest bilateral trade in the world, and Canada is a North American Free Trade Agreement (NAFTA) partner and a NATO ally. Even for Canadians who feel that the US has exploited Canada, or has meddled in its internal affairs, the US remains the devil they know.

In contrast, to most Canadians, China is the unknown. Some Canadians worry that China's increased investment will erode Canadian sovereignty. Canadians worry that Chinese companies, especially in cases of full ownership of Canadian energy extraction, may undermine Canada's labor, human rights, and environmental standards, given Beijing's poor domestic records in these areas, but Chinese SOEs are also becoming more sensitive to their image abroad. To establish themselves in Canada for long-term operations, Chinese firms have strong incentives to conform to labor, environment, and other Canadian norms.

Canadian business' view on Chinese investment

On the other hand, Canadian business generally welcomes investment from China because of the shortage of domestic capital in the resource sector. Yet, similar to the general population, they have their own conditions for the entry of Chinese capital, which was reflected in their responses to the Nexen–CNOOC deal.

Canada has the largest reserves in the world in which private companies are free to invest – more than half of the global total – and yet the sector lacks capital, people, and pipes (*Economist,* 2012), particularly because most of Canada's reserves are in oil sands, which involve mega-projects with very significant regulatory, capital, commercial, and environmental components. To exploit its hydrocarbons (including oil sands), Canada needs capital investment of US\$50–60 billion a year (Alberta Oil, 2013). According to David Collyer of the Canadian Association of Petroleum Producers (CAPP), such sums are 'far more than Canadian capital markets can raise' (Bennet Jones, 2014), a view shared by Canadian Natural Resource Minister Joe Oliver (Alberta Oil, 2013).

After PM Harper announced changes in foreign investment rules simultaneously with the approval of the Nexen–CNOOC deal, business leaders began to worry about the flow of capital and production. Glen Schmidt, president and CEO of Laricina Energy, worried that 'no money from China' would force Canadian firms to turn to costly and time-consuming strategies to get investment funds (Alberta Oil, 2013). In fact, a quarterly survey conducted by Gandalf Group and KPMG reveals that 42 percent of Canadian business leaders believed the government should have allowed the unconditional acquisition of Nexen by CNOOC (Gandalf Group, 2012). Moreover, Chinese investment by firms with less 'state' background would be more welcome, given the overall shortage of capital.

The major concern of executives who would have wanted more conditions on the Nexen deal is to 'ensure the operation (of Nexen) is not determined by Chinese state interests' (Gandalf Group, 2012: 7). Among the remaining 58 percent of executives, four-fifths thought the deal should be approved if Chinese state interests did not determine its operations, if Canadian jobs are protected, and if reciprocal access to Chinese markets is forthcoming (Gandalf Group, 2012). Similar concerns were expressed by the CEO of Canadian Natural, who said that foreign investment is good for the energy industry and the country, but there should be limits to what Ottawa is prepared to approve given the strategic importance of the oil sands. Suncor CEO Steve Williams, who felt that there was no reason for Ottawa to block the CNOOC deal, recognized that the government would have to grapple with the bigger issue of where to draw the line on foreign investment in key sectors, and Cenovus CEO Brian Ferguson argued that Canada needed domestic champions in the oil industry (*Globe and Mail,* 2012). Finally, John Manley, president and CEO of the Canadian Council of Chief Executives was probably speaking for all Canadian business executives in the oil sector when he said:

> It goes without saying that Canadian investors ought to be afforded the same access to China that Chinese investors are afforded to Canada....Unfortunately, this is not the case today. This lack of openness is an obvious source of frustration for Canadian investors, particularly given the recent dramatic increase in Chinese investment in Canada'.
>
> (Canadian Council of Chief Executives, 2011)

Canada's energy ties with the US

In 2001, the National Energy Policy Development Group, which was constituted by then US President Bush, reported on its findings in a lengthy policy analysis, replete with numerous suggestions as to how to enhance US energy security and efficiency, and decrease costs for US citizens. In that report, directed by Vice President Dick Cheney, the continued development of Canada's oil sands 'can be a pillar of sustained North American energy and economic security', (National Energy Policy, 2001: 134), making Canada's tar-sands oil highly important for US energy security. At the time the report was written, Vice President Cheney reportedly said that the US did not mind if Canada sold oil to China as long as the Canadians remembered that 'it is our oil'.[1]

Around 2006, when China demonstrated a strong interest in Alberta's Tar Sands, Canadian officials stationed in Washington DC asked the Americans how they would feel if Canada sold oil to China. Reports were that although the US did not actively oppose Canadian energy sales to China, US officials were 'wary' of the situation and were monitoring it. At that time, however, concerns about pariahs trumped worried about allies who might be edging closer to China because US State Department officials were pleased because Canada offered China an alternative energy source from Iran, Venezuela, and Sudan (Zweig personal communication, 2010).

Yet Canada remains the largest foreign supplier of energy to the US, well above Mexico and Saudi Arabia (Chapter 1, Table 1.1). Further inroads by Chinese firms into the Canadian energy market are likely to worry certain elements within the US body politic. More remarkably, in terms of Canadian dependence on the US market, PM Harper, in his speech in Guangzhou in 2012 emphasized that 'currently, 99 percent of Canada's energy exports go to one country – the United States' (Harper, 2012).

So, does China's involvement in Canada's energy sector affect Canadian–US energy relations? First, exporting oil to China will be possible only on a small scale in the foreseeable future, given the limited pipeline infrastructure on the Canadian west coast. The potential for large-scale supply exists only if Enbridge's Gateway pipeline gets the regulatory approval required for its construction, which will not be easy. Second, much of China's global oil production is not shipped to China, but is sold on the world market. As of 2015, Syncrude production was flowing south to the US and Sinopec's 9 percent ownership was not changing this arrangement. China certainly remains interested in a pipeline that ships oil from Alberta to the Canadian West coast, precisely because there is no large-capacity, direct pipeline from Alberta to the west coast. Kinder Morgan's TMX Loop project, which was completed in 2008, links existing pipelines from Alberta to a port in southern Vancouver. Its shipping capacity is only 300,000 barrels per day, while Enbridge's new Gateway pipeline would have an additional 550,000 barrels per day capacity, but it will not be functional for several years. Moreover, there is a great deal of opposition within British Columbia, and especially in Vancouver, to the building of a pipeline through the region that would bring few local benefits to the region but enormous environmental risks.

Finally, where is the red line for Chinese investment in Canada's energy sector? In the past few years, Canada has become more confident in believing that the country has the necessary regulatory framework in place to cope with increased Chinese investment. The current Canadian national discourse is more focused on whether investment from China will provide social and economic benefit to Canada and on the environmental impact of pipelines running to the west coast and prolonged large-scale extraction of oil sands. In the following sections, we will look at two projects that suggest the triangular nature of China's energy diplomacy.

Sino–Canadian energy links increase support in the US for the Keystone XL pipeline

A major environmental and energy issue linking the US, Canada and China is the Keystone XL pipeline project, which could transport synthetic oil sands and crude oil from Alberta to Texas; however, the project includes a controversial routing over the Ogallala Aquifer in Nebraska and the risk of water pollution mobilized environmental forces in the US to challenge the program and eventually convince President Obama to block the project.

Proponents, on the other hand, claimed that if this oil was not going to US via the pipeline, it would be likely to go east to China by tanker (Foxnews, 2012). After all, China was likely to make a very significant bid for oil coming out the Tar Sands (Rubin, 2012). Rick Perry, the Texas governor and failed presidential candidate, echoed this view by arguing that blocking the pipeline would result in the oil ending up in China. 'Obama wants us to believe he is for jobs, economic opportunity and greater energy security, and his Keystone decision does help meet those goals – for the People's Republic of China' (*The Guardian*, 2012).

The CNOOC–Nexen deal

A recent manifestation of the triangle involved an effort by a Chinese company to purchase Nexen, an important energy company in Canada. On 23 July 2012, CNOOC made a generous offer of US$15.1 billion for Nexen, Canada's 10th biggest oil company, which has oil sands operations in Alberta, shale gas in British Columbia, and extensive exploration and production holdings in the North Sea, Gulf of Mexico, and offshore West Africa (Reuters, 2012a); however, although the purchase price was 60 percent higher than the extant value of the shares, the deal was not easy, involving many aspects of the triangular model. These include a pushback from Canadian society and the political opposition, Canadian government support for exporting resources, complex Sino–Canadian relations, and publicly expressed US Congressional concerns about Chinese incursions into its energy security.

The Canadian view

Canadian stakeholders and PM Harper saw the CNOOC–Nexen deal as a way to welcome Chinese FDI into Canada's resource-based economy that did not significantly challenge Canada's national security because of 'its scale and its importance in the Canadian context (which is not big enough in terms of national security)'. He saw the deal as beneficial for the Canadian economy (Mayeda, 2012).

The sale also tied in with the Canada–China Economic Complementarities Study, the Sino–Canadian joint study that underpinned PM Harper's shift in his China policy. The section on the environment, energy, and natural resources, emphasized that,

> With a vast energy resource endowment and growing oil production, Canada can strengthen its position as a reliable, politically stable and competitive energy supplier to China. For example, Canada's oil exports to China are currently limited due to infrastructure constraints, but the potential construction of pipelines to the Canadian west coast would open up the Chinese market for Canadian oil and natural gas producers....Over the coming decades, massive investments will be required to further develop Canada's natural resources potential. China's growing investment interest in Canada's natural resources is adding to the diversity of domestic and foreign funding sources available to finance Canadian natural resources projects.
>
> (Economic Partnership Working Group, 2012: 4)

But although Nexen shareholders backed the deal, opposition grew among politicians and the general public in Canada. According to a survey by Abacus Data, 69 percent of Canadians opposed the deal, while only 8 percent approved, with 23 percent uncertain. Among the opponents, 58 percent cited the fact that 'Nexen operates in one of Canada's core strategic industries, and a foreign company should not have control of such an important resource' (BBC, 2012).

The fact that the purchaser is an SOE, which is linked to the Chinese state, complicates the issue. Ray Boisvert, a former deputy director at the Canadian Security Intelligence Service, told an audience in Ottawa in October 2012 that 'State-owned enterprises have the same marching orders or essentially the same mandate or mission' as China's intelligence services, which is to serve the interests of the party and the state. In his view, China remains a 'persistent and aggressive' perpetrator of espionage activities that could threaten Canadian interests (McCarthy, 2012).

The opposition NDP declared the economic bid a political issue precisely because the deal involved a Chinese SOE (Scoffield and Levitz, 2012). Paul Boothe, a former deputy minister in Industry Canada, questioned whether Canada should allow SOEs to invest in the country, whether it should insist on diversification of SOE investment, and whether it should limit state-led investment from China (Scoffield and Levitz, 2012). According to Scoffield and Levitz (2012), 'Mr. Harper's looming decision is not merely one of looking at the economic benefits, or deciding how to handle state-owned enterprises. It also has to pre-empt a public backlash'.

Nevertheless, in December 2012, the Canadian government approved the CNOOC–Nexen deal, giving the SOE a stake in Canada's largest oil-sands project and the biggest position in the Buzzard oil field in the UK's North Sea. But although allowing the takeover, PM Harper said that Canada would not approve state-owned companies taking controlling interests in any more oil-sands projects, except in 'exceptional circumstances'. Unfortunately for Canada, the result of this public posture has been a major freeze on foreign investment in the entire Tar Sands sector (Ivison, 2013).

The view from the US

One complication to the CNOOC–Nexen deal arose because 10 percent of Nexen's assets are in US territorial waters in the Gulf of Mexico where it operates deep-water drilling wells. Some US Congressmen insisted that the takeover be approved by the Committee on Foreign Investment in the United States (CFIUS), an US interagency board that reviews economic deals for their national security implications. House Minority Leader Nancy Pelosi expected a thorough review by CFIUS. According to Pelosi's spokesperson, 'This deal prompts great concern about the Chinese government's continued attempts to use its state-owned enterprises to acquire global energy resources' (Reuters, 2012b). Senator John Hoeven, representing North Dakota, a major oil-producing state in the US, worried that the US would find itself buying Canadian oil back from the Chinese, rather than working 'with our closest friend and ally, Canada' (Reuters, 2012b). The most direct security-related challenge came from US Senator James Inhofe, a Republican from Oklahoma, who was a key opponent of the Unocal sale in 2005. 'I have serious national security concerns with the Chinese government, acting through one of its corporations, purchasing a company that will give it control over significant US oil and gas resources' (Orol, 2012). In fact, Inhofe has long argued that the Department of Defense, rather than Treasury, should serve as the chair of CFIUS, which would increase the likelihood that any energy purchases by Chinese SOEs in the US would be rejected on security grounds.

US Senator Charles Schumer of New York, however, emphasized the trade component of the deal, rather than its national security implications, which he saw as limited. His letter to Treasury Secretary Tim Geithner, who chaired CFIUS, called for a 'long-term' perspective, emphasizing that the approval of the CNOOC–Nexen deal, the largest foreign acquisition by a Chinese company, could be used to pressure China to allow more foreign investment into China itself (Kirchgaessner, 2012). He felt that the deal offered the US rare leverage over China and the approval should be conditional on China allowing more FDI (Reuters, 2012a). Massachusetts Democrat Ed Markey, a ranking member of the House Committee on Natural Resources, saw it as a massive transfer of wealth from the American people to the Chinese government. He believes that CNOOC should at least pay royalties to the US taxpayer on all oil produced on or near US shores (*Wall Street Journal*, 2012).

Finally, one cynical observer of the deal in the US said that the core of opposition 'is about China', and as one insider put it, 'I don't know if the issues would be the same if it was Exxon-Nexen. I don't think it's about foreign investment' (Mayeda, 2012). The US Congress was cautious about any security threat from China in recent years, and it therefore blocked the CNOOC–Unocal deal in 2005 and challenged the Chinese technology company, Huawei, in 2012 from doing business in the US market by citing national security concerns. CNOOC–Nexen in the US was bipartisan, although the parties' suggestions about how to resolve the issue diverged.

David Goldwyn, a former State Department energy official, was confident that the Keystone XL pipeline would be approved, especially considering future Middle East turmoil. 'I think it would be a huge waste of a great opportunity to provide supply security. We don't often get the choice of where we can get our oil from. In this case we get to choose Canada. That's an opportunity we shouldn't miss', he said in an interview. He saw no threat from Chinese inroads into Canada because there is more than enough oil for all concerned. By investing to boost Canadian production, the Chinese 'are growing the pie to meet their own demand. That's a whole lot better than mopping up supply from the existing pie and creating competition for resources' (Gillies, 2011). Similarly, Republican Congressman Randi Forbes, who had led congressional opposition to the failed Unocal bid, said that although he did not like the CNOOC–Nexen deal, he felt that he could not do much about it because it is was a Canadian company. 'Whatever we would do would simply be talking in the wind because we don't have any legal authority to stop this action' (Reuters, 2012b). Although the Senators were concerned about CNOOC's bid for a Canadian oil company, 'they stopped short of saying that the US government should try to do something to stop it' (Reuters, 2012b).

US officials are indeed wary that Canadian–China energy deals will undermine US interests. Alberta Premier Ed Stelmach said US government officials had raised concerns about a pipeline to the Pacific in terms of 'Well, are you still going to be able to supply us?'. Moreover, William Cohen, Secretary of Defense under President Clinton, said that any Chinese–Canadian oil partnership must be done 'with some diplomacy and care' (that isn't) 'a threat to the United States.

Canada can do whatever it wants', Cohen said in Toronto after a public debate about whether China will dominate the twenty-first century, but 'Canada knows it has a very close and vital relationship with the United States. I'm sure there will be discussions'. These comments infuriated Eddie Goldenberg, Chief of Staff to former PM Jean Chretien, who retorted to Cohen's statement that Canada should care less if some US officials are leery about Canada selling oil to China. 'We're not the 51st state. It's not the business of the United States to decide where Canada sells its resources' (Gillies, 2011).

The truth is that if the US allowed Tar Sands oil to flow southward, the Chinese energy companies could be more than willing to ship the oil to the US. Chinese oil majors often pump oil overseas and sell it in third markets, rather than ship it back to China. Moreover, the Gateway Pipeline from the Tar Sands to Canada's west coast is under threat because residents and government officials in British Columbia worry about the environmental impact of a pipeline running close to the city of Vancouver. So, if CNOOC bought Nexus, it would still need a US outlet for its oil, and without the Keystone deal, that direction of flow is at risk.

Conclusion

Facts run against the assumptions of those who perceive China's 'Going Out' strategy as predatory behavior (Frieberg, 2006), or those who are concerned that China's presence in Canada's energy sector may deprive the US of its supplies. The dragon has returned to Canada, but cautiously. Beijing has become more sensitive to the political, economic, social, and environmental conditions of investments in Canada, often settling for minority positions in their equity and joint venture agreements. Most importantly, at present, crude produced by Chinese capital in Canada is only flowing south to the US market, helping to secure US energy supply.

Nevertheless, the triangle is in play in North America and helps explain the overall dynamic of China's efforts to buy into this market. The US is concerned about China's actions because it relies so heavily on Canadian oil imports and because, in a time of crisis, the oil in the land of the US's closest ally is a critical source of US energy security. Sovereignty is also an issue because Canada's oil resources are Canada's to sell, regardless of the views of US politicians. Finally, Canada needs investment for its energy resources, and Chinese companies are extremely active worldwide searching for energy. So, Canada needs investors and China is hungry to invest.

Still, as members of a democratic society closely integrated into the West, Canadians worry about Chinese SOEs buying significant shares of their natural resource companies. Canada does not need US pressure to make it wary; pushback remains a natural response to these type of purchases until Chinese goals and interests become more transparent and less state directed. Yet, US pressure is clearly part of the game influencing Sino–Canadian ties.

Note

1 The statement might be apocryphal, but it does reflect the views of the Vice President.

References

Alberta Oil, 2013. Will Canada's new foreign investment rules kill oil sands development? 1 March. Available online at: www.albertaoilmagazine.com/2013/03/foreign-invest-ment-oilsands/ (accessed 2 April 2015).

BBC, 2012. CNOOC's Nexen bid: Shareholders approve $15.1bn deal. 21 September. Available online at: www.bbc.co.uk/news/business-19671219 (accessed 2 April 2015).

Bennett Jones, 2014. The Canadian oil sands: a backgrounder. Available online at: www.bennettjones.com/OilSands/ (accessed 2 April 2015).

Bloomberg, 2004. Noranda ends exclusive talks with China Minmetals. *Bloomberg,* 16 November. Available online at: www.bloomberg.com/apps/news?pid=newsar-chive&sid=aSoDroH3ZY5Q&refer=canada (accessed 2 April 2015).

Burney, Derek H. and Ackhurst, Kevin, 2011. Canadian protectionism? Political and legal considerations for foreign investment in Canada, Norton Rose Fulbright, 10 August. Available online at: www.mondaq.com/canada/x/142238/Inward+Foreign+Investment/Canadian+Protectionism+Political+and+Legal+Considerations+for+Foreign+Invest-ment+in+Canada (accessed 1 April 2015).

Canadian Council of Chief Executives, 2011. Reciprocity is key in the Canada–China rela-tionship, Manley says. 21 November. Available online at: www.ceocouncil.ca/news-item/reciprocity-is-key-in-the-canada-china-relationship-manley-says (accessed 2 April 2015).

Costen, Whelan, 2004. MP's should have say – sale of Noranda Inc. 19 October. Avail-able online at: www.vivelecanada.ca/article/144732396-mp-s-should-have-say-sale-of-noranda-inc (accessed 9 May 2015).

Cummins, Chip, and MacDonald, Alistair, 2012. Resource-rich Canada looks to China for growth. *Reuters,* May 13.

Dobby, Christine, 2012. Athabasca to sell oil sands interest to PetroChina for $680M. *Financial Times,* January 3.

Economic Partnership Working Group, 2012. Canada–China Economic Complementarities Study, clean energy section. Available online at: www.international.gc.ca/trade-agree-ments-accords-commerciaux/agr-acc/china-chine/study-comp-etude.aspx?view=d (accessed 1 April 2015).

Economist, The, 2012. The sands of grime. *The Economist,* 17 November. Retrieved from: http://www.economist.com.hk/news/business/21566686-become-energy-giant-canada-needs-capital-people-and-pipes-sands-grime (accessed 1 April 2015).

Foxnews, 2012. After Obama blocks Keystone Pipeline, China readies $15.1 billion Canadian oil deal. 28 July. Available online at: http://nation.foxnews.com/canadi-an-oil/2012/07/28/after-obama-blocks-keystone-pipeline-china-readies-151-billion-canadian-oil-deal (accessed 2 April 2015).

Frieberg, Aaron L., 2006. 'Going Out:' China's pursuit of natural resources and implica-tions for the PRC's grand strategy. *NBR Analysis,* 17 (3).

Gandalf Group, 2012. The 28th quarterly C-Suite survey: foreign takeovers and national energy strategy. 1 October. Available online at: www.gandalfgroup.ca/downloads/2012/Q3/C-Suite%20SeptOct%202012%20TC.pdf (accessed 2 April 2015).

Gillies, Rob, 2011. China eyes Canada's oil sands. *Associated Press*, 27 June. Avail-able online at: www.huffingtonpost.com/2011/06/27/china-canada-oil-sands-alber-ta-tar_n_885032.html (accessed 1 April 2015).

Globe and Mail, The, 2012. Protect Canadian ownership of oil sands firms, executives urge. *The Globe and Mail,* 25 September. Available online at: www.theglobeandmail.

com/news/national/protect-canadian-ownership-of-oil-sands-firms-executives-urge/article4568562/ (accessed 15 May 2015).

Grant, Michael, 2012. *Fear the dragon? Chinese foreign direct investment in Canada: trade, investment policy, and international cooperation.* Ottawa, ON: The Conference Board of Canada.

Guardian, The, 2012. Keystone pipeline opponents fight on with petition challenge. *The Guardian,* 13 February. Available online at: www.guardian.co.uk/environment/2012/feb/13/keystone-pipeline-petition (accessed 2 April 2015).

Harper, Stephen, 2012. PM delivers address in Guangzhou, China. 10 February. Available online at: www.pm.gc.ca/eng/media.asp?category=2&id=4671 (accessed 2 April 2015).

Ivison, John, 2013. Foreign investment in the oil sands have dropped off a cliff since Nexen takeover. *The National Post,* October 1, p. 1.

Kirchgaessner, Stephanie, 2012. US senator challenges CNOOC–Nexen deal, *Financial Times,* 27 July.

Liepert, 2011. Alberta Minister of Energy Ron Liepert, Keynote Speech, Canada–Asia Energy Cooperation Conference and the 7th Canada–China Energy & Environment Forum. Calgary, 8 September.

Litvak, Isaiah A., 2006. China Minmetals Corporation and Noranda Inc, Case Study 906M13, Richard Ivey School of Business, The University of Western Ontario. Available online at: http://hbr.org/product/china-minmetals-corporation-and-noranda-inc/an/906M13-PDF-ENG (accessed 2 April 2015).

Mayeda, Andrew, 2012. Executives urge Harper to approve CNOOC's Nexen bid. *Bloomberg,* 27 September.

McCarthy, Shawn, 2012. CNOOC's bid for Nexen fuels security concerns, *The Globe and Mail,* 17 October. Available online at: /www.theglobeandmail.com/globe-investor/cnoocs-bid-for-nexen-fuels-security-concerns/article4619564/ (accessed 15 May 2015).

National Energy Policy, 2001. Report of the National Energy Policy Development Group, May. Available online at: www.wtrg.com/EnergyReport/National-Energy-Policy.pdf (accessed 31 March 2015).

New York Times, 2004. Chinese plan to buy bigger miner stirs Canadians Left and Right. *New York Times,* 29 October.

New York Times, 2005. Noranda spurns China and buys up Falconbridge. *New York Times,* 11 March. Available online at: www.nytimes.com/2005/03/10/business/worldbusiness/10iht-noranda.html (accessed 2 April 2015).

Orol, Ronald D., 2012. CNOOC–Nexen US opposition become bipartisan. *Market Watch,* 6 August. Available online at: www.marketwatch.com/story/cnoon-nexen-us-opposition-becomes-bipartisan-2012-08-06 (accessed 2 April 2015).

Reuters, 2012a. CNOOC to buy Nexen for $15.1 billion in China's largest foreign deal. *Reuters,* 23 July. Available online at: http://mobile.reuters.com/article/article/idUS-BRE86M0CF20120723?irpc=962 (accessed 2 April 2015).

Reuters, 2012b. Republicans say China oil deal highlights US inaction. *Reuters,* 27 July. Available online at: http://in.reuters.com/article/2012/07/26/usa-congress-CNOOC–idINL2E8IQAXQ20120726 (accessed 2 April 2015).

Rubin, Jeff, 2012. On Keystone, environmentalists lose by winning. *Bloomberg News.* Available online at: www.bloomberg.com/news/2012-09-25/on-keystone-environmentalists-lose-by-winning.html (accessed 2 April 2015).

Scoffield, Heather and Levitz, Stephanie, 2012. CNOOC–Nexen decision pivotal for Harper agenda. *CTV News,* 4 November. Available online at: www.ctvnews.ca/politics/cnooc-nexen-decision-pivotal-for-harper-agenda-1.1023980 (accessed 15 May 2015).

US Energy Information Administration (EIA), 2014. International energy statistics. Available online at: www.eia.gov/cfapps/ipdbproject/IEDIndex3.cfm?tid=3&pid=3&aid=6 (accessed 2 April 2015).

Wall Street Journal, 2012. Markey urges conditional block of CNOOC–Nexen deal. *Wall Street Journal,* 30 July. Available online at: www.wsj.com/articles/SB10000872396390 44422690457755921178037486 8 (accessed 15 May 2015).

Wenran, Jiang, 2009. Seeking a strategic vision for Canada–China relations, *The International Journal,* 64 (4): 891–909.

Wenran, Jiang, 2012. China's demand for energy and its impact on Canada–US relations. Original paper submitted for this volume, December 2012.

Zhang, Kenny, and Victor Chen, 2011. Growing and diversifying Chinese investment in Canada: 2000–2010. *Asia Pacific and Globalization Review,* 1 (1): 37–54.

Part III
America's neutrals

7 Angolan agency and Chinese and US oil politics, 1975–2014[1]

Alex Vines

Introduction

China and the US are the two leading importers of Angolan oil, although US imports of Angolan crude are dramatically declining. Angola is today a key player in Africa's oil industry as a major supplier and exporter and the newest member of Organization of the Petroleum Exporting Countries (OPEC), producing around 1.7 million barrels per day. Between 2004 and 2007, Angola posted the highest global increase in oil output (ahead of Russia, Azerbaijan, Brazil, Libya and Kazakhstan, among others).

China and the US delayed establishing diplomatic relations for some years after Angolan independence in 1975 but did not suffer in terms of access to Angola's oil. Angola's political elite has adeptly protected its oil production to extract 'rents' to lubricate its neo-patrimonial system, and triangulated China, the US and Big Oil. According to Soares de Oliveira (2007: 595), 'During Angola's long civil war of 1976–2000, the state oil company, Sonangol was 'an island of competence' as most other institutions imploded through attrition and mismanagement while Sonangol played on oil 'neutrality' while the Cold War raged on. Since the end of the Civil War, this pattern has continued, as China and the US have found that elite politics, under the guise of Angolan sovereignty, coupled with technical pragmatism, has framed Angola's oil decision making'.

Background

For both China and the US, oil has driven relations with Angola over the past decade. For Angola, the name of the game has been partnership diversification. Africa currently supplies about 12 per cent of the world's oil (a 17 per cent increase in the last 10 years), boasts significant untapped reserves and has surpassed the Middle East as the largest regional supplier of crude oil to the US. Individually, Nigeria is the sixth largest crude oil supplier to the US, Angola is the ninth and Algeria is the eleventh, although this situation is changing due to shale oil and gas production in the US. Still, Africa offers the US a way to diversify oil supplies away from Middle East oil and also access to new gas reserves. It is not only the US that is vying to access African oil; China's continued economic growth, key to the country's stability and the Communist Party's legitimacy, requires substantial imports of energy, minerals and other materials. According to China's 2003 'National Energy Strategy and Policy', 'oil is the key factor in the creation of public wealth, and also one kind of most important commodity influencing the global political pattern, economic order and military operations'. Moreover, after the US Congress blocked the sale of Unocal in 2005 to the China National Offshore Oil Corporation (CNOOC), China's leaders concluded that competing on the open market was too risky. Thereafter, they intensified pressure on Chinese companies to diversify their energy suppliers to spread the risk.

China became a net importer of oil in 1993, and in 2003 became the world's second largest consumer of petroleum products in the world (and third largest importer) surpassed only by the US. Although 55 per cent of African oil and gas went to Europe and the US in recent years and only 16 per cent to China, this situation is changing. China is projected to surpass the US in 2015. In 2010, China received around 30 per cent of its imported crude oil from Africa, with one senior official in Beijing stating that the country aims to increase this figure to 40 per cent in the next 5–10 years.

In 2010, Angola was the second largest supplier of crude to China after Saudi Arabia, with China representing 45 per cent (790,000 barrels per day) of total Angolan exports, compared with the US, which imported 23 per cent (400,000 barrels per day) of Angolan oil. Although Chinese oil diplomacy towards Angola has one decade of history, oil has driven US policy since Angola's independence. For example, in a 1975 National Security Council meeting, senior US officials discussed which faction to support in Angola's civil war in anticipation of the Portuguese withdrawal from the country. Secretary of State James Schlesinger suggested that the US 'might wish to encourage the disintegration of Angola. Cabinda [Province] in the clutches of Mobutu [President of Zaire] would mean far greater security of the petroleum resources' (Ganesan and Vines, 2004: 313). Although the US did not follow this strategy, the enclave of Cabinda remains a significant source of Angola's oil.

Today, US planners still worry about the predictable access of Angolan oil onto the international market, and particularly about China's role in that process. A December 2011 report entitled 'African Security in Strategic Perspective', commissioned for the United States Africa Command, worried that:

> A deeper China–Angola relationship, forged through stronger economic links, could render Angola contractually bound to export a rising portion of oil to China. While oil is a fungible resource, the dramatic increase in Angolan exports to China could eventually come at the expense of its longstanding priority market in the US This would weaken the US–Angola energy security relationship over time, even if US companies, such as Chevron and ExxonMobil, remain active investors in the country.
>
> (Eurasia Group, 2011: 72)

A US Army War College report on the national security implications for the US and China of Africa's booming oil and gas exploration and production concluded that:

> Because of a domestic boom in shale oil and gas in the United States, our nation's energy imports from Africa have been falling rapidly in recent years, raising the key strategic issue of whether Africa matters as much to US energy security as it once did...while Africa may be becoming less important for US energy security, it is becoming more important for broader US national security. This is for a variety of reasons, such as the extraordinary trade and investment opportunities that the rapidly growing continent represents, including the need for $2.1 trillion in oil and gas sector investments between now and 2035 to realize its potential.
>
> (Brown, 2013: xiii)

As Figure 7.1 shows, since 2008, although US imports of Angolan oil have indeed declined dramatically, exports to China have risen dramatically, and exports to Taiwan and India have also increased.

China and Angola

China's growing role in Angola has generated debate and speculation, particularly in the US (Campos and Vines, 2008). But, after the civil war, when reconstruction became the government's top priority, China, which had established diplomatic relations in 1983, significantly assisted these efforts by kick-starting over 100 projects in energy, water, health, education, telecommunications, fisheries and public works. Why such extensive involvement? For Angola and China, relations are both pragmatic and strategic. Thus, speaking during Prime Minister Wen Jiabao's 2006 visit, Angolan President dos Santos described bilateral relations as 'mutually advantageous' and 'pragmatic' with no 'political preconditions' (*Jornal de Angola,* 2006). To put it succinctly, 'China needs natural resources and Angola wants development' (Angola Press News Agency, 2007).

By 2009, China had loaned Angola at least US$13.4 billion, while some estimated loans of US$19.7 billion (*Reuters,* 2011). In 2009/2010, China signed another US$10 billion in credit lines (African Development Bank, 2011). In return, China's Sinopec group obtained oil equity through a joint venture.

Bilateral trade has grown spectacularly. Bilateral trade in the 1990s ranged between US$150 million and US$700 million, and in 2000 it exceeded US$1.8 billion. By the end of 2005, it had reached US$6.9 billion. By 2013, bilateral trade of US$36 billion made Angola China's second largest trading partner in Africa. Most of this trade has been in oil exports, while official Chinese imports remain smaller and consist mostly of food products and consumer goods.

The US has been a leader in the import of Angolan oil for a long time, but after 2002, China's imports increased seven-fold, whilst US imports increased only 3.5 times (*Portugal News,* 2007). By 2013, Angolan crude exports in 2013 represented over 14 per cent of China's total oil imports, allowing Angola to run

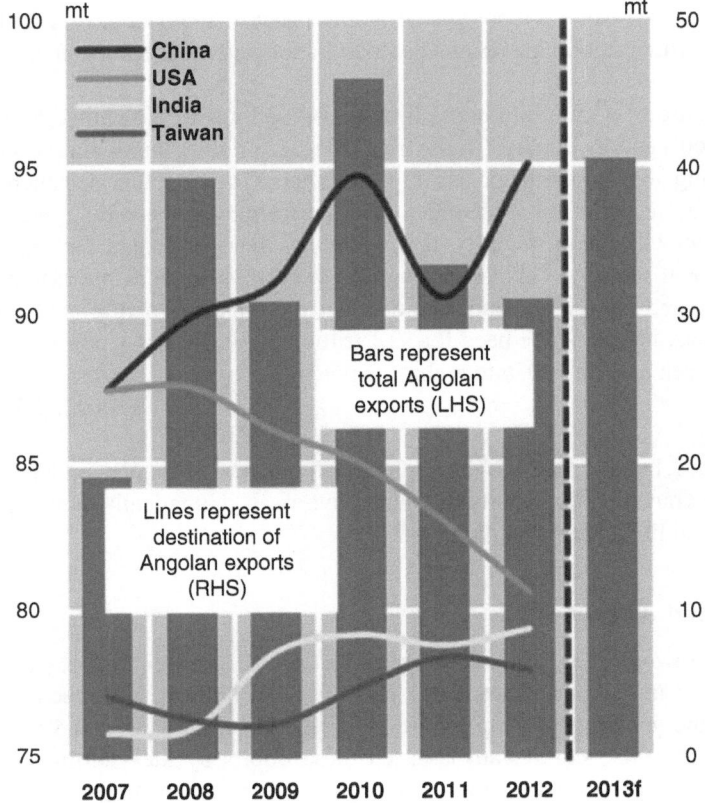

Figure 7.1 Angolan crude oil exports.

Source: Clarkson Research Services Limited, 2013.

a large trade surplus with China. Yet, the increase of infrastructural projects and the greater competitiveness of Chinese exports, compared to European exports to Angola, will allow significant penetration of Chinese products into Angola over the next few years.

China has backed up its oil imports with loans for construction and Chinese construction sites are now spread across the country. According to Angolan officials, 258,391 Chinese nationals held Angolan work visas in 2011 (SME, 2012), while Chinese Premier Li Keqiang, during his May 2014 visit to Angola, acknowledged that 260,000 Chinese nationals were working in Angola (out of one million in Africa). In fact, some estimates put the number of Chinese workers in Angola at 400,000 (Corkin, 2013: 109).

Political and diplomatic relations

China became involved in Angola during the anti-colonial struggle, giving support at various junctures to the country's three major liberation movements. Yet most support went to the National Union for the Total Independence of Angola (UNITA), with only fleeting support for the Popular Movement for the Liberation of Angola (MPLA) and the National Front for the Liberation of Angola (FNLA). Thus with the victory of the MPLA in 1975, China initially refused to formally recognize Angola's independence and the MPLA-led government until 1983, when it distanced itself from the UNITA rebels. Yet by the 1990s, Angola became China's second largest trading partner in Africa (overtaking South Africa by the end of the decade), mostly due to defence cooperation.[1] In October 1998, President dos Santos visited China for the first time, seeking to 'expand bilateral ties'; thereafter, cooperation between the two countries has involved frequent bilateral visits by important officials. In 2010, China established a 'strategic partnership' with Angola and China's then Vice-Premier, Xi Jinping, who visited Luanda in March 2011, claiming that his visit marked a new phase in Sino–Angolan relations. During Premier Li's visit in May 2014, the first visit by a Chinese premier in 8 years, new agreements in construction and energy infrastructure were signed.

'Angola mode': oil-backed loans for infrastructure

Once the war ended in 2002, Sino–Angolan relations quickly shifted focus from defence and security to economic development. The bulk of Chinese financial assistance in Angola is reserved for key public investment projects in infrastructure, telecommunications and agro-businesses under the Angolan Government's National Reconstruction Programme (Campos and Vines, 2008).

The China Construction Bank (CCB) and China's Export–Import Bank (Ex–Im Bank) provided the first funding for infrastructure development in 2002. In fact, according to China's submissions to the UN Registry on Conventional Arms, China transferred no major military equipment to Angola after 2006.

Financial relations between China and Angola grew in late 2003, when a 'framework agreement' for new economic and commercial cooperation was formally signed by Angola's Ministry of Finance and the Chinese Ministry of Trade. In March, 2004, China's Ex–Im Bank pledged the first US$2 billion oil-backed loan to Angola to fund the reconstruction of shattered infrastructure throughout the country. The loan, payable over 12 years at a deeply concessional interest rate, was divided into two phases, with $1 billon assigned to each.[3]

Ex–Im Bank is using this deal structure more and more – what the World Bank calls the 'Angola mode' or 'resources for infrastructure' – whereby repayment for loans for infrastructure development is made in natural resources. This approach has a long history that originates with Japan. In the case of Ex–Im Bank, the arrangement is used for countries that cannot provide adequate financial guarantees and allows them to package together natural resource exploitation and infrastructure development. According to the World Bank, these loans offer on average an interest rate of 3.6 per cent, a grace period of 4 years and a maturity of 12 years (World Bank). The only unique thing about the 'Angola mode' is that Chinese engagement has been quick and loans have been large. Angolans argue that, over time, China's investment in the Democratic Republic of the Congo will probably become more significant than its investment in Angola.

These loans operate like a current account. When ordered by the Ministry of Finance, disbursements are made by Ex–Im Bank directly into the accounts of the contractors. Repayment starts as soon as a project is completed. If a project is not undertaken, no repayment is made. Revenue from oil sold under this arrangement is deposited into an escrow account from which the exact amount towards servicing the debt is then deducted. The government of Angola is free to use the difference at its own discretion.

Chinese government officials believe that oil-backed loans are the most beneficial arrangement because they offer the greatest security and have regularly indicated this preference to their Angolan counterparts. The Angolans, however, seem to continue to want to move away from the 'Angola mode' because these loans are costly and repayment depends heavily upon international oil prices. China is clearly seeking to secure more oil concessions, but at the same time is under pressure to provide better local content provisions in contracts for its companies (Interview with Chinese Officials, Luanda, 4 March 2009).

Chinese oil companies

China has shown great interest in Angola's extractive industries (Vines *et al.,* 2009). Following the opening of China's first credit line to Angola in March 2004, China Petrochemical Corp, better known as Sinopec Group, acquired its first stake in Angola's oil industry, 50 per cent of the BP-operated Block 18. Sonangol Sinopec International (SSI) is a joint-venture majority, owned by Sinopec with Angola's National Oil Company, Sonangol, and Hong Kong-based Beiya (now Dayuan) International Development. Although BP's former license partner Shell had agreed to sell its stake to India's state-owned Oil and Natural Gas Corporation (ONGC), the Chinese in their first involvement in the Angolan oil industry side-lined ONGC with an offer that media sources estimated at US$725 million (China Institute, 2006). The company reportedly spent an additional $1.5 billion to develop and explore the block (International Herald Tribune, 2006).

The Angolan state oil company also entered a long-term uplift agreement to supply Unipec, which Africa Energy Intelligence estimated could result in Sinopec (as the parent company) lifting up to 100,000 barrels per day (Africa Energy Intelligence, 2005). Additionally, the two parties signed a Memorandum of Understanding (MoU) to jointly study plans for the exploration of the shallow offshore Blocks 3(05) and 3(05A) (previously known as Block 3(80) – see earlier) that had been withdrawn from Total in late 2004. Later that year, Sonangol agreed that China Sonangol International Holding Limited (CSIH) would acquire the 25 per cent stake. CSIH does not have any Sinopec participation but the CSIH stake was handed over to SSI (where Sinopec holds a 55 per cent interest) in 2007.

Since 2004, China has obtained equity partnerships in Angolan deepwater oil blocks through Sinopec's majority in SSI and in shallow-water blocks through the CSIH, a joint venture between Sonangol and Hong Kong-based private business interests.

China has continued to purchase equity oil in Angola. In December 2011, Angola finally formalized agreements with several international oil companies for 11 exploration blocks in the country's first sub-salt licensing round. Sonangol had indicated early on that these complex pre-salt blocks would require technical ability and financial backing. BP, Total and ConocoPhillips operate two blocks each, and Repsol, ENI and Cobalt International Energy operate one each.

Minority stakeholders included CSIH, which in the lengthy negotiations seems to have been used as a stalking horse to up the signature bonus commitments. Exxon Mobil was initially thought to be set to take equity in Blocks 38, 39, 40 but did not appear in the final line-up. Perhaps a combination of inflated prices and reputation concerns related to CSIH convinced Exxon to walk away. Despite this, a joint venture between Sonangol and Sinopec purchased Marathon's 10 per cent stake in Block 31 in June 2013 (Ma, 2013).

As shown in Table 7.1, the fortunes of the SSI and CSIH joint ventures appear to be declining. In April 2014, a joint venture between the independent Genel and White Rose Energy Ventures acquired a 15 per cent share in the Statoil-operated Block 38 for a price of US$59 million. Following the acquisition, Statoil retains a 55 per cent stake, while the remaining 30 per cent is held by Sonangol P&P. Sonangol has approved this deal, which was also facilitated by CSIH's inability to pay its way and increasing reputation anxiety by Statoil about CSIH's beneficial ownership.

Inter-Chinese rivalry

One cannot really speak of China as a unified actor in Angola due to the increasing number of Chinese interests and actors in Sino–Angolan oil relations as Chinese national oil companies (NOCs) actively compete with each other abroad. Sinopec Corporation's Chairman of the Board of Directors, Dr Su Shulin, visited Luanda in April 2008. A delegation from Zhen Hua Oil (an affiliate of NORINCO) followed in March 2008 to sign a MoU with Sonangol (Africa Energy Intelligence, 2008). This competition is manifested in the 'Marathon saga'. In July 2009, Marathon International Petroleum, Angola Block 32 Ltd, a subsidiary of Marathon Oil, entered into an agreement with CNOOC and Sinopec, whereby the Chinese companies were to purchase a 20 per cent interest in a production-sharing contract and joint operating agreement in Block 32 for US$1.3 billion, effective from 1 January 2009. Marathon Oil was to retain a 10 per cent interest. The companies expected to close the deal by year-end 2009, subject to government and regulatory approvals. Unusually, China National Petroleum Corporation (CNPC) made a separate bid. In the end, CNOOC and Sinopec both lost out, as did CNPC because Sonangol exercised its right of first refusal and offered the stake to CSIH. This is not the first time oil politics have not gone smoothly for Chinese NOCs.

Table 7.1 China's exploration and production assets in Angola[1]

Block (s)	Company	Year acquired	Percent acquired	Partners (%)
31	CSIH (from TEPA)	2010	5	BP [OP] (26.67), Sonangol (45), Statoil (13.33), Marathon (10)
32	CSIH (from Marathon)	2010	20	Total [OP] (30), Sonangol P&P (35), Marathon (10), Galp (5)
15 (06)	SSI	2006	25	ENI Angola EXPL. [OP] (35), Sonangol P&P (15), TOTAL (15), Falcon Oil (5), STATOIL (5)
17 (06)	SSI	2006	27.5	TOTAL [OP] (30), Sonangol P&P (20), Somoil (10), Falcon Oil (5), Acrep SA (5), Partex (2.5)
18 (06)	SSI	2006	40	Petrobras [OP] (30), Sonangol P&P (20), Falcon Oil (5), Geminas (5)
3/05 and 3/05A	CSIH (from SSI in 2007)	2005	25	Sonangol P&P [OP] (25), Ajoco (20), ENI Angola EXPL. (12), SOMOIL (10), NAFTGAS (4), Ina-naftaplin (4)
18	SSI	2004	50	BP [OP] (50)
38/11	CSIH	2011	15	Statoil [OP] (50), Sonangol P&P (30)

Source: Author-compiled from industry sources, 2014.

Notes:

1 CSIH held shares in Blocks 19/11 (40), 20/11 (30) and 36/11 (50) in 2011, and Block 38/11 in 2014. Over 2012–14, these shares shifted to other stakeholders, indicating a decline in CSIH's presence in the pre-salt blocks.

Chinese negotiations with a domestic oil refinery collapsed in 2007 when Sonangol decided to manage the project on its own.

This was the first time Chinese companies openly competed amongst themselves for Angolan concessions, with the biggest winner being CSIH. Similarly, efforts by CNOOC in 2011 to partner with other Asian NOCs, such as Indonesia's Pertamina, for Angolan oil concessions have yet to succeed; however, in future, CSIH and SSI's liquidity and reputation issues could handicap development of lucrative oil blocks.

The China International Fund, the GRN and local politics

From its genesis, oil-backed borrowing became an effective tool for the Angolan presidency to control spending priorities, bypassing the inefficient financial system. As a result, the Gabinete de Reconstrução Nacional (GRN) became the epicentre of this exercise of ultimate power.

The GRN was set up in 2005 to manage large investment projects and ensure rapid reconstruction of infrastructure prior to national elections. The initial head was a military adviser to the president, General Helder Vieira Dias 'Kopelipa'. The China International Fund Ltd (CIF), a private Hong Kong-based institution, provided the loans to undertake these projects.

A senior government official close to the president valued GRN's projects at US$10 billion, while in April 2007, the World Bank published an Angolan Ministry of Finance estimate of the loan as $9.8 billion at Libor plus 1.5 per cent (World Bank, 2007). The US Department of State's '2008 Investment Climate Statement – Angola' estimates the CIF loan at between US$2.9 billion and US$9 billion (US Department of State, 2008). In addition to the uncertainty regarding the exact figures, how these funds were allocated across projects is also unclear.

The opaqueness of the CIF loans began with the first loan in March 2004; thereafter, CIF's opacity attracted renewed media attention. In March 2007, Chinese construction company Hangxiao Steel Structure Co Ltd was investigated by the China Securities Regulatory Commission (CSRC) for stock-price rigging in deals related to Angola (*Xinhua Online*, 2007), but on 17 October 2007, the Ministry of Finance denied any misuse of Chinese funds. The Ministry also published details of the lines of credit managed by the ministry (*Jornal de Angola*, 2007). Unlike projects undertaken by the Ministry of Finance, how the GRN managed its funds remains unclear.

Putting China into context

Since the early 1990s, China's thirst for energy transformed it from self-sufficiency into the world's second largest oil importer, making it vulnerable to global market shocks and increasingly reliant on its allies in Luanda (Alves, 2012: 106). From Angola's perspective, China provides the funding for strategic post-conflict infrastructure projects, which is rarely forthcoming from Western donors. Moreover, China offers better conditions than commercial loans, lower interest rates and longer repayment time compared to non-Chinese credit lines, secured in 2004, which demand higher guarantees of oil, no grace period and higher interest rates.

More importantly, China provided financing when concessional funding was not available to Angola. Recurrent episodes of hyperinflation and stabilization had prevented any lasting accord with the International Monetary Fund (IMF), without which the World Bank could only provide emergency and humanitarian aid. Moreover, the IMF insisted on a more serious level of transparency in the use of resource wealth. China, however, provided a new model of cooperation based on credit lines, which contrasts with Western aid attached to conditionality.

Despite the importance of the loans and the enormous number of Chinese construction workers in Angola in 2011, China's influence in Angola is often overstated by Western media. According to Horta, (2011: 20) 'one would expect...that Angola would fall under greater Chinese control', but 'while China has gained an impressive economic presence in Angola, its political and diplomatic influence is growing weaker by the day and its soft power is rather weak'.

Similarly, although Chinese loans for infrastructure are numerous, Angola's developmental needs are much larger. Addressing the OPEC summit in November 2007, President dos Santos stated that, 'Angola needs over $20 billion to guarantee its reconstruction (dos Santos, 2007). The Angolan government is therefore borrowing from other traditional commercial partners, such as Portugal and Brazil, who remain fully engaged in Angola's post-war reconstruction. Also, rejecting any rigid adherence to dogma, President dos Santos's 2008 address at New Year to the diplomatic corps stressed government plans to reinforce bilateral and commercial relations with other countries (i.e., other than China):

> globalization naturally makes us see the need to diversify international relations and to accept the principle of competition, which has in a dynamic world, replaced the petrified concept of zones of influence that used to characterize the world.
>
> (Vines *et al.*, 2009: 56)

This view resonates well with Angola's 'neutral position' in Zweig's model. Other than publicly stating its desire to diversify its international and economic relations, Angola's oil production relations have changed little. Western firms still dominate Angola's oil landscape with Chevron, ExxonMobil, Total, BP and ENI as Sonangol's key partners. Moreover, while Angolan officials are open about their cooperation with China, they are candid about not wanting to depend on any one partner.

For example, the dramatic decline in Angolan exports to the US since 2011 has opened the door for India, which could be a natural partner and could soon overtake the US as an energy importer. In 2010, it signed an MoU, strengthening cooperation in Angola's oil sector; however, Indian companies lack the cash to compete with China for oil concessions. Indian oil companies are also unlikely to obtain any major oil block operatorships or equity because of cash and technical limitations.

Angola is therefore extremely conscious of its sovereignty and refuses to allow any foreign power to dominate it, politically or economically[4] (Corkin, 2013: 130). Angola exports much of its oil to China but skilfully maintains a wide portfolio of partnerships.

US and Angola

US–Angolan relations deepened when Washington recognized Angola's government in 1993. Previously, the US gave UNITA significant covert support, especially in 1974–76, until prohibited by the US Congress's Clark Amendment in 1976. Covert aid resumed after the amendment was repealed in 1985, totalling US$250 million between 1986 and 1991 – the second largest US covert programme, exceeded only by aid to the Afghan mujahidin.

The Clinton Administration made Angola a priority in Africa, providing US$500 million for humanitarian assistance. The US government also supported commercial development through a US$350 million Ex–Im Bank loan and Trade Agency assistance. US Secretary of State Madleine Albright, while visiting Angola in December 1997, announced that Angola supplied the US with 7 per cent of its oil imports, three times what Kuwait supplied just before the Iraqi invasion.

Angola initially viewed President George W Bush Jr's election with trepidation because the Republican Party had previously associated with Savimbi and UNITA; however, by the late 1990s, President dos Santos recognized the 'unipolar moment' and the US's status as the only superpower. In 1999, Angola was the US's second largest site for investment and third largest trading partner in sub-Saharan Africa, with the majority of this trade being oil, which exceeded 750,000 barrels per day. US investment in Angola's petroleum sector was over US$4 billion, making the US Angola's largest trading partner, purchasing 50 per cent of its oil exports. In a speech in February 2002, President Dos Santos stated:

> With significant investment from US companies, our petroleum sector is well established and steadily expanding, American companies including Exxonmobil, ChevronTexaco and Ocean Energy have been and will continue to be valued partners in developing our oil resources. Numerous other US businesses such as Haliburton, Global Marine Drilling, McDermott and Oceaneering International are providing vital support to the oil industry.
>
> Angola already supplies about 5 per cent of the oil imported into the US and we expect this to increase in the very near future. As a non-OPEC oil producing country, we want to contribute to its energy security, thus I am convinced that from this cooperation Angola will have the necessary reciprocal advantages for the resolution of its problems.[5]

A similar message was communicated in May 2004, in a second speech given by President dos Santos in Washington DC stating that:

> As I had the opportunity to state recently in a letter to President Bush, Angola would like to give preferential treatment to the US market in all this process, which is why it extended the oil exploration rights of Chevron–Texaco for another thirty years, and in partnership with that company, Exxon Mobil and others, is implementing a project for the production of liquefied natural gas.
>
> (Ibp USA, 2006)

In that same month, President dos Santos agreed to allow Chevron to publish its signature bonus payment for renegotiated Block 0, which was offshore of Cabinda. This was the first time that Luanda had made information about its revenues from a multinational oil company working in Angola available; however, just over 2 years later, in December 2006, Washington and US oil majors were all caught off-guard by Angola announcing that it was joining OPEC.

Nevertheless, Angola in 2013 was the second-largest trading partner of the US in sub-Saharan Africa (after Nigeria), mainly because of its petroleum exports. US exports to Angola primarily consist of industrial goods and services – such as oilfield equipment, mining equipment, chemicals, aircraft and food.

In 2014, energy security was a prime driver of a more focused US policy on sub-Saharan Africa. The US imports around 20 per cent of its oil from Africa, only slightly less than it does from the Middle East, which accounts for around 22 per cent (Energy Information Administration [EIA], 2013). The size of the Angola lobby in the US is partly attributable to this fact; however, in comparison to the amount of national resources, political and physical, that are invested in the Middle East, Africa has received only minimal attention in Washington to date – a situation that is likely to change gradually.

African politicians have a history of skilful diplomacy and playing great power suitors off each other in order to secure special concessions for themselves. A key reason why African leaders support US policies has been a pragmatic calculation of self-interest based on the acceptance, however reluctant, of a unipolar world order in which the US stood as the world's only superpower. China's dramatic increase in commercial and diplomatic investment across Africa since the turn of the century is changing this view, particularly as commercial and cultural links between African states and China grow. For Europe and the US, the growing presence in Africa of 'emerging powers', such as China, has pushed African issues higher up the policy agenda (Lyman and Morrison, 2006).

Unlike China and a number of European countries, US firms, outside the oil sector, have lagged behind others in seeking new markets in Africa. Although US investment in Africa outside the energy sector has remained almost static over the past decade, its cumulative level of investments in sub-Saharan Africa as a whole has not yet been displaced by Asian investments, and the US remains Africa's largest trading partner.

This relative lack of private sector investment undermines US policy effectiveness on the continent. It means that US relations with African countries lack the complex business links that are so essential for cementing political ties, and that contributes to a more nuanced and sympathetic African view of US intentions and values.

US companies dominate Angolan oil

Angolan oil is plentiful and accessible and is the type of high quality crude that the US needs. By 2011, the US imported around 3 per cent of its oil from Angola, which accounted for 14 per cent of Angola's total oil exports. As of 2008, Block 15, located offshore of Soyo, provides 30 per cent of Angola's crude oil production. Exxon Mobil, through its subsidiary Esso, is the operator, with a 40 per cent share. In 2005, Block 15's second major subfield, Kizomba B, came on line, producing about 250,000 barrels per day. Chevron operates Block 0, offshore of Cabinda, which provides about 20 per cent of Angola's crude oil production. In 2007, Block 0 had a total production of 370,000 barrels per day, and drilling activity continues at a high level. Chevron also operates Angola's first deepwater section to go into production, Block 14, which started pumping in January 2000 and produced 105,000 barrels per day in 2006. Despite the Chinese walking away from the project, plans for a refinery in Lobito, with a processing capacity of 200,000 barrels per day, have come to fruition. The US firm KBR has been selected to do the front-end engineering and design work, while the UK's Standard Chartered Bank will provide financial consultation on the project. Sonangol announced that commissioning would begin in 2017 or 2018.

The corporate oil mergers of 1999–2001 had the effect of consolidating major oil company assets in Angola (Fina and ELF within TFE, Amoca with BP, Exxon and Mobil, Chevron and Texaco); however, by 2000 there were obvious exceptions, with the Chinese, Indians and several significant independent actors remaining untouched by the mergers. The position of companies, particularly from the US, in Angola's upstream therefore reflected competitiveness, pragmatic strategy and geopolitical considerations. Even during the Cold War, a time when the US supported the UNITA rebels in Angola, US oil companies such as Chevron, Texaco and Exxon maintained a successful presence in Angola and enjoyed the support of the government in Luanda.

Angola is increasingly shifting from being an important source of crude oil and gas exports for the US to being of interest in helping to maintain peace and security in central Africa. US international oil companies (IOCs) continue to dominate offshore Angolan oil production, and US strategic planners are not particularly worried by China because they believe Angola's leaders will seek to diversify their partnerships globally.

Trying to reinvigorate bilateral relations

During her 11-day visit to Africa in August 2009 (the longest ever to Africa by a US Secretary of State), Hillary Clinton visited seven African countries, including Angola. Along with South Africa and Nigeria, Clinton's choice of Angola highlighted one of the administration's key priorities in Africa – the establishment of good relations with what is likely to become the three 'anchor states' of sub-Saharan Africa.

Hillary Clinton's mission in Angola was to build a strategic partnership between the two countries and followed a visit by the Angolan foreign minister to Washington in May 2009. In Angola, Clinton spoke of the two countries as 'partners, friends and allies' (*Angola Press,* 2009). When the US established diplomatic relations with Angola in 1993 during the Clinton Administration, the relationship was initially warm (see Partricio, 1997). Following the collapse of the Lusaka peace process and the resumption of war in the latter half of 1998, the Clinton Administration backed away from an active involvement in the search for peace in Angola. Instead, a concerted effort was made to strengthen the bilateral relationship with the Government of Angola through the establishment of the US–Angola Bilateral Consultative Commission, which met formally three times during 1999–2000 to discuss a wide range of issues.

The decision to visit Angola on Clinton's maiden visit to Africa was seen in Luanda as proof that the US wanted to work constructively with the government of José Eduardo dos Santos. As a country in transition, Secretary Clinton's approach to Angola was seen as pragmatic. She simultaneously called for good governance and presidential elections, while also declaring that the US was 'satisfied' with Angolan efforts so far and would encourage further efforts to promote transparency. Furthermore, Clinton stated that she was convinced that presidential elections would occur 'in due course' upon ratification of the new constitution.

The US and Angola sought to resuscitate their bilateral consultative commission. In November 2009 in Washington DC, the US and Angola launched meetings under a new framework for sustained bilateral engagement, with working groups on energy and security cooperation. This was followed in July 2010 by Secretary Clinton signing an agreement in Washington for a Strategic Partnership Dialogue with Angola's foreign minister. In May 2014, US Secretary of State John Kerry visited Luanda briefly and congratulated Angola on its diplomacy as the chair of the International Conference of Great Lakes region. Kerry also said that the US and Angola would begin an 'energy dialogue' about increasing economic cooperation in the areas of energy, technology, infrastructure and agriculture. The US Commerce Department also planned to open a commercial service office in Angola soon, according to Kerry. Although US officials hoped this would open a new chapter in bilateral ties between the two countries, this has not happened and President dos Santos did not attend the US–Africa leaders' summit in Washington DC in August 2014. Still, in early 2012, Sonangol withdrew from a natural gas project in Iran after the tightening of US-led sanctions on that country.

Good governance and corruption

Like China, the US relationship with Angola will continue to experience challenges. Good governance and anti-corruption limitations in Angola are major issues for the US and its companies. This was highlighted by a banking crisis in Washington DC in late 2010, when Angola and 16 other African missions faced US bank account closures because of probity issues. The banking crisis had been triggered by a US Senate investigation subcommittee, which in a February 2010 report cited Angola 'for an ongoing corruption problem, weak anti-money laundering controls and a cash intensive banking system' (All Africa, 2010). Both the Bank of America and HSBC closed their Angolan embassy accounts, resulting in problems issuing visas, cancellation of the Angolan national day in Washington DC and the recall of the Angolan ambassador in Washington to Luanda for consultations. Citibank also closed its accounts for the Angolan Central Bank and other Angolan government entities.

This issue became a high priority for the Obama Administration, with the State Department confirming it was engaged in efforts to assist. In May 2011, Chris McMullen, the then US ambassador to Angola, acknowledged that the banking crisis represented a significant fissure in US–Angola relations. Further, the 'Lugar–Cardin' Provision in the US Dodd–Frank Act of 2010 that requires all US-listed/registered extractive industry companies to disclose payments to governments has also made some senior Angolan officials feel increasingly uncomfortable with the US.

Such pragmatism needs to be placed within a broad, consistent, positive vision of the democratic values and prosperity agenda that the US seeks if it is to remain a leading voice for reform-minded engagement in sub-Saharan Africa in the face of ever-growing global competition. The risk is that a plethora of diverse US approaches to specific countries will undermine the central message of the need for better governance across the region and overwhelm the policymaking process in Washington.

Despite Washington's concerns over the transparency of the Angolan oil sector, the actions of some US companies have been controversial. In 2012, Houston-based Cobalt International Energy became the subject of an investigation by the Department of Justice and the Securities and Exchange Commission (SEC). Three key figures in the Luanda government have confirmed that they held previously concealed shares in Nazaki, an Angolan oil company linked to Cobalt (Burgis, 2013). As of October 2014, the corruption investigation remained unresolved, and in August 2014 the US regulator issued a Wells Notice, indicating that the SEC believes it has gathered sufficient evidence of wrongdoing potentially to bring charges. In the same month, the state-owned *Jornal de Angola* announced that Nazaki would no longer be allowed to keep its equity, and Cobalt shortly afterwards in a regulatory filing stated that two Angolan companies – Nazaki and Alper Oil – had transferred their interests in an offshore oil venture operated by Cobalt to Sonangol (Economist Intelligence Unit, 2014).

These developments signal that, as with China's Sonangol, the government is trying to be seen as more transparent. It is also a pragmatic assessment that Block 21 needs significant financial investment, and concerns over corruption have clearly dissuaded a number of supermajors from considering buying Cobalt, despite reports of impressive oil discoveries in its Block 21. As Sinopec discovered through SSI and CSIH, Cobalt is finding that opaque joint ventures can have a long-term impact on business.

It is impossible to discuss current US–Angola ties without considering the former's booming domestic energy sector. The United States' domestic crude oil production increased by almost 54 per cent between 2011 and 2014; consequently, American demand for Angolan oil plummeted by more than 70 per cent over the same period (EIA). The US shale gas revolution has also impacted Angola LNG, a Soyo-based liquefied natural gas (LNG) plant co-owned by Sonangol, Chevron, ENI, BP and Total, which was constructed at a cost of US$10 billion and with the initial intention of exporting LNG to the lucrative US market.

The sudden disappearance of the US market for Angolan gas was unexpected and has forced a complete rethink of where Angola's LNG will be exported to. Angola LNG's first cargo was successfully delivered to Brazil's Petrobras in June 2013, and global demand is expected to remain stable for the next few years at least, owing to Chinese growth and Japan's post-Fukushima energy needs (Wallace, 2013). However, with the US no longer an export option and the current boom of LNG plant construction around the world, the viability of Angola LNG in the long run is far from certain and is not helped by ongoing technical problems at the LNG plant.

The sharp decline in US demand for Angola's oil has meant that Luanda is increasingly reliant on Beijing and its other Asian partners, with India close to overtaking the US as the second largest buyer of Angolan crude.

Conclusion

Both China and the US have increasingly found that good governance issues have complicated their relationships with Angola, which have impacted on their oil relationship. To this day, Sonangol enjoys the political backing of the presidency. As discussed earlier, a joint CNOOC/Sinopec bid for Marathon Oil's 20 per cent of Block 32 in Angola was obstructed by Sonangol exercising its pre-emption rights. Western and Asian oil companies are finding that African states are calling the shots and that joint venture partnerships, including with their rivals or host national oil companies, are the future. This is not resource nationalism; rather it is host government strategic pragmatism and diversification. Whereas Europe's big oil companies are seeking partnerships with their Chinese counterparts in Africa and elsewhere, such steps have yet to happen significantly with US firms. It may be just a question of time because the oil landscape is changing and there is room for triangularization on mutual challenges, including obtaining sufficient oil production and international best practice for the global economy.

This does not mean that Chinese oil companies will simply replace Western ones. Although the Chinese government wanted its NOCs to prosper in Angola, since the early 2000s they have been forced by Angolan politics to enter into inefficient and risky joint-ventures, such as through CSIH. It is China's Ex–Im Bank that has been successful, facilitating the entry and continued growth of Chinese construction companies and whose credit loans have often been oil-backed (Corkin, 2011). It is oil provided as collateral for loans rather than oil concessions or equity that China has gained from Angola. This is unlikely to change because Western oil companies have significant technological advantage and China entered late in the game when much of the shallow water pickings have already been taken.

There are few signs of triangulation between the US and China in Angola over oil. Technology and skills for ultra-deep or pre-salt exploration and production guarantees US oil companies a major footprint in Angola for many years to come, even if the amount of oil exported to the US dramatically declines due to shale gas. Chinese capital, as we saw in the financing of BP's effort in Block 18, will continue to play a role, but what is unclear is how much equity oil Angola will lock into future infrastructure and trade deals with China.[6] With over 40 per cent of Angolan oil exports going to China in 2013, this is a serious question less for the US than for other emerging economies thirsty for diversified oil supplies.

Notes

1 This chapter draws on Campos and Vines, 2008; Vines, 2011 and Vines and Cargill, 2010. Daragh Neville provided research assistance for this paper.
2 In the mid-1990s, when low commodity prices squeezed Angola, Taiwan offered significant incentives to switch recognition to Taipei albeit without success. Still, Taiwan imported 6 per cent of Angolan oil exports in 2010.
3 The terms of the loan are Libor plus a spread of 1.5 per cent, with a grace period of up to 3 years.
4 Manuel Vicente, then President of Sonangol criticized the Chinese in the Angolan media, claiming that 'we can't construct a refinery just to make products for China'. This would suggest some resistance on the strategy of locking in supplies through long-term contracts, which China has applied elsewhere. However, according to Chang Hexi, the Chinese Economic Counsellor in Luanda, negotiations over the refinery were deliberately obstructed by the Chinese negotiators because they were not genuinely interested in the deal (Centre for Chinese Studies, 2007).
5 Text of speech on file with author.
6 This also includes a 2009 deal between Trafigura and DT Group, which, through a set of front companies set up in Angola, the Bahamas and Singapore, controls one of the largest oil swap arrangements in the world, much of it for China (Weiss, 2014).

References

Africa Energy Intelligence (2005) The Angola–China connection. *Africa Energy Intelligence*, 27 July 2005.

Africa Energy Intelligence (2008) A controversial figure pops up in Beijing. *Africa Energy Intelligence*, 18 June 2008.

African Development Bank (2011) Angola – 2011–2015 Country Strategy Paper and 2010 Country Portfolio Performance Review. Abidjan: African Development Bank.

All Africa (2010) Angola and 16 other African missions facing U.S. bank account closures, 16 November. Available online at: http://allafrica.com/stories/201011160010. html (accessed 8 April 2015).

Alves, A.C. (2012) 'Taming the dragon: China's oil interests in Angola', in: M. Power and A.C. Alves (eds.), *China and Angola: A Marriage of Convenience?* Cape Town: Pambazuka Press, pp. 105–23.

Angola Press (2009) Chefe de estado fala de cooperação com Hillary Clinton. *Angola Press,* 11 August. Available online at: www.portalangop.co.ao/angola/ pt_pt/noticias/politica/2009/7/33/Chefe-Estado-angolano-fala-cooperacao-com-Hillary-Clinton,9bf3a1df-5bcc-4815-8be1-24857e8e7532.html (accessed 30 January 2014).

Angola Press News Agency (2007) Angolan leader addresses OPEC summit in Saudi Arabia, 19 November 2007.

Brown, D. (2013) *Africa's Booming Oil and Natural Gas Exploration and Production: National Security Implications for the United States and China.* Carlisle, United States Army War College Press.

Burgis, T. (2013) Cobalt's returns from Angolan venture raise wider concerns. *The Financial Times,* November 20. Available online at: www.ft.com/cms/s/0/36e28cf6-4bb5-11e3-a02f-00144feabdc0.html#axzz2rsUayY7 (accessed 30 January 2014).

Campos, I. and Vines, A. (2008) Angola: a pragmatic partnership. Working paper presented at CSIS Conference Prospects for Improving US–China–Africa Cooperation, December 5, Center for Strategic and International Studies.

Centre for Chinese Studies (2007) *China's engagement of Africa: preliminary scoping of Africa case studies*. Stellenbosch: Stellenbosch University.

China Institute (2006) *Sinopec beats ONGC, gets Angola block*. China Institute, University of Alberta, 14 July 2006.

China's National Energy Strategy and Policy 2000–2020 (2003) Available online at: http://csep.efchina.org/report/2006102695218495.5347708042717.pdf/Draft_Natl_E_Plan0311.pdf (accessed 8 April 2015).

Clarkson Research Services Limited (2013) Crude in West Africa: all augurs well for Angola. Available online at: www.clarksons.net/sin2010/markets/Feature.aspx?news_id=34669 (accessed 30 January 2014).

Corkin, L. (2011) China and Angola: strategic partnership or marriage of convenience?, *CMI Angola Brief*, 1 (1): 1–4.

Corkin, L. (2013) *Uncovering African agency: Angola's management of China's credit lines*, Farnham: Ashgate.

Economist Intelligence Unit (2014) EIU Angola Country Report, 3 September. Available online at: www.eiu.com (accessed 9 April 2015).

dos Santos, J.E. (2007) OPEC Summit, Riyadh, Kingdom of Saudi Arabia, 17 November 2007.

Energy Information Administration (EIA) (2013) *Petroleum and other liquids*. Available online at: www.eia.gov/petroleum/ (accessed 30 January 2014).

Eurasia Group (2011) African security in strategic perspective, 22 December. New York, NY: Eurasia Group.

Ganesan, A. and Vines, A. (2004) 'Engine of war: resources, greed and the predatory state', in Human Rights Watch (ed.), *Human Rights Watch World Report 2004*, New York, NY: Human Rights Watch.

Horta, L. (2011) 'China's waning influence in Angola,' *Diplomatic Courier*, 26 August 2011.

Ibp USA (2006) *Angola President Jose Eduardo dos Santos Handbook*. Washington, DC: International Business Publications.

International Herald Tribune (2006) China buys more Angolan crude than Saudi. *International Herald Tribune*, 19 March.

Jornal de Angola (2006) PR defende cooperação constituva com a China. *Jornal de Angola*, 21 June 2006.

Jornal de Angola (2007) Governo nega mau uso dos créditos da China, communicado do Ministério das Financa. *Journal de Angola*, 18 October 2007.

Lyman, P. and Morrison, J.S. (2006) *More than humanitarianism: a strategic US approach to Angola*. New York, NY: Council on Foreign Relations.

Ma, W. (2013) Sinopec's joint venture raises stake in Angola deepwater block, *World Oil*, 24 June. Available online at: www.worldoil.com/news/2013/6/24/sinopec39s-jv-raises-stake-in-angola-deepwater-block (accessed 9 April 2015).

Partricio, J. (1997) *Angola – EUA: Os caminhos do bom-senso*. Lisbon: Publicações Dom Quixote.

Portugal News (2007) Angola exporta 29.9 bilioes de dolares em petroleo, *Portugal News*, 8 August.

Reuters (2011) China lends Angola $15bn but creates few jobs. *Reuters*, 6 March 2011.

SME (2012) Signed agreement of police cooperation between Angola and China. 25 April. Available online at: www.sme.ao/index.php?option=com_content&view=-article&id=653:rubricado-acordo-de-cooperacao-policial-entre-angola-e-china&catid=50:noticias&Itemid=122&lang=en (accessed 30 January 2014).

Soares de Oliveira, R. (2007) Business success, Angolan style: postcolonial politics and the rise and rise of Sonangol. *Journal of Modern African Studies,* 45, 4: 594–619.

US Department of State (2008) 2008 Investment Climate Statement – Angola. Washington, DC: US Department of State.

Vines, A. and Cargill, T. (2010) Sub-Saharan Africa: providing strategic vision or fire-fighting? In: R. Niblett, (ed.), *America and a Changed World: A Question of Leadership.* London: Wiley-Blackwell, 49–71.

Vines, A., Wong, L., Weimer, M. and Campos, I. (2009) *Thirst for African oil: Asian oil companies in Nigeria and Angola.* London: Chatham House.

Wallace, R. (2013) Fukushima fallout fuels LNG demand. *The Australian,* 15 November. Available online at: www.theaustralian.com.au/news/features/fukushima-fallout-fuels-lng-demand/story-e6frgabx-1226759922652 (accessed 30 January 2014).

Weiss, M. (2014) The 750 million dollar man. *Foreign Policy,* 13 February. Available online at: www.foreignpolicy.com/articles/2014/02/12/the_750_million_dollar_man_trafigura_angola_general_dino (accessed 28 April 2015).

World Bank (n.d.) Building bridges: China's growing role as infrastructure financier for sub-Saharan Africa, executive summary. Available online at: http://siteresources.worldbank.org/INTAFRICA/Resources/BB_Final_Exec_summary_English_July08_Wo-Embg.pdf (accessed 9 April 2015).

World Bank (2007) Annex 7 – non-concessional borrowing of Angola since 2004. In: International Development Association interim strategy note for the Republic of Angola, *Report No. 39394-AO,* 26 April 2007. Available online at: http://documents.worldbank.org/curated/en/2007/04/7584493/angola-interim-strategy-note (accessed 30 January 2014).

Xinhua Online (2007) Hangxiao Steel Structure defends handling $4.4bn Angola contracts. *Xinhua Online,* 27 March.

8 Triangular or parallel?

China's relations with Nigeria in the context of the US's ties with Abuja

Ian Taylor

Nigeria is China's third major export destination in Africa after South Africa and Angola. Currently, China is Nigeria's largest source of imports and third major trade partner (*Vanguard,* 2015). To gauge the exponential rise in Sino–Nigerian trade relations, consider that in 1998 trade volume was US$384 million, and by 2014 the trade volume between the two countries had reached US$18.1 billion (*Ventures Africa,* 19 February 2015). The bilateral trade volume represents over 30 per cent of China's total trade with the whole of West Africa. Signifying the importance of Nigeria to China, Abuja signed a Memorandum of Understanding (MoU) with Beijing in January 2006 on the Establishment of a Strategic Partnership, becoming the first African country to sign such an agreement. Other anecdotal points back up the importance that China places on Nigeria: China's first scheduled direct flight to Africa was between Beijing and Lagos via Dubai (inaugurated in 2006) and the first newspaper in Chinese in West Africa was started in Nigeria in 2005.

Chinese imports of Nigerian crude petroleum have been increasing year-on-year and although Nigeria remains relatively insignificant as an importer for China, the Chinese market is seen by the government in Abuja as extremely important. Indeed, as Nigerian exports to the US have collapsed (see Table 8.1), the government in Abuja has been desperately looking to alternative markets, with China one of the prioritised destinations (*This Day* (Lagos), 3 October 2015).

Nigeria is an emerging strategic partner for China, and Beijing is investing heavily in both commercial and political terms in the country. These efforts have important implications for the US, which for many years saw Nigeria as an important player in the region, because it has extensive trade links with Abuja and has often cast Nigeria as an ally of some import on the continent (Falola and Heaton, 2008). As late as 2010, Johnnie Carson, US Assistant Secretary for African Affairs, when he met with international oil industry actors in Lagos in February 2010, stressed that:

> Nigeria [was] the most important country in Africa for the US due to the size of its population; presence of hydro-carbons; peacekeeping role in ECOWAS, especially in Sierra Leone and Liberia; its seat on the United Nations Security Council; [and] the strength and size of its financial markets'.
>
> (cited in WikiLeaks, 2010a)

Table 8.1 Nigerian exports of crude petroleum to China (US millions)

2000	2001	2002	2003	2004	2005	2006	2007	2008	2009	2010	2011	2012
231	138.5	75	28	359	440	195	463	240	620	717	731	818

Source: Observatory of Economic Complexity, 2015.

Nigeria is the seventh largest Muslim country in the world, with an Islamic population that will eclipse Egypt by 2015; however, when examining Sino–Nigerian ties in the context of the US, we must tame our hyperbole. This chapter analyses contemporary Sino–Nigerian relations and how the dynamics of Sino–American rivalry – or at least competition – plays out within a framework of relations and networks. In doing so, it contextualises some of the problems that both external actors must manage in developing linkages with the African continent's most populous nation and it also analyses some of the problems Chinese actors have experienced in Nigeria. As will be shown, Washington DC is not overly concerned by China's rise in the country; the US's economic ties with Abuja are declining and there is little perception that US interests are being threatened by China. Indeed, given Washington's precipitous waning of interest in African oil, China is pushing at an open door in Nigeria and is unlikely to face much opposition from the US. Before discussing such issues, however, the political economy of Nigeria and the environment that external actors, both Chinese and American, encounter needs to be reviewed because ultimately it is this environment that makes Nigeria a rather difficult partner.

The Nigerian context

Regrettably, Nigeria is a byword for corruption and malgovernance (see Lewis, 1996; Maier, 2000; Cunliffe-Jones, 2010; Okonjo-Iweala, 2012; Adebanwi and Obadare, 2013). The country currently produces 10 per cent of the oil consumed by the US, and currently holds roughly half of the Gulf of Guinea's oil reserves (Neumann, 2004). Yet a World Bank report estimated in 2005 that as much as 80 per cent of Nigeria's oil revenues benefited just 1 per cent of the country's population (cited in *Daily Mail*, 9 August 2013). Although the Niger Delta region produces 90 per cent of Nigeria's oil and over 75 per cent of the country's export earnings, very little of the wealth has been seen by residents in the Delta (Avuru, 2005). Since independence in 1960, only one Nigerian head of state – the current president, Goodluck Jonathan of Bayelsa State – has had origins in any one of the oil-producing states. In fact, 'the northern predominantly Hausa region has benefited in a disproportionate manner from oil resources, contributing to grievances by the rest of the country and ongoing instability' (White and Taylor, 2001: 333).

Due to the low level of governance and the misallocation of revenues, Nigeria's social indicators are appalling. In 2014, life expectancy remained very low at 51 years (lower than in Somalia) and Nigeria's Human Development Index-score of 0.459 puts Nigeria at position 156 out of 169 in the country rankings. In 2014,

84.5 of Nigerians live on less than US$2 a day, i.e. they live in absolute poverty as defined by international institutions. Typically, 'despite the improvements in fiscal management, budgets [are] not implemented as stated, funds [are] impounded by the President, and extra-budgetary spending continue[s]' (Gillies, 2007: 576).

At the same time, the media in Nigeria is extremely active and vibrant, and it acts as a watchdog on malgovernance. Although by no means perfect, it tries to keep track of particularly serious issues, such as allegations of corruption against top government officials. The media and wider civil society are also highly critical of foreign actors seen to be exploiting Nigeria. This surveillance and reporting makes Nigeria a very different place for the Chinese to operate in than Angola (Jibo and Okoosi-Simbine, 2003; see Chapter 7, this volume).

Despite these efforts by civil society to disclose corruption, well-connected political insiders steal approximately over 100,000 barrels of oil per day, worth circa US$1.46 billion a year. Government budgets are routinely estimated on projected income, based on assumptions that are wildly below the actual revenues collected (Nwankwo, 2002). In 2013, projected revenue was US$80 billion from oil exports, but even by mid-year US$7 billion had gone 'missing' (*Bloomberg News*, 1 November 2013). It is anyone's guess where the surplus finances from oil sales go – certainly not into government coffers or to broad mass of Nigeria's citizens (*allAfrica*, 2004). Indeed:

> Some Western diplomats estimate that Nigeria lost a minimum average of US$4 billion to US$8 billion per year to corruption over the eight years of the Obasanjo administration [1999–2007]. That figure would equal between 4.25% and 9.5% of Nigeria's total GDP in 2006. To put those numbers in perspective, a loss of 9.5% of the United States' GDP to corruption in 2006 would have translated into $1.25 trillion in stolen funds or $222 billion (GBP 108.6 billion) in the case of the United Kingdom's economy.
>
> (*Daily Trust*, 2007)

In December 2013, a letter from the governor of the Central Bank of Nigeria to the Nigerian president suggested that the Nigerian National Petroleum Corporation (NNPC), the state oil company, had failed to account for nearly US$50 billion in crude oil sold between January 2012 and July 2013, amounting to 76 per cent of crude oil sold by the NNPC and almost equal to the federal budget expenditure for both years (*Financial Times*, 2013).

Like most oil-rich states in Africa, those who control Nigeria's government are corrupt, self-serving and uninterested in promoting broad-based development in the country (see Peel, 2009). Clientelism, patronage and graft are central to the whole political economy of the country (Smith, 2007). The current Minister of Petroleum Resources, Deiziani Allison-Madueke is under investigation by the House of Representatives Committee for embezzling US$309 of public funds to charter private jets for herself. This is on top of a Petroleum Revenue Special Task Force report that revealed non-payment of US$167 million in oil block signature bonuses by concessionaires, non-payment of a reported US$3.027 billion in royalties on crude oil and gas, and that over US$2.6 billion was paid as subsidy in 2012 to fake petroleum marketers (*Premium Times* (Abuja), 28 January 2014). When President Jonathan nominated Alison-Madueketo as Secretary-General of the Organization of Petroleum Exporting Countries (OPEC), the organisation was forced to extend the tenure of the incumbent, Abdullah al-Badri of Iraq, until June 2015 in an unprecedented move clearly linked to the raft of allegations and scandals around the Nigerian minister.

What exists in Nigeria is a longstanding form of 'pirate capitalism' (Kelechukwu, 2011), brought about by how oil intersects with the extant polity, in which the state is the major source of wealth and fortune for the ruling elite, but where the ordinary Nigerian misses out (Schatz, 1984). Due to such serious inequalities, violence has now become a norm in the oil-rich Delta, where militants from the Movement for the Emancipation of the Niger Delta (MEND) routinely attack the oil industry, through kidnappings, assassinations and car bombings (Okonta, 2008). Meanwhile, in the north of the country whole swathes are off-limits to state forces, and the Islamist organisation Boko Haram has been fighting since 2001 to put in place an austere form of Islam in Nigeria. This has led to 10,000 deaths between 2001 and 2013 (Campbell, 2013: 139). Endemic instability led Johnnie Carson to remark privately that 'it is possible that Nigeria could be a future Pakistan' (cited in WikiLeaks, 2010a: 1).

Interest in Nigeria's oil by outsiders will exacerbate problems if the current political economy of the country remains. Energy is the main focus of interest in Nigeria, driving Nigeria–China–US relations and pulling Beijing and Washington into the Gulf of Guinea, a subject we now turn to.

The West African context

Oil provides the foundation for commerce and industry, the means for transportation and the ability to wage war. It is *the* prize to capture (Yergin, 1991). In this context, talk of a 'scramble for Africa's oil' reflects the nineteenth century's 'Scramble for Africa' (Klare and Volman, 2006a). West Africa has now emerged as a hugely important source of oil, affecting Nigeria's relations with both China and the US. According to the US Department of Energy, the combined oil output by all African producers is projected to rise by 91 per cent between 2002 and 2025 (from 8.6 to 16.4 million barrels per day) (Klare and Volman, 2006a: 611). Within oil circles – at least until recently – there was growing excitement about the 'alluring global source of energy in Africa' (Shaxson, 2007: 24), with Africa identified as the 'final frontier' in the quest for global oil supplies (Klare and Volman, 2006b). Indeed, until the light tight oil (LTO) and unconventional gas revolution (see later), it was predicted that the US would source 25 per cent of its oil imports from Africa by 2015 (Ghazvinian, 2007: 8). Certainly, until quite recently it was being asserted that the Gulf of Guinea was of major strategic relevance in global energy politics (Alao, 2007: 168).

Although not solely a race between Chinese and US corporations, dynamics in West Africa were heavily influenced by actors from these two states (Frynas and Paulo, 2007: 230). Policy analysts in Beijing saw the broader, global political milieu as intrinsically linked to Chinese energy security and felt that China was vulnerable until and unless it could diversify its oil sourcing and secure greater access to the world's oil supplies (Taylor, 2006: 937). Between 2002 and 2025, Chinese energy consumption will rise by 153 per cent, and China is currently the second largest consumer of oil globally, after the US (Xu Conglin and Bell, 2013).

From the US perspective, the 'war on terrorism and preparations for war against Iraq...enormously increased the strategic value of West African oil reserves' (Ellis, 2003: 135). The high level of interest from such major importers certainly raised the level of competition over Africa's oil. Although corporations headquartered in other states, for example in Britain, Brazil, France, India and Malaysia, played important roles in the scramble, it was the ostensible Sino–American competition for oil on the continent that grabbed the most headlines.

Unlike the colonial scramble for Africa, however, African agency has been far more present with important implications for these relationships. Many African governments are quite proactive, and Nigeria is no exception. Although the nineteenth century scramble 'was driven and dictated by European colonial interests', today 'African leaders act in the role of decision-makers' (ibid: 135). The government's ability to negotiate favourable contracts is considerable, as many oil-rich African governments skilfully play the oil game – albeit for the benefit of the incumbent elites.

The Chinese entry into Nigeria

Nigeria's China policy was hinged upon the government developing a solid relationship with Beijing across the gamut of economic and political policies. This milieu has attracted numerous Chinese companies to Nigeria, particularly in the construction, oil, telecommunications and pharmaceuticals fields. The level of reported commercial investment into Nigeria is astonishing. In 2006, China announced plans to invest US$267 million in the first-phase of the Lekki Free Trade Zone (FTZ) in Lagos (*World Finance,* 4 September 2014). Approved by the Chinese government, the Lekki FTZ is 'the first of its kind the Chinese government has ever built abroad' (*People's Daily* (Beijing), 13 May 2006). This Sino–Nigerian joint venture links three Chinese companies and the Lagos state government. The first-phase of this truly gigantic project covers 15 square kilometres, and Phases II and III aim to cover 150 square kilometres with a total investment of US$5 billion. It will focus on heavy industry manufacturing, chemicals, petroleum processing, pharmaceuticals, cars, logistics, import/ export businesses, a deep-water port, tourism, real estate, education, banking and finance, amongst others. Whether or not Nigeria will ever see a 150 square kilometre FTZ in Lagos with 300,000 Nigerians employed is a matter of speculation, but the proposed project suggests that China has large-scale ambitions and a deep interest in the country. It is important to note that in 2005 Sino–Nigerian ties were officially upgraded to a strategic partnership, and yet development of the FTZ will have to confront the growing concern about Chinese work practices. According to a Nigerian report:

> Chinese companies are notorious for their tendency to bring in their own workers as opposed to hiring locally. Local content has no meaning to the Chinese-run companies. This policy does not in any way address issues of unemployment in the host nations. Safety standards within their industries are another area of concern. The fire incident at a Chinese-owned industry in Ikorodu Town, Lagos State revealed that it was standard practice to lock the workers in while on duty. In this particular case, this policy hindered the workers' escape route from the fire and resulted in many of them losing their lives.
>
> (*This Day* (Abuja), 15 February 2007)

Given the well-known corrupt tendencies of Nigerian government officials, safety inspectors were probably bribed to look the other way, but genuine issues remain that have led to demands that the Federal Government 'should terminate every contract with the Chinese government and severe all Sino–Nigerian trade relationships because of China's refusal to follow internationally accepted best practices and due process', according to Shehu Sani, President of the Civil Rights Congress (*Daily Trust* (Abuja), 2009). Nigerian trade union officials apparently told David Goldwyn, the visiting Coordinator for International Energy Affairs, that they did not want the Chinese in the Nigerian oil sector. 'The Chinese are here and that is a huge problem!' one official was quoted as saying, mentioning a list of the worst five countries to work for that included the Chinese (WikiLeaks, 2010b: 2). The unionists unfavourably compared the Chinese to the Americans, noting that when ExxonMobil 'wrongfully fired a worker', the union applied pressure through the US steel workers and the employee in question was given a choice of being re-hired or compensated – the Nigerian chose the latter. But 'If [we have] a problem with a Chinese company…who can [we talk] to?' (WikiLeaks, 2010b).

Politically explosive is the flooding of Nigeria's markets with cheap Chinese products. As much of Nigerian public opinion sees it, these goods undermine local commercial operations and put Nigerians out of work. Textiles and garments account for 15 per cent of China's total exports to Nigeria:

> The main culprit behind the Nigerian debacle is the Chinese invasion of Nigerian markets…Nigerian fabrics are more costly to the Nigerian than the Chinese fabrics, and no amount of national fervour can salvage the Nigerian textile factories from their present crisis.
>
> (Institute of Development and Education for Africa, 2005)

Such realities provoke political outbursts, such as when Mallam Sanusi Lamido Sanusi, the Governor of the Central Bank of Nigeria, warned in March 2013 that:

China takes from us primary goods and sells us manufactured ones. This was also the essence of colonialism. China is no longer a fellow underdeveloped economy. China is the second biggest economy in the world, an economic giant capable of the same forms of exploitation as the West. China is a major contributor to the de-industrialization of Africa and thus African underdevelopment.

(*This Day*, 13 March 2013)

China is not necessarily totally responsible for this decline. Although the dynamics behind this are not necessarily the fault of the Chinese, Beijing's reputation suffers in light of the vitality of Nigerian civil society. Currently,

Nigeria's media scene is one of the most vibrant in Africa. State radio and TV have near-national coverage and operate at federal and regional levels. All 36 states run at least one radio network and a TV station. There are hundreds of radio stations and terrestrial TV networks, as well as cable and direct-to-home satellite offerings. There are more than 100 national and local press titles.

(*BBC News*, 12 September 2013)

Unlike other African oil producers where Chinese companies have a strong presence, such as Angola, Equatorial Guinea and Sudan, the volatility of Nigeria's polity is reflected in its media (Ojo, 2003). Consequently the multitude of media actors closely scrutinises all foreign operations (including the Chinese) by making Beijing quite comfortable.

Nigeria's oil and China

Nigeria is Africa's leading oil producer; 80 per cent of government revenue and 95 per cent of export income comes from oil (*Bloomberg News*, 1 November 2013). Nigerian crude oil is high quality and has a light, low sulphur grade known as sweet, which is valued because of its high gasoline content and relatively cheap processing outlay. Estimates are that Nigeria contains 176 trillion cubic feet of natural gas reserves, mainly from onshore fields and the Niger Delta, making of the largest holder of natural gas proven reserves in Africa (Nwaogaidu, 2013: 163). However, because of the lack of a gas infrastructure, 75 per cent of associated gas is flared and 12 per cent re-injected.[1] Nigeria's government set a target of zero flaring by 2010, which was (predictably) shifted to 2012; in 2013 the Minister of Petroleum Resources stated that gas flaring would be reduced to 2 per cent by 2014 (*Reuters*, March 2014).

Nigeria's oil industry depends greatly on joint-venture operations, composing 95 per cent of Nigeria's crude production. The very first oil and gas operations in 1956 began with the then Shell D'Arcy. Shell, an Anglo–Dutch corporation, continued its dominance until Nigeria joined the OPEC in 1971, whereupon the Nigerian state began to take a more active role in the country's oil resources through the aforementioned NNPC. Unlike in other OPEC nations, where a state-owned national oil company often took direct control of production, Nigerian multinationals were permitted to carry on operations, but under Joint Operating Agreements. After OPEC membership, new players in the Nigerian oil industry signed various joint ventures with the state, including Gulf Oil, Texaco, Elf Petroleum and Mobil, in addition to Shell. Joint ventures account for approximately 95 per cent of all crude oil output and the main players are:

- Shell Petroleum Development Company of Nigeria Limited (SPDC): a joint venture operated by Royal Dutch Shell (30 per cent) and NNPC (55 per cent), with minority interests held by TotalFinaElf (10 per cent) and Agip (5 per cent). Shell formerly operated alongside British Petroleum as Shell–BP, but BP has since sold all of its Nigerian concessions.
- Chevron Nigeria Limited (CNL): a joint venture between NNPC (60 per cent) and the US's Chevron (40 per cent), historically the second largest producer in Nigeria.
- Mobil Producing Nigeria Unlimited (MPNU): a joint venture between NNPC (60 per cent) and Exxon–Mobil (USA) (40 per cent).
- Nigerian Agip Oil Company Limited (NAOC): a joint venture owned by NNPC (60 per cent), Agip (20 per cent) and ConocoPhillips (USA) (20 per cent).
- Total Petroleum Nigeria Limited (TPNL): a joint venture between NNPC (60 per cent) and Total (France).
- NNPC Texaco–Chevron Joint Venture: a joint venture between NNPC (60 per cent), Texaco (US) (20 per cent) and Chevron (US) (20 per cent) (Nigerian Investment Promotion Commission, 2015).

Shell, the colossus that has long dominated Nigeria's oil industry, has put much of its Niger Delta activities up for sale, citing oil theft, which was costing the company US$700 million (*Daily Independent*, 12 November 2013).

For many years China was absent from Nigeria's oil industry; however, this situation changed in the early 2000s through a mix of diplomacy, sweetener deals (often unrelated to the actual oil industry) and a close relationship with the then president of Nigeria, Olusegun Obasanjo. According to Obasanjo's biographer:

[Obasanjo] had long been exasperated by the international oil companies that were eager to exploit Nigeria's oil and gas but refused to invest in enterprises, such as oil-refining or in infrastructural development that were not essential to their own core businesses. During 2004, as dramatic economic growth in China, India and South Korea created an international scramble for oil, [Obasanjo] offer[ed] Asian national oil companies 'rights of first refusal' to explore and develop oil blocks in return for undertakings to invest capital and technology in Nigerian infrastructure.

(Iliffe, 2011: 275)

In October 2004, the government announced that Nigeria needed an annual investment of US$10 billion to reach the target of 40 billion barrels of reserve by 2010. Consequently, China National Offshore Oil Corporation (CNOOC) signed an agreement with the Nigerian government to locate upstream oil and gas assets that might be incorporated into downstream projects. In December 2004, Sinopec, the state-owned Chinese oil company, and NNPC signed an agreement to develop Oil Mining Lease (OML) 64 and 66 in the Niger Delta. Within OML 64, five exploration wells were drilled, one of which came across hydrocarbon resources. OML 66 has been far more successful, having drilled 18 exploration wells and 12 encountering hydrocarbon resources (*China Daily* (Beijing), 9 December 2004). Sinopec also has a contract with the Nigerian Petroleum Development Company (NPDC) and Italy's Eni to develop the Okono and Okpoho fields, which have combined reserves of 500 million barrels. Later, in July 2005, CNOOC and NNPC signed a US$800 million contract that guaranteed 30,000 barrels per day to China over a five-year period, to be reviewed every year (*Pambazuka News,* 14 December 2006).

Building on such developments, in April 2006 the Nigerian government offered China four oil exploration licences in exchange for US$4 billion in infrastructure investments. The two countries then signed seven development agreements granting Abuja export credits worth US$500 million (*The Vanguard* (Lagos), 27 April 2006). China agreed to repair the Kaduna Refining and Petrochemicals Company whilst undertaking other investment projects, such as building a hydropower plant in the Mambila, Plateau State, in return for exercising the 'right of first refusal' on oil blocs. Several of the oil blocs on offer were relinquished by previous operators.

CNOOC took over the commitments of a contractor of the deepwater block, OPL 246, which had been assigned earlier to South Atlantic Petroleum Limited, a company owned by former Nigerian Defence Minister, Theophilus Danjuma. Illustrating that politics, patronage and oil are inextricably linked in Nigeria, Danjuma immediately went to court to negate the deal, claiming that his company's acreage had been revoked for political reasons. The Federal High Court in Lagos later judged in favour of Obasanjo's government. Consequently CNOOC bought a 45 per cent working interest in OPL 246, paying US$2.3 billion plus an adjustment of US$424 million for other expenses. The US$2.3 billion will finance the NNPC's 50 per cent equity stake in OPL 246 and, in return, CNOOC will have a 70 per cent share in profits from the field, while the NNPC takes the remaining 30 per cent. CNOOC agreed to refund the US$600 million already spent by French company Total in developing the field (*This Day* (Abuja), 21 April 2006).

However, Sino–Nigerian relations were not as smooth as initially thought and were highly dependent on Obasanjo, who left office in 2007. In March 2007, therefore, the Nigerian government announced that it was considering reviewing its plans vis-à-vis the Kaduna refinery on the grounds that Chinese promises to invest in the refinery had not materialized. The Director General of the Bureau of Public Enterprises, Irene Chigbue, stated that the plan to get CNOOC to manage the 110,000-barrels-per-day Kaduna refinery 'had run into hitches as the CNOC have not been forthcoming with the takeover plans' (*This Day* (Abuja), 6 March 2007).

In the run-up to Obasanjo finally retiring (after an aborted attempt at changing the constitution), much of the apparently 'robust' Sino–Nigerian ties started unravelling:

> When Obasanjo left office…nothing of [the plethora of] plans was yet visible on the ground, but the difficulties were increasingly obvious. The viability of the projects…was often uncertain. The agreements had no binding provisions regulating the relative timing of the oilfield and infrastructure developments…The infrastructural estimates were often inflated and the financial details opaque, disputed and burdensome to Nigeria. Suspicions proliferated of kickbacks to finance Obasanjo's political schemes and line his agents' pockets. The Asian firms realised that they had plunged into a political and business environment they did not understand.
>
> (Iliffe, 2011: 276)

The failure of the Kaduna oil refinery project was a setback for the Chinese government's Nigeria policy, requiring China to significantly re-evaluate how best to conduct business. One Nigerian analyst identified two factors that stalled China's infrastructure for resources deal in Nigeria:

> First, the policy failed because of the interest of Nigerian elites, who felt implementing the deal would cut them off from profit from crude oil sales on the international market. It is also significant to note that China's offer when deploying its infrastructure for resources [fell] below the prevailing market price.
>
> (Umejei, 2013)

Second, Western oil interests in Nigeria fiercely opposed the deal – the 'Chinese foray into Nigeria was halted by powerful Western elements, because at a point in 2008/9, Nigeria became the last gambit in the global chess game' (Umejei, 2013).

In May 2010, Nigeria's state-run oil firm NNPC and China State Construction Engineering Corporation (CSCEC) signed a US$23 billion deal to seek financing and credits from Chinese authorities and banks to build three refineries and a fuel complex in Nigeria (*BBC News*, 14 May 2010). This announcement was greeted with media suspicion:

> The recent $23 billion loan for building three new refineries should concern most Nigerians. To build these refineries, Nigeria will create almost as much debt as the $30 billion that was eliminated during President Obasanjo's tenure in office…One thing we tend to be poor at is capacity building. Who will run four massive new plants? Us or the Chinese? If it is the Chinese, as is likely, how will technology transfer take place?...Many feel that the rush by politicians to sign these agreements and large loans was driven less by thoughtful consideration of Nigeria's needs than by their desire for the large commissions and bribes that are rumoured to have to accompanied them. In the end, this arrangement could be hugely ruinous for Nigeria…Vast loans

are being taken out with abandon by reckless, greedy politicians who have learned little from the past.

(Vanguard (Lagos), 27 April 2006)

Other problems in Sino–Nigerian relations around the oil industry also undermine the win–win situation about which Beijing likes to boast. These problems may be seen as an alternative site of 'pushback' against Chinese influence, in particular the security situation surrounding the oil industry in the Niger Delta is increasingly problematic. Following President Hu Jintao's April 2006 state visit to Nigeria and the signing of a US$4 billion infrastructure investment deal, Nigerian militants from MEND warned Chinese companies to 'stay well clear' of the Niger Delta or risk facing attack. MEND also claimed responsibility for a car bomb attack near the port town of Warri, stating that the blast was 'a warning against Chinese expansion in the region', adding that, 'the Chinese government by investing in stolen crude places its citizens in our line of fire' *(BBC News*, 30 April 2006).

In early 2006, MEND militias began attacking oil installations and kidnapping foreign oil workers, leading to a 20 per cent reduction in Nigeria's oil production. Nigeria turned to China for military supplies to protect the oil fields after it claimed that Washington had been tardy in its response to the decreasing security situation in the Delta (Taylor, 2007). Washington's reluctance reflected anxiety over widespread corruption within and human rights violations by Nigeria's security forces who stand accused of

> politically motivated killings; the use of lethal force against suspected criminals and hostage-seizing militants in the Niger Delta; beatings and even torture of suspects, detainees, and convicts; and extortion of civilians, as well as child labour and prostitution, and human trafficking.
>
> (Grimmett, 2006: 22)

China appeared unconcerned, although this had costs to Beijing's reputation: 'When America balked at supplying Nigeria's trigger-happy military, China offered dozens of patrol boats. "They are impossible. They just don't care what we or anyone else says", complained a member of one Dutch human rights advocacy group' *(The Guardian* (London), 28 March 2006). Nigerian elites are less troubled. As one report puts it:

> Nigeria is reportedly seeking to buy naval patrol boats from China to help protect its Niger Delta rigs from rebel attack, at a time when traditional allies are nervous of sending more weapons into an already volatile region. One Niger Delta state governor, reacting to concerns over attacks on Shell's facilities and rumours the firm might even pull out of the region, grinned and [said] 'If the Brits don't want the oil, we'll sell it to the Chinese'.
>
> *(Agence-France Presse* (Lagos), 26 April 2006)

This bravado, however, has been short lived.

Sino–US competition?

China's ties with Nigeria in the early 2000s generated a great deal of excitement within Nigeria, encouraged by Beijing's willingness to work closely with the Obasanjo regime. This buzz coincided with an upsurge of interest by Washington in West Africa due to its status as an important energy provider. During this period talk emerged of a strategic competition between China and the US. Thus in 2007, Frynas and Paulo (2007) saw a rivalry over the West spread out, specifically between the US and China.

It would be tempting to cast Chinese entry into Nigeria as stimulating a major competition; however, Chinese actors have often succeeded because Western companies either have not been interested in bidding for a particular project or have not played ball with the Nigerian government and its demands, whilst the Chinese have been prepared to do so. For example, in discussions with the US embassy in Abuja, NNPC officials admitted that the Chinese were, for now, their favourite partners. As one Nigerian official stated, 'We know what had happened in the Sudan and Chad and we know enough about them to know where we want them and where we don't' (WikiLeaks, 2010b: 1). Nevertheless, these same officials appraised Chinese oil companies' attempts to obtain deep water oil mining leases, stating that 'Shell Nigeria had opened the door for the Chinese by resisting [Abuja's] efforts to pass the proposed Petroleum Industry Bill (PIB) and telling the National Assembly that the "Nigerian oil industry would be dead" if the PIB passed,[2] so they brought in the Chinese' (WikiLeaks, 2009).

The Americans, however, have not been concerned about China's entry into Nigeria. While briefing US diplomats in Lagos in 2010, US Assistant Secretary for African Affairs, Johnnie Carson, stated that, 'The United States does not consider China a military, security or intelligence threat', although 'China is a very aggressive and pernicious economic competitor with no morals' (WikiLeaks, 2010c: 2). In fact, Carson asserted that Washington would only start worrying if Beijing crosses some 'trip wires' – if China was developing blue water navy, signing military base agreements, training armies, or developing intelligence operations. Carson's tone indicated that the administration did not believe Beijing had done any of these.

Rather than panicking about China's rise, Carson asserted that 'the influence of the United States has increased in Africa…The United States' reputation is stable and its popularity is the highest in Africa compared to anywhere else in the world. Obama has helped to increase that influence' (ibid.). In his testimony before the Senate Committee on Foreign Relations Subcommittee on African Affair in Washington DC in February, 2010, Carson did not even mention China. In reality, many people have overstated the Chinese economic and political influence in Nigeria. Beijing's role peaked under Olusegun Obasanjo between 1999 and 2007, but rapidly dissipated. As John Campbell (2010: 3) of the Council on Foreign Relations put it:

> Despite China's escalating energy requirements, its attempts to expand its energy relationship with Nigeria has largely failed. Obasanjo, seeking an alternative to the big Western oil companies, sought to secure Chinese

Table 8.2 US imports from Nigeria (in US$ millions)

2002	2003	2004	2005	2006	2007	2008	2009	2010	2011	2012	2013
5.9	10.4	16.2	24.2	27.8	32.7	38.1	19.1	30.5	33.8	19.0	11.5

Source: United States Census Bureau: Foreign Trade, January 2013.

investment in big infrastructure projects in return for oil block concessions he would grant on highly favourable terms, known as 'oil for infrastructure' projects...Both sides made numerous dramatic announcements, including promises that the Chinese would rebuild the railway system and construct power plants and refineries.

However, many of these deals 'lacked transparency and were short on details' (ibid.). Nigeria's exports to the US have fallen precipitously (see Table 8.2). Why this is so hinges on one simple fact: over 99 per cent of US imports from Nigeria are made up of oil. In 2009, Nigeria was the fifth largest foreign oil supplier to Washington; by 2013, however, Nigeria had fallen to eighth. The following section discusses why.

The 'energy revolution'

In the past few years, forecasts for growth in US oil production and oil reserves have changed radically, springing the development of LTO – both crude oil and condensate in all tight formations, including shale basins. Currently, the global oil industry is considering LTO's potential and its likely implications for global oil supply and demand. This is important because over the past decade, as noted earlier, the unparalleled expansion in demand from non-OECD countries, such as China and India, compelled them to seek new oil sources worldwide, including Africa. This expansion in demand was at the root of the so-called (and possibly short-lived) new 'scramble for Africa'. Recently, oil companies have started to develop unconventional hydrocarbons, successfully bringing to the market several large and under-exploited oil and natural gas liquid resources. US shale/tight oil, Canadian tar sands, Venezuela's extra-heavy oil and Brazil's pre-salt oil are the main examples, shelving concerns about 'peak oil' and energy shortages that had propelled oil prices upwards. With little doubt that unconventional resources through the 'energy revolution' will satisfy global demand, the debate now revolves around the speed and price at which these resources can be extracted.

The energy revolution derives from technological innovations in horizontal drilling and hydraulic fracturing (or 'fracking'). The US Energy Department forecasts that the US production of crude and other liquid hydrocarbons will average around 11.4 million barrels per day by 2014, which would place the US just below Saudi Arabia's expected output for 2013 of 11.6 million barrels per day (*US Today*, 2012). Several forecasts put US production at between 13 and 15 million barrels per day by 2020, with the International Energy Agency (IEA) suggesting that the US may supplant Saudi Arabia as the world's largest producer (IEA, 2013). At this juncture, Saudi oil remains cheaper to tap than tight oil because LTO needs a price above US$70 per barrel to be profitable (break-even prices of most tight oil are in the range of US$40–60 per barrel); however, the recent turnaround in the US production of crude oil is extraordinary and will have major implications for Africa (and the world). As the US meets more of its current and future demand for oil from indigenous supplies, imports from its traditional suppliers will fall. This decline has been happening in dramatic fashion in Africa, with countries such as Congo-Brazzaville dropping from 26.27 million barrels to 7.2 million barrels between 2010 and 2013. Similarly Equatorial Guinea dropped from 21.06 million barrels to 6 million barrels during the same time period (see Table 8.3).

In its *Medium-Term Oil Market Report* for 2013, the IEA asserted that LTO and other aspects of the energy revolution will act as a 'supply shock' to the global oil market that will be 'as transformative to the [energy] market over the next five years as was the rise of Chinese demands in the last 15 years' (IEA, 2013). Exports of Nigerian oil to the US almost halved between 2011 and 2012, with a continued decline in 2013. In late 2014 Nigeria become the first country to completely stop selling oil to the US due to the impact of the shale revolution, 'an astounding reversal as the African nation was only four years ago one of the top-5 oil suppliers to America' (*Financial Times*, 2 October 2014).

Consequently, Nigeria has experienced difficulties finding alternative destinations for its crude and has had to cut prices. At the start of 2013, weak demand

Table 8.3 US imports of total crude oil and products from Africa, 2010–2013 (in thousands of barrels)

	2010	*2011*	*2012*	*2013*
Algeria	186,019	130,723	88,487	42,014
Angola	143,512	126,259	85,335	78,672
Cameroon	19,728	13,921	12,356	1,225
Chad	11,312	18,473	11,004	23,944
Congo-Brazzaville	26,276	19,275	11,341	7,293
DRC	3,225	3,999	137	276
Equatorial Guinea	21,063	8,500	15,100	6,073
Gabon	17,022	12,557	15,886	8,993
Ghana	215	3,832	313	993
Ivory Coast	3,560	1,450	1,634	87
Liberia	–	20	1	3
Libya	25,595	5,542	22,281	21,407
Nigeria	373,297	298,732	161,558	102,611

Source: US Energy Information Administration, various dates.

forced Nigeria to sell some cargoes of its oil below the official selling price, with a loss of US$380,000 on a typical cargo (*Wall Street Journal*, 6 March 2013).

Such a scenario is very serious for Nigeria. Addressing the Nigeria Economist's Group Summit in May 2013, Diezani Alison-Madueke (2013: 1), Minister of Petroleum Resources, forecast a rather pessimistic outlook:

> US dependence on oil imports is expected to continue declining over the next 10 years reaching a share of about 43% of total oil consumption by 2020 from 67% in 2005…Between 2007 and 2011, US shale gas share of total gas supply increased from 8% to 32%; consequently pipeline & LNG import share of total gas supply declined from 16% and 3% in 2007 to 12% & 1% respectively. As a result of shale gas production, it is projected that US will become a net exporter of natural gas in the year 2020. This is already evident in the decline of Nigeria's LNG exports to the US from 12% in 2007 to 1% in 2011.

Accordingly, 'unprecedented growth in USA gas reserves inevitably *eliminates* [the] USA as a destination for Nigerian gas' (Alison-Madueke, 2013: 1, emphasis added). In addition, the growth in gas reserves helps re-establish the US as a major producer of industries such as petrochemicals and fertilizers, and in effect slashing the market options for such products from Nigeria.

The trend of declining crude oil imports into the US continues, with a particularly sharp decline in imports from Nigeria. This was due to the idling in late 2011 of two refineries in the Philadelphia area (ConocoPhillips Trainer refinery and Sunoco's Marcus Hook refinery) (US Energy Information Administration, 2012). Both were significant buyers of Nigerian crude and proficient to run light-sweet crude oils. Problematically for Nigeria, the price of Nigeria's crude oil is high relative to what is now being produced in the US. Moreover, US refineries are encouraged to switch from imported crude to inland, domestically produced crude.

Given that crude oil comprised virtually all of the US's imports from Nigeria, Abuja faces a *massive* problem in the future. Edward Morse, Head of Commodities Research at Citigroup Global Markets, had predicted that 'sometime before mid-2014, the US and Canada will stop importing crude from West Africa altogether' (Philips, 2013). As noted, this has already happened with regard to Nigeria. Nigerian elites, however, hope that China and other Asian economies will replace the US as Abuja's main destination. Finance Minister Ngozi Okonjo-Iweala stated in 2013 that the gap in demand for the nation's oil created by the US would hopefully be filled by China and India: 'China is desirous of increasing its lifting of Nigerian oil…US demand for Nigerian oil has fallen, India and China have taken up the slack' (*African Business*, 29 August 2013). Chinese consumption, however, is not a permanent solution to Nigeria's woes because, according to the US Energy Information Administration, China has an estimated 1,275 trillion cubic feet, or 36 trillion cubic meters, of technically recoverable shale-gas reserves, more than Canada and the US combined (*Wall Street Journal*, 23 July 2012).

In addition, Chinese responses to the fall in oil prices are skewing the global oil market and undermining the ability of a country like Nigeria to predict future revenues. When oil prices fell in 2014, China rushed to buy and stockpile in order to increase energy security. This saw a 9.5 per cent growth in China's crude-oil imports in 2014 (*Reuters*, 27 January 2015). This purchasing has made the task of assessing the true Chinese demand extremely difficult because the Chinese actual demand for oil and how this affects prices is now unclear. Working out how much China may buy is fundamental to any serious price forecasting, particularly because the global slump in oil has curtailed previous concerns over the outlook for demand. As China's economy slows – pulling down growth in oil demand – and the stockpiling increases, guessing Chinese demand in the future is at best sketchy. Beijing's goal is to have 90 days of oil in reserve supply; strategic reserves at present are between 40–50 days (*Wall Street Journal* (New York), 25 February 2015). China astutely bought when prices were low, therefore saving the country billions of dollars, but this is not good news for the suppliers – Nigeria included – and a battle between oil-producing nations for the Chinese market will only intensify. In this game, China holds all the cards.

Conclusion

Despite all the hype in the mid-to-late 2000s about US–China rivalry for Nigeria's oil, this clash has not materialized. US interest in Nigeria's crude, which reflected US strategic interest in the Gulf of Guinea, peaked in 2006–8, and has since dwindled rapidly. China's needs, on the other hand, which were the target of resistance by Western interests and largely criticized by local civil society and the post-Obasanjo regime, may yet prove a short-term saviour of Nigeria's oil sector. Indeed, the irony here should not be ignored.

Today, however, the main challenge for both China and the US has always been confronting and negotiating the reality of Nigeria's political economy. No amount of high-sounding rhetoric from Beijing about fraternal ties and mutual benefits can hide or escape this fact. Nigeria's politics have long been an open scramble for power in which elites compete to control the state in order to capture the mega-benefits associated with the country's enormous oil revenues. 'New' actors, such as China, must navigate their relations with Nigeria, but this environment, as other actors have found, is inherently unstable and dangerous where long-term guarantees mean little. Massive profits may be accrued quickly, but violence and perpetual crises within Nigeria's polity mean that engagements pursued by external actors are always highly vulnerable (Hill, 2012).

China as with all other actors in Africa, needs stability and security for its investments to flourish and for its connections with the continent to be coherent. China's heavy involvement in fixed infrastructure assets means that Chinese companies cannot stand aloof from the very real problems that characterise Nigeria. Ultimately, rather than competing with the West, China must converge with US policy aims in Nigeria, even if this is so far unacknowledged by Beijing, otherwise the dynamics of Nigerian politics might lead Chinese investors to retreat to the oil economy, endangering investments in the non-oil sector. China would then become just another actor in Nigeria's enclave economy – a possibility, but wholly against Beijing's rhetoric on mutual benefits and win–win situations.

For the US, Nigeria is becoming less and less important as the energy revolution in North America (and elsewhere) negates Washington's need for Nigerian oil. Given that, for all practical purposes, oil is the only thing that Nigeria exported to the US, US–Nigerian trade relations are heading towards a dead-end. Taking into account Nigeria's notorious governance record, China's rather hands-off approach to governance matters may well run afoul of its own interests. Governance, peace and security are crucial to Nigeria, as is reducing poverty and building infrastructure. It is in the latter where China may play its role and where the US must engage Beijing in identified areas where mutual interests converge, assuming that Washington will retain some interest in Nigeria. As Johnnie Carson remarked, 'Nigeria cannot afford to implode or run aground' (WikiLeaks, 2010a: 1). Although the US's economic interests in Nigeria are in a state of abrupt decline, such a situation would not be in the interests of either Beijing *or* Washington DC.

Notes

1 Re-injection is where gas is returned to the reservoir via 'injection wells', awaiting future markets or, alternatively, to repressurize the gas field, serving to improve extraction rates from adjacent wells.
2 The PIB seeks to create a transparent administrative system, amending the Petroleum Profit Tax Administration, which currently treats information relating to company profits as confidential. The PIB also clarifies procedures for bidding processes and the retention of licences and leases and also simplifies collection of revenues by emphasising rents and royalties and less on taxes from petroleum. Furthermore, the PIB will compel companies to employ Nigerians in their employments and contract awards, especially on community related development projects.

References

Adebanwi, W. and Obadare, E. (2013) *Democracy and prebendalism in Nigeria: critical interpretations.* Basingstoke: Palgrave Macmillan.

African Business (2013) Jonathan makes 'significant' trip to China. *African Business*, 29 August. Available online at: http://africanbusinessmagazine.com/africa-within/countryfiles/jonathan-makes-significant-trip-to-china/ (accessed 2 April 2015).

Agence-France Presse (Lagos) (2006) Chinese oil safari hits Nigeria. *Agence-France Presse* (Lagos), 26 April. Available online at: www.terradaily.com/reports/Chinese_Oil_Safari_Hits_Nigeria.html (accessed 2 April 2015).

Alao, A. (2007) *Natural resources and conflict in Africa: the tragedy of endowment.* Rochester, NY: Rochester University Press.

Alison-Madueke, D. (2013) The Future of Nigeria's Petroleum Industry. Ministerial keynote address by Diezani Alison-Madueke, Honorable Minister Ministry of Petroleum Resources at the Nigeria Economist's Group Summit, Abuja, Nigeria, 24 May.

allAfrica (2004) The Leon H Sullivan Foundation to host Presidents Obasanjo and Clinton. *allAfrica,* 26 October. Available online at: http://allafrica.com/stories/200411050845.html (accessed 2 April 2015).

Avuru, A. (2005) *Politics, economics and the Nigerian petroleum industry.* Lagos: Festac Books.

BBC News (2006) Car blast near Nigeria oil port. *BBC News,* 30 April. Available online at: http://news.bbc.co.uk/1/hi/world/africa/4959210.stm (accessed 2 April 2015).

BBC News (2010) China to build $8bn oil refinery in Nigeria. *BBC News,* 6 July. Available online at: http://news.bbc.co.uk/2/hi/8681814.stm (accessed 2 April 2015).

BBC News (2013) Nigeria profile – media. *BBC News,* 12 September. Available online at: www.bbc.co.uk/news/world-africa-13949549 (accessed 2 April 2015).

Bloomberg News (2013) Nigeria says revenue gap may reach as much as $12 billion, *Bloomberg News,* 1 November. Available online at: www.bloomberg.com/news/articles/2013-10-31/nigeria-finance-ministry-says-revenue-gap-may-reach-12-billion (accessed 2 April 2015).

Campbell, J. (2010) Who's in charge, China or Nigeria? Available online at: www.cfr.org/china/s-charge-china-nigeria/p22383 (accessed 11 March 2015).

Campbell, J. (2013) *Nigeria: dancing on the brink.* Lanham, MD: Rowman & Littlefield.

China Daily (Beijing) (2004) China, Nigeria sign oil development agreement. *China Daily* (Beijing), 19 December. Available online at: www.chinadaily.com/cn (accessed 21 December 2004).

Cunliffe-Jones, P. (2010) *My Nigeria: five decades of independence.* New York, NY: Palgrave Macmillan.

Daily Independent (2013) Uncertainty over Nigeria's daily oil production persists. *Daily Independent,* 12 November. Available online at: http://dailyindependentnig.com/2013/08/uncertainty-over-nigerias-daily-oil-production-persists/ (accessed 2 April 2015).

Daily Mail (2013) A country so corrupt it would be better to burn our aid money. *Daily Mail,* 9 August. Available online at: www.dailymail.co.uk/debate/article-2387359/Nigeria-country-corrupt-better-burn-aid-money.html#ixzz3YJyW9o7q (accessed 2 April 2015).

Daily Trust (2007) Nigeria: when Nollywood partners UNFPA on reproductive health, gender issues. *Daily Trust,* 14 December. Available online at: http://allafrica.com/stories/200712140542.html (accessed 2 April 2015).

Daily Trust (2009) Nigeria: ASUU – branches vote to suspend strike. *Daily Trust,* 8 October. Available online at: http://allafrica.com/stories/200910080437.html (accessed 2 April 2015).

Ellis, S. (2003) Briefing: West Africa and its oil. *African Affairs,* 102 (406): 135–8.

Falola, T. and Heaton, M. (2008) *A history of Nigeria.* Cambridge: Cambridge University Press.

Financial Times (2013) FirstGroup deflects spin-off call. *Financial Times,* 12 December. Available online at: www.ft.com/intl/cms/s/0/da22307c-628c-11e3-bba5-00144feabdc0.html#axzz3V7yaCivh (accessed 2 April 2015).

Financial Times (2014) Victim of shale revolution, Nigeria stops exporting oil to US. *Financial Times,* 2 October. Available online at: http://blogs.ft.com/beyond-brics/2014/10/02/victim-of-shale-revolution-nigeria-stops-exporting-oil-to-us/ (accessed 2 April 2015).

Frynas, J. and Paulo, M. (2007) A new scramble for African Oil? Historical, political, and business perspectives. *African Affairs,* 106 (423): 685–90.

Ghazvinian, J. (2007) *Untapped: the scramble for Africa's oil.* London: Harcourt.

Gillies, A. (2007) Obasanjo, the donor community and reform implementation in Nigeria. *Round Table,* 96 (392): 569–86.

Grimmett, R. (2006) *Conventional arms transfers to developing nations, 1998–2005.* Washington, DC: Library of Congress.

Hill, J. (2012) *Nigeria since independence: forever fragile?* Basingstoke: Palgrave Macmillan.

Iliffe, J. (2011) *Obasanjo: Nigeria and the world.* Oxford: James Currey.

Institute of Development and Education for Africa (IDEA) (2005) The tragedy of African textile industries. *Institute of Development and Education for Africa,* 14 February. Available online at: http://www.africanidea.org/tragedy.html (accessed 2 April 2015).

International Energy Agency (IEA) (2013) *Medium-Term Oil Market Report, 2013.* Paris: International Energy Agency.

Jibo, M. and Okoosi-Simbine, A. (2003) The Nigerian Media: an assessment of its role in achieving transparent and accountable government in the Fourth Republic. *Nordic Journal of African Studies,* 12 (2): 180–95.

Kelechukwu, I. (2011) *Where do we go from 50? Nigeria in perspective.* Accra, Ghana: Hans Publications.

Klare, M. and Volman, D. (2006a) The African 'oil rush' and US National Security. *Third World Quarterly,* 27 (4): 609–28.

Klare, M. and Volman, D. (2006b) America, China and the scramble for Africa's oil. *Review of African Political Economy,* 33 (108): 297–309.

Lewis, P. (1996) From prebendalism to predation: the political economy of decline in Nigeria. *Journal of Modern African Studies,* 34 (1): 79–103.

Maier, K. (2000) *This house has fallen: Nigeria in crisis.* London: Penguin.

Neumann, L. (2004) European policy and energy interests – challenges from the Gulf of Guinea. *In:* R. Traub-Merz and D.A. Yates, eds. *Oil policy in the Gulf of Guinea: security & conflict, economic growth, social development.* Bonn: Friedrich-Ebert-Foundation, 59.

Nigerian Investment Promotion Commission (2015) Available online at: http://nipc.gov.ng (accessed 2 April 2015).

Nwankwo, A. (2002) *Nigeria: the stolen billions.* Enugu, Nigeria: Fourth Dimension.

Nwaogaidu, J.C. (2013) *Globalization and social inequality: an empirical study of Nigerian society.* Münster: Lit Verlag.

Observatory of Economic Complexity (2015) Available online at: https://atlas.media.mit.edu (accessed 2 April 2015).

Ojo, E. (2003) The mass media and the challenges of sustainable democratic values in Nigeria: Possibilities and limitations. *Media Culture Society,* (25) 6: 821–40.

Okonjo-Iweala, N. (2012) *Reforming the unreformable: lessons from Nigeria.* Cambridge, MA: MIT Press.

Okonta, I. (2008) *When citizens revolt: Nigerian elites, big oil and the Ogoni struggle for self-determination.* Trenton, NJ: Africa World Press.

Pambazuka News (2006) Environmental impact: more of the same? *Pambazuka News,* 14 December. Available online at: http://pambazuka.org/en/category/comment/38851 (accessed 2 April 2015).

Peel, M. (2009) *A swamp full of dollars: pipelines and paramilitaries at Nigeria's oil frontier.* London: IB Tauris.

People's Daily (Beijing) (2006) Sino–Nigerian free trade zone set to bolster investment. *People's Daily* (Beijing), 13 May. Available online at: http://news.xinhuanet.com/english/2006-05/12/content_4539373.html (accessed 2 April 2015).

Philips, M. (2013) Falling US oil imports will reshape the world crude market. *Bloomberg Businessweek,* 16 January.

Premium Times (Abuja) (2014) Odinkalu, Falana, others ask EFCC to act on reports on fraud in Nigeria's petroleum industry. *Premium Times* (Abuja), 7 April. Available online at: www.premiumtimesng.com/news/154142-odinkalu-falana-others-ask-efcc-act-reports-fraud-nigerias-petroleum-industry.html (accessed 2 April 2015).

Reuters (2014) Nigeria hopes its gas can keep the lights on. *Reuters,* 21 March. Available online at: http://uk.reuters.com/article/2014/03/21/uk-nigeria-gas-idUKLNEA2K00F20140321 (accessed 2 April 2015).

Reuters (2015) China commodity trade data show winners are scarce: Russell. *Reuters,* 27 January. Available online at: www.reuters.com/article/2015/01/27/column-russell-china-commodities-idUSL4N0V60QI20150127 (accessed 2 April 2015).

Schatz, S. (1984) Pirate capitalism and the inert economy of Nigeria. *Journal of Modern African Studies,* 22 (1): 45–57.

Shaxson, N. (2007) *Poisoned wells: the dirty politics of African oil.* Basingstoke: Palgrave Macmillan.

Smith, D. (2007) *A culture of corruption: everyday deception and popular discontent in Nigeria.* Princeton, NJ: Princeton University Press.

Taylor, I. (2006) China's oil diplomacy in Africa. *International Affairs*, (82) 5: 937–60.

Taylor, I. (2007) Arms sales to Africa: Beijing's reputation at risk. *China Brief,* 7 (7).

The Guardian (2006) China's goldmine. *The Guardian*, 28 March. Available online at: www.theguardian.com/business/2006/mar/28/china.g2 (accessed 2 April 2015).

This Day (Abuja) (2006) NNPC approves China's $2.3 billion stake in OPL 246. *This Day* (Abuja), 21 April. Available online at: http://allafrica.com/stories/200604210439.html (accessed 2 April 2015).

This Day (Abuja) (2007) China – friends or foes. *This Day* (Abuja), 15 February. Available online at: http://allafrica.com/stories/200702160080.html (accessed 2 April 2015).

This Day (Abuja) (2007) FG to review sale of Kaduna refinery. *This Day* (Abuja), 6 March. Available online at: http://allafrica.com/stories/200703060217.html (accessed 2 April 2015).

This Day (2013) Sanusi: China is major contributor to Africa's de-industrialisation. *This Day*, 13 March. Available online at: www.thisdaylive.com/articles/sanusi-china-is-major-contributor-to-africa-s-de-industrialisation/142029/ (accessed 2 April 2015).

This Day (Lagos) (2015) As US shuts its door on Nigeria's oil exports. *This Day* (Lagos), 3 October. Available online at: www.thisdaylive.com/articles/as-us-shuts-its-door-on-nigeria-s-oil-exports/190455/ (accessed 2 April 2015).

Umejei, E. (2013) Why did China's infrastructure for resources deal fail in Nigeria? *African Arguments*, September 2. Available online at: http://africanarguments.org/2013/09/02/why-did-chinas-infrastructure-for-resources-deal-fail-in-nigeria-by-emeka-umejei/ (accessed 2 April 2015).

United States Census Bureau: Foreign Trade (2013) Available online at: www.census.gov/foreign-trade/index.html (accessed 2 April 2015).

US Energy Information Administration (2012) Potential impacts of reductions in refinery activity on northeast petroleum product markets. US Energy Information Administration, 27 February. Available online at: www.eia.gov/analysis/petroleum/nerefining/update/ (accessed 2 April 2015).

US Energy Information Administration (various dates) Available online at: www.eia.gov/ (accessed 2 April 2015).

USA Today (2012) U.S. could surpass Saudi oil output by 2020. *USA Today*, 23 October. Available online at: www.usatoday.com/story/money/business/2012/10/23/us-top-oil-producer/1652937/ (accessed 2 April 2015).

Vanguard (2006) April 27. Available online at: https://pressroom.vanguard.com/content/press_release/VGI_launches_low_cost.html (accessed 2 April 2015).

Vanguard (Lagos) (2006) Nigeria: govt offers China oil blocs for $4 billion. *Vanguard* (Lagos), 27 April. Available online at: http://allafrica.com/stories/200604270208.html (accessed 2 April 2015).

Vanguard (2015) China–Nigeria bi-national relations, productively on high speed – Chinese Consul General. *Vanguard,* 7 February. Available online at: www.vanguard-ngr.com/2015/02/china-nigeria-bi-national-relations-productively-high-speed-chinese-consul-general/ (accessed 2 April 2015).

Ventures Africa (2015) Nigeria, China trade volume rises to $18bn. *Ventures Africa,* 19 February. Available online at: www.ventures-africa.com/2015/02/nigeria-china-trade-volume-rises-to-18bn/ (accessed 2 April 2015).

Wall Street Journal (2012) Shale gas may hold promise For China. *Wall Street Journal,* 23 July 23. Available online at: http://www.wsj.com/articles/SB100008723963904434 37504577544910500662588 (accessed 2 April 2015).

Wall Street Journal (2013) Nigeria bearing brunt of U.S. shale-oil boom. *Wall Street Journal,* 6 March. Available online at: www.wsj.com/articles/SB10001424127887324662404578333981454427720 (accessed 2 April 2015).

White, G. and Taylor, S. (2001) Well-oiled regimes: oil and uncertain transitions in Algeria and Nigeria. *Review of African Political Economy,* 28 (89): 323–24.

WikiLeaks (2009) Chinese oil companies not so welcome in Nigeria's oil patch, 2 December. Available online at: www.wikileaks.org/plusd/cables/09ABUJA2170_a.html (accessed 2 April 2015).

WikiLeaks (2010a) Confidential cable, 'Subject: Chinese Oil Companies not so Welcome in Nigeria's Oil Patch,' 23 February, US Embassy, Abuja, Nigeria.

WikiLeaks (2010b) Confidential cable, 'Subject: African Embassies Suspicious of US–China,' 11 February, US Embassy, Abuja, Nigeria.

WikiLeaks (2010c) Confidential cable, 'Subject: Assistant Secretary Carson Meets Oil Companies in Lagos,' 23 February, US Consul, Lagos, Nigeria.

World Finance (2014) Lekki free zone set to transform Nigeria's fortunes. *World Finance,* 4 September. Available online at: www.worldfinance.com/infrastructure-investment/lekki-free-zone-set-to-transform-nigerias-fortunes (accessed 2 April 2015).

Xu Conglin and Bell, L. (2013) Emerging markets account for all growth in oil use forecast in 2014. *Oil and Gas Journal,* 6 January: 26.

Yergin, D. (1991) *The prize: the epic quest for oil, money and power.* New York, NY: Simon and Schuster.

9 Perspectives and limits on Sino–US competition

The Kazakhstan case study

Sébastien Peyrouse

Introduction

As a hegemonic power, the US has interests worldwide and, even in regions where it is relatively less influential, its symbolic weight remains decisive on the strategic, economic, and cultural levels and as a normative power. For Kazakhstan, which emerged on the international arena in 1991 upon the disappearance of the Soviet Union, the US is a key symbolic and strategic partner, but its importance was overestimated by the Kazakhstani elites in the early 1990s.[1] Today, these elites have a view that is more pragmatic, which ascribes to Washington a more modest place in their foreign policy. This is balanced by the important role of Russia, the growing role of China and the modest role of the European Union (EU), seen potentially as another West.

The main strategic and economic drivers in Kazakhstan comprise a ménage à trois involving Russia, China, and Kazakhstan. To include the US, one must speak of a ménage à quatre, or else refer to the existence of several triangles: the main, driving triangle is the Russia–China–Kazakhstan triangle; the Russia–USA–Kazakhstan triangle is second in importance; and the US–China–Kazakhstan is of least importance. Indeed, there isn't a China–US–Kazakhstan relationship that does not include Russia, which is omnipresent in the agendas of all three countries. On the Kazakhstani side, ambiguity does not weigh on the hierarchy of priorities in foreign policy terms. As President Nursultan Nazarbayev has stated several times, in speaking of his country's so-called 'multi-vectorial rationales', Russia is the first ally, China is the second, and the West is third (Sultanov and Muzaparova, 2005). In addition, because the EU has become Kazakhstan's foremost trading partner – 27 billion euros in 2012, which was 32.4 percent of its total trade (European Commission, 2012) – and because Kazakhstan was elected head of the Organization for Security and Co-operation in Europe (OSCE) for 2010, official Kazakhstani texts have begun to speak more and more clearly of their path towards Europe, rather than of their rapprochement with the US.

Energy diplomacy: the US/China containment of Russia?

With close to 40 billion barrels in proven reserves, Kazakhstan possesses 3.2 percent of total world oil reserves. In 2018–2019, with the exploitation of the gigantic site of Kashagan, it will come to dominate oil production in the Caspian Basin area (GAB International Business Network for Armenia and Caucasus, 2007). This Caspian production cannot replace the Middle East's role as the world's largest oil supplier, but it will play an important role outside the Organization of Petroleum Exporting Countries (OPEC), in particular for neighboring powers such as Russia and China, and potentially also for Japan and the Indian subcontinent. Close to three-quarters of the oil from Kazakhstan is situated in the western regions, which are dominated by three large deposits: Kashagan, Tengiz, and Karachaganak. The Kazakh oil sector seems set for exponential growth following a doubling of production between 2000 and 2008. In 2008, Kazakhstan extracted 70 million tons of oil, and produced 81.8 million tons of oil in 2013 due to the Tengiz and Karachaganak sites. Thereafter, the figures should skyrocket as the exploitation of Kashagan begins. If projections bear out, the country will produce two million barrels per day in 2015 and three million in 2020 (Raballand and Gente, 2007). Particularly well endowed in terms of oil reserves, Kazakhstan is placed 14th among world exporters, but envisages entering into the top ten world exporters by 2050 (Perspectives on Caspian Oil and Gas Development, 2008).

Russia still largely dominates the export routes of Kazakh oil. Prior to 1997, Astana was obliged to export via the only existing oil pipeline from Soviet times, the Atyrau–Samara, which is today controlled by Transneft. Both the West and China now challenge this Russian monopoly. Chinese firms have failed to enter the three main sites in the country (Tengiz, Karachaganak, and Kashagan) and Chinese firms – with some exceptions, such as AktobeMunayGas and KarazhanbasMunay – must therefore specialize in old or remote fields, which are considered technically difficult to exploit. The China National Petroleum Company (CNPC), however, has Beijing's diplomatic and financial support, enabling it to outbid competitors during negotiations, offer complementary good neighbor measures, and accept the authorities' requirement that the state-run company, KazMunayGas, be systematically associated with all activities. These strategies elicit angry reactions from competitors, especially Indian firms who often criticize China's aggressive energy policy which creates market distortion. In less than a decade, Chinese companies have successfully launched themselves into the Kazakhstani market, and by 2010, they were managing about one-quarter of Kazakh production (Peyrouse, 2008). The grand Chinese strategy has been to connect all the acquired fields with the giant Sino–Kazakh pipeline. Now complete, China has the advantage of an oil pipeline of more than 2,200 kilometers in length connecting the shores of the Caspian to the Dostyk–Alashankou border post, with an expected capacity of 20 million tons a year in 2018 (Peyrouse, 2007).

Although this pipeline is the very embodiment of the success of Sino–Kazakh oil cooperation, its dimensions remain modest and it does not resolve the problems that Beijing and Astana face. On the Chinese side, after costly optimization works, the volumes of the pipeline will represent only 10 percent of the country's needs (40 out of 400 million tons), a volume that will not undermine the domination of the Middle East. On the Kazakh side, the volumes transported by the pipeline will also stay modest – out of 80 million tons produced, nearly 25 million tons (nearly 30 percent of Kazakh oil exports) went to China (Demytrie, 2010). In terms of production, the 13 million tons produced by Chinese companies in 2013 represented only one-quarter of Kazakhstan's total production. This share might increase with the agreement signed in 2013 by Kazakhstan's state energy company KazMunaiGas with CNPC, which purchased an 8.3 percent stake in Kazakhstan's Kashagan Caspian offshore project.

Since 2005/06, a new large-scale energy project, the first Chinese transnational gas pipeline, has also been driving relations between Beijing and Astana. The Kazakh reserves (3 trillion cubic meters proven) could help supply the gas pipeline that Beijing negotiated with Turkmenistan in 2006 and would enable Astana to collect transit duty on Turkmen and Uzbek gas sent to China. Inaugurated in December 2009 and upon completion of a new line by the end of 2015, the overall delivery capacity of the Central Asia–China Gas Pipeline will have a capacity of 55 bcm per annum (CNPC, 2009), with Kazakhstan, Uzbekistan, and Turkmenistan each supplying a third (Energy Information Administration, 2008).

As part of this energy diplomacy, the US seeks to end the Russian monopoly that was born of the Soviet heritage. Since the beginning of the 1990s, Washington has supported large international majors such as Chevron and ExxonMobil in their efforts to exploit the two main oil deposits, Tengiz and Kashagan (International Energy Agency, 2008). Today, about two-thirds of oil produced in Kazakhstan is controlled by the main international majors, leaving China and Russia in the minority. The US also supports all the transport projects that bypass Russia and Iran by linking Central Asia, the Caucasus, and Turkey.

In terms of energy geopolitics, the US has succeeded in limiting Russian influence over the exploitation of deposits and is not unduly worried about the weight carried by China, which has not managed to gain access to the main sites. As far as the export routes are concerned, US success is less clear. Transit via Russia therefore continues to make up the majority of Kazakh oil exports. The US policy to promote export routes for oil to the West has only met with limited success for the present. The policy to promote gas export routes, which mainly concerns Turkmenistan, has also proven more complex than was anticipated: as of 2015 neither exports towards the south (the Turkmenistan–Afghanistan–Pakistan–India gas pipeline) nor towards Europe (Nabucco and Southern Corridor) have yet to become a reality.

In this energy diplomacy, the triangularization of foreign policies combines two contradictory trends. US and Chinese strategies that form part of a parallel, which are not coordinated, attempt to contain Russian influence over extraction and export routes, but they do so in two opposite directions: westward for the US (Caucasus–Turkey–Europe) and eastward for China. Washington, however, does not seem to have perceived China's growing control over Kazakh hydrocarbons as a challenge to its interests because Chinese presence is still only complementary and not decisive. As far as Beijing is concerned, it seems more preoccupied by Western competition in the domain of gas (Turkmenistan) than in oil (Kazakhstan). US and Chinese energy policies on Kazakhstan are complementary in their immediate opposition to Russia and potentially competition in terms of the direction of their export routes, a trend that is likely to gather strength in the years to come.

The uranium issue: a growing Russia–China condominium

Not all energy matters are limited to hydrocarbons. For Kazakhstan, uranium is also a strategic component. The country is the leading world producer, ahead of Canada and Australia, and has set up an ambitious development program for the civil nuclear industry between 2010 and 2020 (Laruelle, 2010).

Russia occupies a privileged place in Kazakh uranium for historical reasons that led to this sector being promoted during the Soviet period. In 2006, Moscow and Astana signed an interstate cooperation program in the civil nuclear sector that included several cooperation agreements worth about 10 billion US dollars (Laruelle and Peyrouse, 2010).

But China also needs uranium, chiefly to complete the construction of dozens of nuclear power plants. In the 2000s, Kazatomprom and the Guangdong Nuclear Power Group (CGNPC) signed several cooperation agreements. Kazakhstan is the largest foreign supplier of uranium to China (Kazatomprom, 2008) and has also set itself up in the Chinese market for building nuclear power plants (Pannier, 2011).

On the other hand, the US has not gained a decisive position over what is becoming one of Kazakhstan's international drivers on a par with oil production. Since 2007, Kazatomprom and the US company Westinghouse have consolidated a privileged alliance. Having bought 10 percent of Westinghouse's shares from Toshiba, Kazatomprom now delivers nuclear fuel to the US company; however, compared to Russian and Chinese involvement on the Kazakh uranium market, and to Areva's equally considerable role, Washington does not carry as much weight as might have been expected. In the nuclear sector, the US cannot really influence the Russia–China–Kazakhstan triangle, it can only maintain a marginal role.

China's trade weapon: disproportionate economic weight

Whereas Central Asia in general and Kazakhstan in particular are often viewed as a locus of geopolitical competition between Russia and the US, the peaceful economic competition is actually between China and the EU, with Russia an increasingly less-powerful economic actor and the US still a small one. As in other regions of the world where Beijing is establishing itself, it has multiple objectives that the Chinese authorities see as intrinsically related. First, the People's Republic of China (PRC) wants to consolidate its geopolitical influence in Kazakhstan by creating economically based good neighborhood relations that diffuse potential political tensions. Second, it wishes to contribute to regional development in order to avoid political and social destabilization, which could have domestic consequences in Xinjiang and slow Chinese economic growth. Last, Kazakhstan provides Chinese products with new markets – markets that could open up to Russia, Iran, and Turkey. For landlocked Central Asia, the Chinese economic engine offers the prospect of a new trans-Eurasian corridor, thereby generating a unique historical opportunity (Laruelle and Peyrouse, 2011).

To manage these strategies, China employs multiple instruments in developing both bilateral relations and collective structures such as the Shanghai Cooperation Organization (SCO). Many Western companies consider Central Asia, including Kazakhstan, as a risky environment where investment conditions are unfavorable or unpredictable. Local authorities therefore seek pragmatic foreign partners who are not only undeterred by the political situation but are also capable of investing in large projects, as well as in small- and medium-sized ones. Beijing has played this investment card by opening highways and railroads, improving electrical grids and hydroelectric resources, building pipelines, exploiting precious mineral resources, and developing trade relations.

China is one of the only investors present in Central Asia that attaches importance to the frequently neglected banking sector, which permits the Central Asian republics to pursue large-scale projects. The Bank of China and the Chinese Industrial and Commercial Bank opened branches in Kazakhstan and in 2009 China extended US$10 billion in loans to Kazakhstan, half of which was an Export–Import Bank of China's loan to its counterpart, the Development Bank of Kazakhstan (Energy Daily, 2009).

Trade between China and Central Asia has been booming for almost a decade. Between 2002 and 2003, it increased about 300 percent, going from about US$1 billion per year to more than US$3 billion. An increase of 150 percent followed between 2004 and 2006, with trade reaching more than US$10 billion, according to Central Asian figures (Raballand and Andrésy, 2007) or US$13 billion according to Chinese figures (Raballand and Kaminski, 2007). Before the global economic crisis in 2008, trade between China and Central Asia exceeded US$25 billion, and trade between Russia and Central Asia was US$27 billion. Kazakhstan represents two-thirds of Sino–Central Asian trade and their bilateral trade reached US$20 billion in 2010 and 24 billion in 2012 (European Union, 2009). The gap between Moscow and Beijing is therefore reducing to the advantage of the latter, whose commercial development appears exponential. Taking the shuttle trade and contraband into account, China's economic presence in Kazakhstan is already greater than Russia's (see Table 9.1).

Between 80 and 90 percent of Chinese exports to Kazakhstan consist of finished, diversified goods, for example consumer products, machinery, processed foodstuffs, textiles, shoes, electronic goods, pharmaceutical products, and automobile spare parts (Wu and Chen 2004). On the other hand, more than 85 percent of Central Asian exports to China consist of raw materials, petrol, and ferrous and nonferrous metals (Paramonov and Strokov, 2007). As Central Asia's worldwide exports focus on raw materials (Myant and Drahokoupil, 2009), China's strength

Table 9.1 Kazakhstan's trade with China, Russia and the US in 2012 (in billion euros)

	China	*Russia*	*USA*
Kazakhstan	19,783	16,060	1,891

comes from exporting consumer products at low prices, which suits the low living standards of the Central Asian populations, whereas Russian, Turkish and Iranian, as well as Western, products remain too expensive. China also provides technological goods to Central Asia's middle-class, particularly in Kazakhstan whose consumption patterns are very developed.

This growing Chinese presence benefits Kazakhstan's economy, but it privileges heavy industry, which is in the hands of the oligarchs and powerful clans. Trade has prompted some Chinese or Kazakhstani private enterpreneurship, and joint ventures co-owned with the middle classes; nonetheless, these businesses benefit Kazakhstani corrupted milieus, customs officers, the police, etc. In addition, Kazakhstani public opinion increasingly condemns Chinese economic investment because Chinese firms bring their own equipment and material and do not give work to local enterprises. Leading personnel are mostly Chinese who live in isolation at their place of work, without interacting with the host society, and the few locals employed are often subjected to appalling working conditions (Laruelle and Peyrouse, 2009a). One must wonder, therefore, if the Chinese presence brings development and helps spread know-how and techniques, local training, and interaction with the investing country. Similar questions are being raised about China's involvement in Africa, while in nearby Afghanistan the response is paradoxical indeed.

Russia remains an important trade partner for Kazakhstan, representing about 19 percent of total Kazakh trade (16 billion euros in 2012), which earns it top spot on the import list ahead of China and the EU. Russia remains a dominant economic actor if energy is taken into account, and is a significant player in heavy industry and infrastructure, both of which are old Soviet specializations. Yet Russia is a relatively modest and rather uncompetitive player in terms of small and medium-sized enterprises (SMEs) and new technologies. Despite the dominance of the energy issue, Russia's trade with Kazakhstan involves other important sectors: electricity (maintaining the Soviet grid facilitates common projects), hydroelectricity, construction, telecommunications (mainly mobile telephony), transport (in particular the freight services) and railways (but not the automobile market), banks (Russo–Kazakhstani partnerships are multiplying), the military–industrial complex, and lastly, certain agribusiness sectors (Russia and Kazakhstan are strengthening their cooperation on cereals). The coming into force of a Russia–Belarus–Kazakhstan Customs Union in 2010 has strengthened unification of the Kazakhstani market with its Russian neighbor (which began with the Eurasian Economy Community, created in 2000 upon the initiative of the Kazakhstan government), without putting China's commercial influence into question.

In this context, the US does not appear to Kazakhstan as a trade power, even if the latter has received more than US$14 billion in foreign direct investment (FDI) from Washington since 1993. In 2012, the US was Astana's seventh largest trade partner with trade levels of US$2.3 billion (Office of the US Trade Representative, 2013), which is less than 3 percent of Sino–Kazakh trade. US companies are active in the mechanics, vehicles, electricity, aviation, and medical and optical instruments sectors. They are also interested in the oil and gas industries, with their mastery of cutting edge technologies for deposit exploration, pipeline equipment, techniques of purification and upgrading to meet ecological norms, reviving exhausted fields, laboratory studies, refinery equipment and the sale of oil products (Laruelle and Peyrouse, 2012). US companies also supply excavation and extraction material, and technologies for water decontamination in the minerals sector. Large US companies, such as United Technologies, Sikorsky, Kellogg, and General Electric are active in the promising domains of aircraft technologies, electricity, and railway transport. The US has also become Kazakhstan's main partner in agricultural machines and equipment, and is targeting cereals and bio-fuel processing, as well as agribusiness.

The existing balance in the energy domain is therefore not replicated in the domain of trade in general. Indeed, in terms of trade, Chinese predominance is overwhelming, and this is only set to increase in magnitude in the years to come. The commercial niches available for other countries are declining: Russia will only remain competitive in heavy sectors linked with electricity, heavy machinery, and transport; whereas the US has a high-tech niche that faces competition from the EU, which targets the same sectors (Peyrouse, 2009). For Moscow, Chinese competition is two-fold: on the Central Asian market, but also on its own market because China became Russia's largest trading partner in 2009. China intends to depend on Russia for no more than 20 percent of her energy needs. The Russian energy strategy to 2030 envisages that Asia–Pacific's share of crude oil and oil products exports will rise to 22–25 percent, with exports of natural gas reaching 19–20 percent (Smith, 2010).

Geopolitical strategies and outcomes for Russia, China, and the US in Kazakhstan

The hypothesis of the triangularization of the foreign policies of the US, China, Kazakhstan, and Russia is most pertinent in the strategic domain, where Washington occupies a position that is much larger than its energy and trade influence.

Russia remains the dominant strategic power for Kazakhstan and security issues drive Moscow's renewed presence in the region. The security challenges for Russia in Central Asia are multiple: any destabilization in the weakest (Kyrgyzstan, Tajikistan) or the most dangerous (Uzbekistan) states will have immediate repercussions in Russia, including Islamist infiltration in the Volga–Ural region and the North Caucasus, if not the whole country; an increase in the inflow of drugs reaching the Russian population, which is already widely targeted by drug traffickers; a loss of control over the export networks of hydrocarbons, uranium sites, strategic sites in the military–industrial complex, and electricity power stations; a drop in trade exchanges; a loss of direct access to Afghanistan; and an uncontrollable surge of flows of migrants particularly of refugees. For Moscow, the security of the southern borders of Central Asia is a question of domestic security – not imperialism but pragmatism – the 7,000 kilometers of the Russo–Kazakhstani border in the heart of the steppes is nearly impossible to securitize, forcing Russia to attack these clandestine flows 'downstream', confirming Central Asia's role as a buffer zone for Russia.

During Putin's two mandates (2000–2008), Moscow returned to its status as Kazakhstan's number one partner. Of the numerous institutions within the Commonwealth of Independent States (CIS), only the Anti-Terrorist Center and the Council of Border Guard Agency Commanders are functional, providing Kazakhstani security services and army with training, including annual exercises organized by Moscow (Peyrouse, 2011). The main Russian–Central Asian multilateral military collaborations, the Collective Security Treaty Organization (CSTO), provides for the sale of military material to member countries at Russian domestic market prices, and has revived cooperation amongst the Russian and Central Asian military industrial complexes. Joint military exercises are carried out annually in one of the member countries, which simulate terrorist attacks (called Rubezh) or anti-narcotics operations (Kanal), and permit greater interaction between border guards and other police and military units. New operations were organized along similar lines: arsenals against arms trafficking; Nelegal against illegal immigration; and Proxi against technological criminality (East Time, 2009). Operation Kanal is alleged to have resulted in the seizure of more than 300 tons of drugs and illicit substances in 2008 alone and has become a permanent institution (Interfax, 2009). The Collective Rapid Deployment Force (CRDF) for Central Asia, comprised of about 4,000 persons made up of Kazakh, Kyrgyz, Russian, and Tajik units, is the only trained armed force in Central Asia capable of intervening in real time. Its main target is border securitization in case of violations by terrorist groups (Soyuz, 2009). Through the CSTO, Moscow hopes to weaken the US military partnership in the region and to make itself the necessary intermediary of military relations between the West and the Central Asian regimes (Frost, 2009).

Thanks to several agreements between Astana and Moscow, Russia has become Kazakhstan's primary supplier of defense equipment, including tanks, helicopters, planes, spare parts, and weapons (Nomad.su, 2008). Kazakhstan is the first client of the Kazan helicopter factory and several small and medium tonnage ships are going to be constructed in the Tartar factories of the Russian industrial–military complex at Kazakh request (Slavin, 2004). Astana also considers itself as a future supplier of arms to the other Central Asian states in the medium term (McDermott, 2008a). The second largest domain of cooperation, which assures Russia its supremacy in the military sector, is personnel training. The Soviet legacy in this sector has enabled Moscow to train a majority of Central Asian military personnel (Marat, 2007), for example 700–800 Kazakh officers are enrolled in Russian Federation institutions at any given time. Kazakhstan also constitutes a major element of the Russian defense system, and although Russia has no military base there, since the 1990s Astana has allowed Moscow use several firing ranges in exchange for military material, specialized maintenance, and officer training (Kenzhetayev, 1998). Recent annual, Russian–Kazakhstani bilateral military exercises also focused on drug-trafficking and illegal migrations (Inform.kz, 2008).

Russia is also present through the Shanghai Cooperation Organization (SCO),[2] but Kazkhstanis see the SCO primarily as a tool of Chinese influence and also one of Beijing's main instruments in Central Asia. One should not see this regional institution as a reflection of multilateralism, however, because it is only a discussion platform, while the real political, economic, and strategic decisions are taken in bilateral frameworks where the power differential is fully in China's favor. China's drivers in Central Asia are all linked to a global, long-term view of its security interests. Once the Soviet Union disappeared, China's primary objective was stability on its north and northwest border by addressing the issue of territorial boundaries with Russia, Kazakhstan, Kyrgyzstan, and Tajikistan and seeking confirmation that they would respect the One China discourse. Yet in China's perception, Central Asia is not only a part of the post-Soviet world but also a part of West Asia and an opening toward Afghanistan, Iran, and Turkey. Chinese reevaluation of continental routes therefore must be understood as part of a long-term historical evolution. Moreover, Central Asia is directly linked to domestic issues. The cultural, linguistic, and religious linkages between the Central Asian and Uyghur populations are critical to Xinjiang's economic development and future political stabilization. Beijing's open door policy and Far West Development Program have transformed this landlocked region into a location for significant subsoil resource exploitation and an outpost for advancing Chinese trade in Central Asia, Afghanistan, and Pakistan (Garver, 2006). Finally, Central Asia has also come onto the Chinese radar as a partial solution to two concerns: securing continental energy supplies that are not subject to global geopolitical complications, and also helping China appear as a peaceful rising power able to play the card of multilateralism (Ong, 2005).

The SCO and its precursor, the Shanghai Group, have successfully eased long-standing tensions between Russia and China, put in place cooperative mechanisms for border management, and established a collective narrative on the common threats they face. But having attained this threshold of development and institutionalization, the organization faces new challenges. Despite its security rhetoric, it is relatively inactive in practice and unable to compete with Russian strategic influence. It cannot be compared to the CSTO structure that offers real interactions among general staffs and army corps; trains soldiers in one of the member countries; and carries out arms sales and logics of technological and logistical interoperability (Frost, 2009). SCO activities largely remain declarations of intent because a lack of coordination among member states is evident, the desire to exchange information is restrained, the financial resources are far too few, and bureaucratic structures too weak. The absence of common jurisdictions in most areas and the lack of exchanges on important and related matters considerably weaken the scope for potential action. In economic terms, the SCO has failed to compete with the Eurasian Economic Community of the Customs Union to guide the development of trade relations among member states, which remain primarily bilateral, or to erase Russian and Central Asian fears of the impending 'invasion' of Chinese products. In the coming years, a growing Sino–Russian competition for Central Asian resources – mainly hydrocarbons and potentially uranium, but probably not electricity – could impede SCO energy-related cooperation.

As a mechanism to reinforce confidence, the SCO has been an historical success, but if viewed as an organization that influences Central Asian security, for the time being it is only a 'paper tiger' (Cooley, 2009). The gap between the Organization's narrative about fighting non-traditional threats and its mechanisms to enable collective, or at least concerted, action is immense. The SCO provides no military guarantees in cases of domestic crises, nor does it offer any structure such as a rapid intervention force or a collective troop force (such as that of the Ministry of Emergency Situations in Russia) that is able to intervene in situations such as natural and ecological catastrophes, sudden populations displacements, a flood of refugees, etc. The SCO has never reacted to a large-scale crisis within one of its member states – its silence during the Kyrgyz events of 2010 confirmed this, as does its incapacity to offer anything collective to a state that, albeit a non-member, is as strategic as Afghanistan. The obsession for consensus and for maintaining the status quo has in fact hampered SCO effectiveness and risks delegitimizing it in the future. Since 2008, SCO has entered into a growth crisis – it has not defined any positive long-term goals, has no well-defined priorities, and refuses to discuss divergences in its members' priorities. The SCO remains a reflection of Chinese willingness to support so-called healthy Central Asian order, free from any of the three evils,[3] and devoid of pro-Western forces that might try to destabilize China. Even if SCO partially limits the room the Central Asian states have to maneuver, particularly relative to the West, it provides the established regimes with an ideological framework to shore up their legitimacy on both the domestic and international fronts.

The US needed some time before it came to appreciate Kazakhstan's growing role in the Central Asian strategic equilibrium because throughout the 1990s it remained focused on Uzbekistan. At that time, Washington considered Astana too close to Moscow, as well as a major proponent of Russian influence in Central Asia. Moreover, the US dented any positive image they enjoyed within Kazakhstani public opinion in the 1990s – the choice of Uzbekistan as the strategic partner for the war against terror, despite it being one of the most dictatorial regimes of the region, distorted Washington's political message by promoting an image of a country indifferent to the everyday difficulties of the population and chiefly obsessed with the defense of its own interests (Laumulin, 2008); however, Astana is the only Central Asian capital that has implemented a multi-pronged foreign policy, remained faithful to its so-called 'multi-vectorial' foreign policy for two decades, and avoided the geopolitical subterfuges and last-minute reversals that characterize its Uzbek neighbor. Today, Astana presents itself as one of Washington's loyal allies, united not only by economic interests but also by a strategic and military partnership, which passed the test despite Russia's revival of influence and China's growing role.

The new US strategy in Afghanistan underlies renewed US interest in the region (Kuchins *et al.,* 2009). The Kazakhstani regime, concerned about the efforts of the international coalition in Afghanistan, appreciates the US's renewal of interest. In

November 2010, it allowed US planes to fly across its territory to supply weapons to forces in Afghanistan – Russia had also given consent – which meant that US Air Force cargo jets could fly from Alaska to Afghanistan without refuelling. This reinforced partnership between the US and Kazakhstan has a two-fold impact: Washington can use new arguments for negotiation because this military cooperation is supposed to accompany a renewal of US economic presence, and in return, Kazakhstan has a new means of influence over Washington because the latter cannot dispense with the former's support without imperiling its operations in Afghanistan.

Astana strengthened its strategic weight in American eyes, and symbolically flaunted this gain by hosting the forum of the Euro–Atlantic Security Council in June 2009. Kazakhstan is the most advanced country of Central Asia in terms of cooperation with NATO thanks to its Individual Plan of Action for the Partnership (IPAP) (McDermott, 2008b). Kazakhstan is the only Central Asian state to create a small peacekeeping force that collaborates with NATO under a UN mandate – the Kazbat Battalion that has been upgraded to the Kazbrig Brigade. Astana hopes to obtain 'interoperability' status with NATO in the coming years, despite its privileged partnership with Russia (McDermott, 2009). Kazakhstan is most sensitive to NATO objectives, such as the reform of civil–military relations, which can affect global political questions (Boonstra, 2007), but it is not indifferent to the issue of the reform of the security services, and it also recognizes the importance of regional cooperation – elements that play in its favor. It also partakes in the Action Plan of the Partnership against Terrorism, which exchanges information with NATO members and hosts the annual Steppe Eagle anti-terrorism exercises.[4]

Among Central Asian states, Astana receives the most aid under the US Foreign Military Financing program. Annual military training exercises, called Steppe Eagle, gather together servicemen from Kazakhstan, the US, and the UK. Between 2003 and 2008, the US also aided Kazakhstan in training a rapid reaction unit capable of responding to all kinds of attacks, but mainly anti-terrorist. In 2008, a new 5-year bilateral agreement provided for training Kazakhstani officers to gain experience of US methods. Since the 2000s, about 400 Kazakh officers received such training in US military institutes, including the West Point Military Academy and the National Defense University. Washington is also involved in creating an institute for professional non-commissioned officers. Financial, technical, and training aid from the US to the Kazakh military increased in the 2000s after Astana established its own Caspian fleet. After 2004, the US offered a modernization program for the Kazakh army along several axes: training officers in the military academies of NATO members, particularly in Turkey, Greece, Italy, and Spain; supplying material for radio and radar surveillance capable of monitoring both the surface and the depth of the Caspian; and modernizing the port infrastructures, particularly those of Atyrau (Laruelle and Peyrouse, 2009b). The US is also monitoring the militarization of the Caspian Sea and is working to maintain the balance of the forces; however, the US's main ally in the Caspian region is Azerbaijan because despite Astana's commitment to NATO structures, relations remain complex due to Kazakh authorities being less overtly opposed to

Moscow than those in Baku. Still, the US is focusing on Kazakhstan as its second most important Caspian partner.

Conclusion: is a triangularization China–US–Kazakhstan relevant?

Competition between China, Russia, and the US in Kazakhstan, as well as their influence in and cooperation with this country, requires a balancing act. All three states want a privileged relationship with each of the other two partners; any alliance between two of the three weakens the position of the third. As such, Washington does not want to see a deep strategic alliance between Moscow and Beijing, which would thwart its advances into Eurasian space and its progress on large international issues, such as constraining Iran or North Korea. Russia sees a Sino–American rapprochement and growing economic and financial interdependency between the two as shaping a Sino–American twenty-first century in which Moscow becomes a second-rate actor. Finally, China does not really welcome improved Russo–US relations, or when the US–EU–Russia trio functions well. These tensions give Kazakhstan significant flexibility to play these several actors against each another, and yet although Kazakh authorities appear to have sufficient decision-making autonomy to counter Russian and US objections, their freedom relative to China seems to be declining and its geopolitical balancing – or so-called 'multivectoralism' – has been the key driver of Kazakhstan's foreign policy success.

The US perceives Russia, not China, as its main rival for influence in Kazakhstan. When framing its foreign policy toward this country, Washington takes into account Russia's privileged relations with Astana and criticizes Moscow's strategies for defending its interests in post-Soviet space, but it also advances its pawns prudently, sensitive to Astana's desire for autonomy and balance. China, for its part, remains largely absent from Washington's Kazakhstan agenda; its energy stakes are not formulated in terms of a West versus China axis of competition, but rather a West versus Russia or West versus Russia and China. On the strategic level, the US sees the SCO as an institution that rivals its own interests in the region, but the real competition comes from Moscow-led institutions such as the CSTO. Russia also retains Cold War schemas, focusing on US strategies of containment without measuring the rhythm of China's advance. Beijing, for its part, clearly prefers Russian control over Central Asia to US domination, and can only be happy about the mutual Russo–US 'neutralization', which leaves the field open for its in economic terms.

In Kazakhstan, the US cannot think like a hegemon because the disappearance of its Soviet enemy has not guaranteed it a new sphere of influence. China has clearly positioned itself as a major power in the region without provoking any US reaction, which essentially monitors Russian ambitions in the region and probably even more so after 2014 events in Ukraine. Although Kazakhstan is important for US involvement in the Caspian Basin, post-Soviet space and the Afghan environment, it is not perceived to be vital for US diplomacy and is not worth getting involved if doing so damages relations with China or even Russia. Similarly, China is advancing cautiously into Kazakhstan – it does not go directly against US interests, it shows its respect for Russian supremacy in the region, and it tries not to fuel the anxieties expressed in Kazakhstani public opinion about the Chinese neighbor.

The China–US–Kazakhstan triangle, however, is not without its tensions – politically Beijing and Washington are clearly opposed in terms of the type of regime that they would like to see Astana adopt. The will of the Kazakhstani authorities to acquire interoperability with NATO does not please Beijing, which does not want to see the US settled permanently on its borders. The Chinese authorities are closely monitoring NATO's military activities in the region in order to secure their own strategic autonomy in Xinjiang and Tibet. Also, competiton with the West over the direction of pipelines could exaccerbate tensions in the decades to come.

The China–US–Kazakhstan triangle therefore has limited relevance because other more pertinent triangles are shaping the local balance. Given Russia's ambitions to recover its influence in several former Soviet republics, the equilibrium between the US and China in Kazakhstan will prove crucial for the strategic balance in the years to come. For the time being, Russia's presence diverts attention and acts as a screen: both China and the US see their relations with Kazakhstan essentially through their relation with Moscow, downplaying their potential conflict and delaying it to a future time when Russia will have faded into the background.

Notes

1 The term Kazakhstani is used to speak about the entirety of the country's citizens, without making any national distinction, whereas the term Kazakh is reserved for the ethnic group, which enjoys a small majority. This is a difference that is present in Russian and in Kazakh, as well as in official terminology employed in Kazakhstan.
2 Its members are Russia, China, the four Central Asian states with the exception of Turkmenistan and observer status has been acquired by Afghanistan, India, Pakistan, Iran, and Mongolia.
3 Also called the three extremisms (sange jiduanzhuyi). This ideological drive is sometimes called the 'Shanghai spirit' (see, for example, Oresman, 2003).
4 For more information, see www.nato.int/cps/en/natohq/topics_49598.htm?selectedLocale=en (accessed 12 April 2015).

References

Boonstra, J., 2007. *NATO's role in democratic reform.* FRIDE Working Papers, May. Brussels: FRIDE.
CNPC, 2009. Central Asia–China Gas Pipeline operational. Available online at: www. cnpc.com.cn/en/FlowofnaturalgasfromCentralAsia/FlowofnaturalgasfromCentral-Asia2.shtml (accessed 11 April 2015).
Cooley, A., 2009. *The stagnation of the SCO. Competing agendas and divergent interests in Central Asia.* PONARS Memo No. 85, September. Washinton, DC: PONARS Eurasia.
Demytrie, R., 2010. *Struggle for Central Asian energy riches.* BBC News, 3 June. Available online at: www.bbc.co.uk/news/10175847 (accessed 26 November 2013).
East Time, 2009. *Problemy sotrudnichestva ODKB i SHOS v oblasti bezopasnosti.* 3 July. Available online at: www.easttime.ru/reganalitic/1/212p.html (accessed 5 April 2014).

Energy Daily, 2009. *China loans 10 bln dlrs to Kazakhstan.* 17 April. Available online at: www.energy-daily.com/reports/China_loans_10_bln_dlrs_to_Kazakhstan_state_media_999.html (accessed 11 February 2014).

Energy Information Administration, 2008. *Kazakhstan.* Country Brief Analysis, February. Washington, DC: Energy Information Administration.

European Commission, 2009. European Union. Trade with Kazakhstan. Available online at: http://trade.ec.europa.eu/doclib/docs/2006/september/tradoc_113406.pdf (accessed 12 June 2010).

European Commission, 2012. European Union. Trade in goods with Kazakhstan. Available online at: http://trade.ec.europa.eu/doclib/docs/2012/september/tradoc_113406.pdf (accessed 4 November 2012).

Frost, A., 2009. The Collective Security Treaty Organization, the Shanghai Cooperation Organization, and Russia's strategic goals in Central Asia. *The China and Eurasia Forum Quarterly,* 7 (3): 83–102.

GAB International Business Network for Armenia and Caucasus, 2007. *Caspian energy and transport issues expand into military–political confrontation.* Available online at: www.gab-bn.com/IMG/pdf/Re11-_Caspian_Energy_And_Transport_Issues_Expand_Into_Military-Political_Confrontation._Microsoft_Word.pdf (accessed 10 April 2015).

Garver, J.W., 2006. Development of China's overland transportation links with Central, South-West and South Asia. *The China Quarterly,* 185: 1–22.

Inform.kz, 2008. *Rossiisko-kazaxstankie voennye ucheniia stanut ezhegodnym.* Altai Transgranichnyi. October 7. Available online at: www.altaiinter.info/news/?id=19902 (accessed 15 April 2014).

Interfax, 2009. *Spetsoperatsiia ODKB 'Kanal' poluchila status postoianno deistvui-uchshego proekta.* 5 June. Available online at: www.interfax.by/news/belarus/56181 (accessed 13 April 2014).

International Energy Agency, 2008. *Perspectives on Caspian oil and gas development.* International Energy Agency Working Paper Series, December. Paris: International Energy Agency.

Kazatomprom, 2008. Press Release on Signing of Agreements for Strategic Partnership Between Kazatomprom, CNNC and CGNPC. Kazatomprom, 6 November. Available online at: www.kazatomprom.kz/en/#!/content/press-release-signing-agreements-strategic-partnership-between-kazatomprom-cnnc-and-cgnpc (accessed 10 April 2015).

Kenzhetayev, M., 1998. *Russian–Kazakhstan military and technical cooperation: structure and perceptive. Export of arms.* Available online at: www.armscontrol.ru/atmtc/kazakhstan/article_mtc_kazakhstan.htm (accessed 17 May 2014).

Kuchins, A., Sanderson, T. and Gordon, D., 2009. *The Northern Distribution Network and the Modern Silk Road.* Washington, DC: Center for Strategis and International Studies.

Laruelle, M., 2010. *Kazakhstan as a uranium power: forthcoming successes and challenges.* Central Asia and Caucasus Analyst, 31 March. Available online at: www.cacianalyst.org/?q=node/5296 (accessed 10 April 2015).

Laruelle, M. and Peyrouse, S., 2009a. *China as a neighbor: Central Asian perspectives and strategies.* Central Asia–Caucasus Institute & Silk Road Studies Program.

Laruelle, M. and Peyrouse, S., 2009b. The militarization of the Caspian Sea: 'great games' and 'small games' over the Caspian fleets. *The China and Eurasia Forum Quarterly,* 7 (2): 17–35.

Laruelle, M. and Peyrouse, S., 2010. *L'Asie centrale à l'aune de la mondialisation. Une approche géoéconomique.* Paris: IRIS–Armand Collin.

Laruelle, M. and Peyrouse, S., 2011. *The Chinese Question in Central Asia.* New York, NY: Columbia University Press.

Laruelle, M. and Peyrouse, S., 2012. *Globalizing Central Asia. Geopolitics and the Challenges of Economic Development.* Armonk: M.E. Sharpe, 53.

Laumulin, M., 2008. The US geopolitical experience in Central Asia: success or failure? *In:* M. Esteban and N. de Pedro, eds. *Great powers and Regional Integration in Central Asia. A Local Perspective.* Madrid: Fundacion Alternativas, and Almaty: KIMEP, 53–78.

Marat, E., 2007. Soviet military legacy and regional security cooperation in Central Asia. *The China and Eurasia Forum Quarterly,* 5 (1): 83–114.

McDermott, R.N., 2008a. *Kazakhstan kak potentsialn'yi postavchshik oruzhiia.* Geokz.tv, June 23. Available online at: www.geokz.tv/article.php?aid=5466 (accessed 19 January 2013).

McDermott, R.N., 2008b. United States and NATO military cooperation with Kazakhstan: the need for a new approach. *The Journal of Slavic Military Studies,* 21 (4): 615–41.

McDermott, R.N., 2009. *Kazakhstan's Defense Policy: An Assessment of the Trends.* Carlisle, PA: Strategic Studies Institute, US Army War College.

Myant, M. and Drahokoupil, J., 2009. International integration and the structure of exports in Central Asian republics. *Eurasian Geography and Economics,* 49 (5): 604–22.

Nomad.su, 2008. *Rosoboroneksport ozhidaet zakaz iz Kazakhstana na boevye mashiny podderzhki tankov proizvodstva Uralvagonzavoda.* 23 June. Available online at: www.nomad.su/?a=4-200806230818 (accessed 16 April 2013).

Office of the US Trade Representative (USTR), 2013. Office of the US Trade Representative, Available online at: www.ustr.gov (accessed 21 April 2014).

Ong, R., 2005. China's security interests in Central Asia. *Central Asian Survey,* 24 (4): 425–39.

Oresman, M., 2003. Catching the Shanghai Spirit. *Journal of Social Sciences (Shanghai),* 12, December. Available online at: www.foreignpolicy.com/articles/2004/05/01/catching_the_shanghai_spirit (accessed 10 April 2013).

Pannier, B., 2011. *Kazakh President energized after China trip.* Radio Free Europe/Radio Liberty, February 23. Available online at: www.rferl.org/content/kazakh_president_energized_china_trip/2318634.html (accessed 12 January 2014).

Paramonov, V., and Strokov, A., 2007. *Economic Involvement of Russia and China in Central Asia.* Oxford, UK: Conflict Studies Research Centre, 12.

Perspectives on Caspian and Gas Development, 2008. International Energy Agency Working Paper Series, December, 7–8.

Peyrouse, S., 2007. *The Economic Aspects of the Chinese–Centra Asia Rapprochement.* Silk Road Papers. Washington, DC: Central Asia and Caucasus Institute.

Peyrouse, S., 2008. Chinese economic presence in Kazakhstan: China's resolve and Central Asia's apprehension. *China Perspectives,* 3: 55–75.

Peyrouse, S., 2009. Business and trade relationship between the EU and Central Asia. EU–Central Asia Monitoring Working Paper 1. Available online at: www.eucentralasia.eu/fileadmin/user_upload/PDF/Working_Papers/WP-No.1.pdf (accessed 19 March 2013).

Peyrouse, S., 2011. Russia–Central Asia: advances and shortcomings of the military partnership. In *Central Asian Security Trends: Views from Europe and Russia,* 1–34. Carlisle, PA: Strategic Studies Institute, U.S. Army War College.

Raballand G. and Andrésy, A., 2007. Why should trade between central Asia and China continue to Expand?, *Asia Europe Journal,* 5(2): 235–52.

Raballand, G. and Gente, R., 2007. Oil in the Caspian Basin: facts and figures. *In:* B. Najman, G. Raballand, and R. Pomfret, ed. *The Economics and Politics of Oil in the Caspian Basin. The Redistribution of Oil Revenues in Azerbaijan and Kazakhstan.* London: Routedge, 9–29.

Raballand, G. and Kaminski, B., 2007. La Déferlante économique chinoise et ses conséquences en Asie centrale. *Monde chinois,* 11: 129–34.

Slavin, M., 2004. *Na Kaspii postroiat novyi flot? [Is a new fleet being constructed on the Caspian?].* Sootechestvennik, March. Available online at: www.russedina.ru/?id=5926 (accessed 2 September 2011).

Smith, M.A., 2010. *The Russo–Chinese Energy Relationship.* Swindon: Defence Academy of the United Kingdom.

Soyuz, 2009. *Krasnaia Zvezda.* Available online at: www.postkomsg.com/news/ bezopasnost/179497/ (accessed 10 April 2015).

Sultanov, B.K. and Muzaparova, L.M. (eds) 2005. *Stanovlenie vneshnei politiki Kazakhstana. Istoriia, dostizheniia, vzgliad na budushchee.* Almaty: IWEP.

Wu, H.-L. and Chen, C.-H., 2004. The prospects for regional economic integration between China and the five Central Asian countries. *Europe–Asia Studies,* 56 (7): 1069–70.

10 The impact of Brazil's expanding hydrocarbon reserves on its relations with the US and the PRC

Susana Moreira

In 2007, Petrobras announced the discovery of the Tupi oil field off the coast of Brazil. It was said to contain 5–8 billion barrels of sweet oil, making it the largest discovery in the Americas since 1976, and the most promising find anywhere since Kazakhstan's Kashagan field in 2000. Tupi increased Brazil's oil reserves by 50 percent and demonstrated the viability of exploration in the pre-salt layer – consisting of oil reservoirs under a shifting layer of salt at depths of between 16,400 and 22,950 feet. Subsequent successful wells have led to estimates for the total size of Brazil's pre-salt reserves ranging from 20 to 110 billion barrels.

This chapter looks at the rapid growth of Brazil's oil reserves and what this means for Brazil and its relations with the US and China, the world's two largest oil consumers. First, it looks at how the world's two top consumers of oil are reacting to Brazil's expanding oil reserves and its changing oil fortunes. Second, it compares their reactions and how these reactions have affected Brazil's behavior. Finally, the chapter concludes with an assessment of how Brazil's oil wealth influences Brazil–US–China relations.

Introduction

Brazil's bountiful natural resources have drawn the attention of outsiders for centuries. The Portuguese were the first to reach its shores but several other countries followed. In the twentieth century, however, foreign access to Brazil's wealth was curtailed. Starting in the 1930s, Brazil assumed greater control over its national resources and sought to diversify the economy away from the production and export of primary products. In the early 2000s, however, Brazil again experienced a significant increase in commodity exports thanks to China's booming demand. A country that had been largely ignored by Brazil quickly became the engine for its export growth. In fact, China was already on track to surpass the US as Brazil's largest trading partner before Brazil uncovered major offshore oil reserves.

Up to that point, the US had accepted Chinese inroads into Brazil. After the discovery of Brazil's newfound oil wealth in the country's offshore, would the US be tempted to undercut China's efforts? Many in the US Congress were already concerned with China's expanding control of overseas oil assets. Fully aware of this concern, would China still attempt to secure access to Brazil's reserves?

This chapter looks at these and other questions raised by the very large oil reserves found in Brazil to determine the impact of this offshore wealth on the triangular relationship between Brazil, the US, and the People's Republic of China (PRC).

From the outside looking in: the US and PRC react to Brazil's burgeoning oil reserves

Brazil's expanding hydrocarbon wealth drew the attention of the US as well as China, the world's largest consumers of oil. Representatives of both countries have repeatedly declared their determination to reduce their exposure to oil price, or volatility of supplies from the Middle East, by securing access to alternative energy sources whether domestically (unconventional oil and gas) or abroad. The Brazilian pre-salt oil fields could become one of these alternative sources, providing China and/or the US with a reliable source of high quality crude; however, tapping these resources will not be easy due to significant technological hurdles to drilling in ultra-deep extreme conditions. Brasilia's decisions on how to develop the area also affect the ability of the US and China to participate in these exciting new finds, the so-called "Blue Amazon". In the end, access will require an effort by both countries to reconcile their energy security imperatives with Brazil's strategic interests and policy decisions.

Pre-salt: an opportunity to expand US–Brazilian cooperation

Since US–Brazil relations were upgraded to a "strategic dialogue" in 2005, energy has become a key element in the bilateral relationship. In the context of sustained oil price increases, both the US and Brazil turned to biofuels as an affordable, clean, and sustainable alternative. Recognizing the benefits of leveraging their collective private and public resources to develop a global market for biofuels, the US and Brazil signed a Memorandum of Understanding (MOU) to Advance Cooperation on Biofuels in March of 2007. The MOU was seen as proof of the US's deepening strategic relationship with Brazil and a testament to Brazil's growing importance as a global energy player.

Only eight months after the bilateral MOU on biofuels was signed, Petrobras announced the oil discoveries in the Tupi pre-salt fields. For the US, the vast offshore reserves meant that more oil could be available more cheaply for all players in the market. This opportunity was particularly important at a time of rapidly rising oil prices and declining production from traditional suppliers, such as Mexico and Venezuela. Actual access to the reserves would ultimately depend on Brazil's willingness to share the wealth and its ability to muster and manage the needed resources (technical and financial) successfully to develop the massive deep-sea fields.

Fully aware that any direct attempts to influence Brazil's strategy for the pre-salt fields could be perceived as interference and possibly backfire (due to deep suspicions of US intentions in Brazil), Washington addressed the issue indirectly.

Table 10.1 China's and the US's share of Brazil's oil exports, 2003–2012 (1,000 barrels)

	Brazil's total oil exports	US's share of Brazil's total oil exports (percent)	China's share of Brazil's total oil exports (percent)
2003	88,246	15	1
2004	84,252	13	8
2005	100,190	16	13
2006	134,336	29	12
2007	153,813	35	10
2008	158,110	31	13
2009	191,859	26	14
2010	230,492	24	25
2011	220,649	27	23
2012	200,528	28	23

Source: Elaborated with data from ANP, 2014.

Its strategy was to bring the pre-salt discussion under the umbrella of the pre-existing strategic, bilateral energy relationship. Building on the success of the biofuels MOU, Washington presented the pre-salt finds as a "golden opportunity" for both countries to combine their resources (private and public) to fulfill their shared interest in bringing the oil online and bolstering Brazilian oil exports to the US (Langevin 2010; see also Table 10.1). "Such a partnership could also offer additional strategic benefits in terms of joint E&P [exploration and production] efforts around the globe" (Cable 08RIODEJANEIRO165 2008). Additionally, it could be an opportunity to improve bilateral relations through cooperation in an area of mutual interest.

Washington deemed positive engagement of the Brazilian government and institutions at the early stages of exploration as critical. Not surprisingly, within the first year of the Tupi announcement, US diplomatic representatives reached out to key energy entities in Brazil. To allay fears that the US did not respect Brazil's sovereignty, US representatives assumed a passive posture in the meetings, enquiring, instead of imposing, what public resources the US could offer to assist Brazil with E&P of the pre-salt fields. They also tried to determine the concerns Brazil might have with a US presence in the pre-salt area and how to address those concerns. As expected, the Brazilian representatives appreciated this approach and these offers of help, generating goodwill. Some even advanced suggestions as to how to deal with US mistakes, including the reactivation of the US Fourth Fleet shortly after the announcement of the Tupi discoveries, which fueled fears among many Brazilians that the US would use its military might to control Brazilian reserves (Cable 08BRASILIA1356 2008).

In parallel, the US government organized three high-profile visits to expand energy cooperation with Brazil and to call public attention to these joint efforts. The visits were intended to demonstrate to Brazilians that the US is a strong economic partner that should be taken into consideration as Brazil charts its "course in developing their newly discovered oil resources and possibly play a larger role in ensuring global energy security" (Cable 08BRASILIA1325 2008). Brazil's initial reaction was encouraging. According to a diplomatic cable, Brazil expressed "interest in having US companies involved in the exploitation of Brazil's oil reserves as well as in receiving high-level US visitors with the intention of developing closer bilateral ties" (Cable 08BRASILIA1325 2008). For the US oil industry this was good news. Many US companies had missed out on the pre-salt discoveries and were trying to catch-up. Like all listed oil companies, US firms' market capitalization and future viability depend on having access to proven reserves, such as Brazil's.

The US government further strengthened its positive engagement of Brazil's key energy players when it responded positively to Petrobras' request for external funding. Overwhelmed by the investment demands of developing the pre-salt finds, Petrobras had decided to reach out to national and foreign financial institutions, as well as export credit agencies of countries that supplied it with equipment (such as the US) or imported its oil (such as China). Shortly after Petrobras secured a US$10 billion loan from China, the US Export–Import Bank (Ex–Im Bank) approved a 10-year, US$2 billion preliminary line of credit to secure the purchase of US goods and services by Petrobras. The agreement stated that the Ex–Im Bank's final commitment could be increased above US$2 billion. Petrobras purchases financed by the Ex–Im Bank were expected to create and maintain over 507,000 US jobs (Export–Import Bank of the United States 2008).

Yet, Brazil was determined to deal with the pre-salt finds on its own terms. Brasilia introduced a new oil regime, designed to allow the Brazilian government and Petrobras to take maximum advantage of the finds. Although the US government closely monitored these developments, it did not intervene in this "very sensitive sovereignty issue, featuring minefields of bureaucratic infighting" (Cable 08RIODEJANEIRO91 2008; Cable 09RIODEJANEIRO288 2009). Private US companies, however, did attempt to intervene by joining other IOCs to lobby Brazil's Congress to enact key amendments. Still, US firms had to tread carefully because these lobbying efforts coincided with a deterioration in relations between the Obama and Lula administrations that greatly undermined the progress that had been made in previous months (Cable 09RIODEJANEIRO369 2009).

US–Brazil relations had been rapidly expanding, as evidenced by increased numbers of joint exercises between their militaries; joint promotion of biofuels development in Africa and Latin America; trilateral cooperation in Africa and Haiti on such issues as health, food security, and institutional strengthening; and an innovative agreement to fight racial and ethnic discrimination. However, the countries' independent foreign policies led to several disputes in late 2009 and 2010, the most conspicuous being disagreement over how to handle Iran's nuclear program. Other regional issues included Washington's willingness to accept the election results in Honduras and the US agreement to use military bases in Colombia (both of which Brazil opposed). In addition, trade issues (from intellectual property and pharmaceuticals to US subsidies for cotton growers and ethanol producers) complicated the bilateral agenda, reinforcing the low level of trust many Brazilians have towards the US (Cable 09BRASILIA1113 2009; Cable 09BRASILIA671 2009; Cable 09BRASILIA1041 2009; Cable 09BRASILIA1411 2009).

Amidst this tumultuous environment, the US Ex–Im Bank granted the first loan installment to Petrobras, (Export–Import Bank of the United States 2010) even as President Lula signed into law the new oil regime, much to the chagrin of US oil companies. As one executive said: "the opportunity set in Brazil has been reduced, there's no question of that. It's really limited now, frankly, to the present licenses that you hold" (Fick 2009). Nevertheless, due to the growing instability and declining access to reserves in other markets, US companies did not show any intention of leaving the Brazilian market. The US government therefore did not express a public view on the new, legal oil regime, except for general comments invoking the importance of transparency and good governance.

Shortly thereafter, Dilma Rousseff was elected Brazil's new President and indicated that building stronger relations with the US would be a top priority of her administration. The US government responded positively to this policy shift, sending Secretary of State Hillary Clinton to Rousseff's inauguration in January 2011. US Treasury Secretary Tim Geithner visited Brazil in February and a few days later Clinton welcomed Brazil's Foreign Minister Antonio Patriota in Washington (*International Business Times* 2011) Interestingly, the pre-salt oil wealth was not mentioned in the official remarks that followed the meeting.

The pre-salt finds were, however, front and center during President Obama's visit to Brazil in March 2011 (Oppenheimer 2011; State Department 2011). During his first visit to the country, Obama welcomed the potential for a new, stable source of energy, referring to the unrest in the Middle East and North Africa. He said the US hoped to participate in the environmentally responsible and technologically advanced development of the massive pre-salt reserves, and emphasized that the US wanted to be one of Brazil's best customers (Meckler 2011; White House 2011a; Yapp 2011). These public statements departed from the US government's initial approach to the pre-salt finds, which had favored more indirect or private forms of engagement. This shift in Washington's policy was reinforced by the decision to make oil and gas the number one priority of the new bilateral Strategic Energy Dialogue between the US and Brazil (White House 2011b).

The year 2011 marked a new and more productive phase of US–Brazil relations. Rousseff's decision to distance herself from Lula's foreign policy was paramount in this shift, as was Washington's openness to this new posture. Still, several contentious issues and a lingering mutual distrust persisted in 2012. First, during Rousseff's January 2012 visit to Cuba, the Brazilian President criticized the human rights abuses perpetrated in the US base of Guantanamo, while ignoring the human rights violations in Cuba (*MercoPress* 2012). Second, the White House did not concede "state visit" status to Rousseff's April 2012 visit to the US, disappointing Brasilia because the White House had granted such an honor to other emerging countries, namely China in 2011, Mexico in 2010, and India in 2009 (*Exame* 2012). Several commercial developments also added to bilateral tensions in 2012.

Notwithstanding, after a 2-hour private meeting with President Obama during her April 2012 trip to the US, President Rousseff declared that oil and gas production was "a tremendous opportunity for further cooperation," with the US supplying equipment and know-how and purchasing some of the oil and gas (Romero and Calmes 2012). Shortly after President Rousseff's US visit, Brazil's Ministry of Mines and Energy and the US Department of Energy organized a workshop on unconventional hydrocarbons (shale gas) as part of the Strategic Energy Dialogue. Also in 2012, both sides agreed to use the US–Brazil Commercial Dialogue to enhance the private sector's role in strategic bilateral energy discussions.

In March 2013, the second meeting of the US–Brazil Strategic Energy Dialogue was held in Brazil. Both governments recognized advances in all five areas of this bilateral cooperation – oil and gas, biofuels, renewable energy and energy efficiency, civil nuclear energy generation, and smart grid and transmission systems. With respect to oil and gas, two joint technical workshops – one on offshore E&P and one on unconventional energy – had been organized in 2012 to facilitate information sharing and exchange of best practices between both governments. Their success prompted the Brazilian government to propose a third workshop for 2013, again on unconventionals, to promote interaction between Brazilian and US companies and investment in the area in Brazil.

These high-level energy contacts were one of many signs pointing to warmer ties between the US and Brazil in 2013. President Rousseff was scheduled to go to the US in October 2013 for a highly anticipated state visit – trade was booming (reaching US$100 billion) and investment flows were expanding. The tide quickly turned, however, when it was made public in July 2013 that the US National Security Agency (NSA) was spying on Brazilians, including President Rousseff and her aides, and companies such as Petrobras. President Rousseff postponed her trip to the US, despite an August visit to Brazil by Secretary of State John Kerry and a phone call by President Obama. In the months that followed, US–Brazil relations stalled once again.

After a full year, President Rousseff finally indicated that she was prepared to move on, so the Obama government quickly sent Vice-President Biden to Brazil in June 2014. Meetings with President Rousseff and Vice-President Temer helped rebuild trust and refocused the relationship on shared interests, such as energy. During the visit, Vice-President Biden stated that: "put simply, the potential for US–Brazil energy cooperation is great." Speaking about oil and gas, he added:

> The United States and Brazil have many lessons learned to share in the regulation and environmental management of hydrocarbons development, be it in the offshore environment or in the development of shale gas. Efficient, safe, and reliable production of oil and gas benefits us all, which is why oil and gas development will remain an important part of our bilateral engagement.
>
> (Lores 2014)

Despite Vice-President Biden's optimistic comments, developing Brazil's new oil wealth has lost some appeal for both the US government and US oil companies. Consistently difficult diplomatic relations, very high Brazilian government take/involvement in the pre-salt and, most importantly, a decline in US oil imports (including from Brazil) thanks to the shale oil and gas revolution, drive this shift (Table 10.1) and all influenced the US's level of interest. For example, there was no US oil company that bid at the Libra pre-salt oil auction in 2013, even though they have a long history in the country and the expertise and financial means to do it.

PRC: securing access to Brazil's oil wealth

Since they forged diplomatic ties in 1974, China and Brazil have steadily promoted bilateral ties, although only modest progress has been made. The thirtieth anniversary of bilateral relations in 2004 was marked by a series of events that seemed to point to a new era of fast-deepening ties, including an exchange of presidential visits, accompanied by large business delegations (personal communication with Head of CASS Department, Beijing, August 14, 2009). Several agreements were signed and promises of massive investment and booming trade flows were made. Public and private mechanisms for dialogue were established in the form of the China–Brazil High-level Coordination and Cooperation Committee (COSBAN) and the China–Brazil Business Council (CBBC). Petrobras and other large Brazilian companies and institutions, such as Varig and BOVESPA, opened offices in China.

However, these bilateral mechanisms for dialogue – COSBAN, CBBC, and the China–Brazil Strategic Dialogue (2007) – produced few results because meetings were constantly delayed and proceedings marred by red tape. Although bilateral trade greatly expanded, it generated as many misgivings as assurances in Brazil. Brazil's economy benefited from the successive, large trade surpluses it amassed with China, but stark disparities exist between China and Brazil's trade performance. In 2006, China quickly surpassed the US to become Brazil's largest trade partner, but Brazil only represents two percent of China's total trade. Furthermore, iron ore, soybeans, and crude oil account for 82 percent of Brazilian exports to China, while the top two Chinese exports to Brazil are machinery and organic chemical products, accounting for 73 percent of Brazil imports from China (China–Brazil Business Council 2014a).

In contrast to trade, Chinese investment in Brazil remained insignificant, despite the great expectations following President Hu Jintao's 2004 visit (Banco Central 2011). Faced with Brazil's complex investment environment, several Chinese projects fell through, while the few that succeeded often required prolonged government support and a strong commitment by Chinese companies and their Brazilian counterparts.

The Gasene gas pipeline project is one such example (Petrobras 2006). The proposed US$1.3 billion project was the object of a 2004 MOU between Sinopec and Petrobras, signed shortly after both companies agreed for joint oil exploration, production, refining, product sales, petrochemicals, and pipelines (*The Economist* 2007). Later that year, the Petrobras–Sinopec partnership to build Gasene became bogged down by the funders' inability to agree on the terms of the loan. China's Ex–Im Bank insisted that a large share of labor, services and goods be procured in China (personal communication with BNDES International Division Representative, Rio de Janeiro, July 23, 2009). BNDES, the Brazilian Development Bank, in turn, refused those terms pointing to the 2004 MOU that states:

> Petrobras, under the guidance of the Brazilian government, requires that Brazilian construction companies and material/equipment and services suppliers to be part of the Engineering Procurement and Construction (EPC) contract, resulting in a minimum Brazilian content of 75 percent of the project.
>
> (Itamaraty 2004: Part 4)

Not waiting for the resolution of this conflict, Petrobras started the project in March 2006, resorting to provisional loans (*Dow Jones Newswires* 2010; *ICIS Chemical Business* 2006). Construction finally began in June 2006, almost a year behind schedule. Sinopec's Brazilian subsidiary, Sinopec International Petroleum Service do Brasil Ltda (est. 2005), headed the construction, which turned out to be heavily dependent on Brazilian sub-contractors because the company had difficulty dealing with Brazil's work visas, labor rules and red tape (personal communication with Sinopec International Representative, Rio de Janeiro June 10, 2010). As negotiations with the Ex–Im Bank were at a standstill, Petrobras cancelled Sinopec's contract for the second phase of Gasene, GASCAC (Sinopec Brasil 2011). Faced with the imminent loss of one of Sinopec's largest overseas service contracts, the company's leadership mobilized its government contacts. As a result, the China Development Bank (CDB) was brought in to replace the Ex–Im Bank in the negotiations. In December 2007, CDB finally inked an agreement with BNDES on lending the latter US$750 million to support GASCAC (Petrobras 2007; *SinoCast China Business Daily News* 2008). BNDES passed on the loan to Petrobras, which in turn revised its stance and contracted Sinopec to run the project (personal communication with Sinopec Brazil Representative, Rio de Janeiro June 11, 2010).

The timing of the CDB–BNDES agreement could not have been better. The first pre-salt discoveries had just been announced. Based on Sinopec's experience in Yemen and Saudi Arabia, where successful service projects had secured access to upstream investment opportunities, Chinese national oil companies (NOCs) hoped that Gasene would yield the same results (Kong 2010).

In late 2008, Jose Sergio Gabrielli, CEO of Petrobras, was touring the world looking for funding for Petrobras's development of the pre-salt reserves; however, the world economic crisis had depressed oil prices and dramatically reduced investors' willingness to invest. He spontaneously decided to meet with the president of the China Development Bank (CDB), Chen Yuan, in Beijing. To his surprise, after a short presentation on the pre-salt finds, Chen asked how much money Petrobras needed. Caught off-guard, Gabrielli came up with a number: US$10 billion. Chen promptly agreed to supply Petrobras with the funds.

This deal became public during Xi Jinping's visit to Brazil in February 2009. Shortly after a bilateral cooperation agreement on energy and mining was signed, Petrobras announced a series of MOUs with Chinese entities: one with CDB regarding the US$10 billion loan, one with China National Petroleum Company (CNPC) to export 40–60,000 barrels per day of crude oil, and one with Sinopec to supply it with 60–100,000 barrels per day (2009–2010) (Fiori 2009). By then Sinopec had become Brazil's major oil importer in China and a major driver of the rapid expansion of Brazilian oil exports to China. Since 2003, oil exports had risen from 228,000 tons in 2000 to 8.1 million tons in 2010, accounting for 14 percent of total bilateral trade (China–Brazil Business Council 2014b; Table 10.1).

In the following months, the two parties negotiated the details of the US$10 billion deal (personal communication with Petrobras Representative, Rio de Janeiro September 1, 2009). The contentious issues were the collateral for the loan and the rules for procurement. In both instances, CDB prevailed. Petrobras reluctantly agreed to secure the loan with the revenue it earns from its supply contract with Unipec – a Sinopec subsidiary. Unipec pays for the oil – purchased at market prices – through an account that Petrobras holds at the CDB. As part of the deal, Petrobras must maintain the equivalent of six months of interest payments on the loan at that account, reducing the uncertainty for CDB but greatly limiting Petrobras' access to the money (Downs 2011: 46–8).

The CDB also ensured that whenever possible, Petrobras would use US$3 billion to purchase Chinese goods and services. CDB actually wanted an outright commitment from Petrobras to buy Chinese machinery and equipment, but Brazilian law required a bidding process and a minimum of 30 percent local content for the industry.

The US$10 billion loan deal was concluded during President Lula's second state visit to China. Three documents were signed: a loan agreement between the CDB and Petrobras, a 10-year crude oil delivery contract for Petrobras to supply Unipec with 150,000 barrels per day in the first year and 200,000 barrels per day thereafter, and a new MOU between Sinopec and Petrobras to cooperate in exploration, refining, petrochemicals, and the supply of goods and services. During the visit, President Lula and President Hu Jintao also signed the Joint Action Plan (2010–2014) to provide guidance to the Brazil–China Strategic Partnership. Article Six of the Joint Action Plan deals exclusively with energy and mining and gives special relevance to oil E&P and trade activities.

The US$10 billion loan deal was the highlight of President Lula's trip (*Xinhua* 2009) because it proved that Petrobras could secure the funding needed to explore the pre-salt finds independently, strengthening his case that Petrobras should be the sole operator of the yet unlicensed, pre-salt areas. Furthermore, it showed that China and Brazil could cooperate, despite many MOUs that had amounted to little (*O Globo* 2009).

Motivated by China's relentless demand for oil, Chinese NOCs moved beyond loan-for-oil contracts to secure equity in the pre-salt reserves. As the new oil regime forbid further foreign ownership of equity in the pre-salt fields, Chinese NOCs were limited to already licensed pre-salt blocks that were not affected by the rule changes. Fortunately for the Chinese NOCs, upstream capital costs and spending had fallen following the global financial crisis, making it cheaper for Chinese NOCs to invest in upstream projects as they encountered less competition from other investors. Also, the appreciation of the Renminbi made overseas assets cheaper for NOCs; however, rather than trying to buy controlling stakes, Chinese NOCs looked for minority stakes and partnerships with oil companies that had oil equity and better long-term relations with the host country.

Chinese NOCs also benefited from Chinese government support. In April 2010, during his second visit to Brazil, President Hu underscored the opportunities for cooperation in energy and infrastructure. Brazilian government representatives publicly welcomed the participation of Chinese oil companies in the pre-salt development, leading Petrobras, Sinopec, and CDB to sign an agreement to assess mutually beneficial opportunities in E&P, downstream, and services. Following this agreement, Petrobras finally fulfilled its May 2009 promise of granting Sinopec access to two blocks in Para-Maranhao, strengthening energy cooperation between the two countries. For a 20 percent stake in each block, Sinopec agreed to fund the costly drilling operations, provide equipment and machinery, and help deal with the massive environmental and logistical challenges (Colitt 2010; Pamplona 2010).

Many details of the Para-Maranhao deal are unknown, including the amount of funding provided by Sinopec, the block's oil reserves and whether the oil produced at the blocks will be shipped to China. This lack of transparency that preserves the interests of Sinopec and Petrobras worries industry experts because it undermines earlier efforts to develop a highly competitive, efficient, and transparent oil sector in Brazil.

Shortly after Sinopec secured stakes in the Para-Maranhao blocks, Sinochem obtained a Chinese NOC's first equity stake in Brazil's deep-sea offshore fields (Chatwynd 2010). The May 2010 sale of a 40 percent stake of the Peregrino oilfield, owned by Norway's Statoil, concluded a hotly contested auction that attracted offers from several companies, including China National Offshore Oil Corporation (CNOOC), Sinopec, Sinochem, Total, Shell, and ExxonMobil. For Sinochem, this deal secured direct access to oil and to Statoil's management and technological expertise (Johnson and Ward 2010; Laroi and Bhatia 2010), but although Sinochem's success fitted Beijing's strategy to secure sources of oil, the auction that preceded it worried Chinese energy policymakers because it underscored the lack of coordination among China's NOCs.

Several months later, Sinopec secured access to Brazil's pre-salt reserves. In October 2010, Repsol completed a capital increase of US$7.1 billion subscribed by Sinopec in exchange for a 40 percent stake in a joint venture (JV): Repsol Sinopec Brazil. The JV, valued at US$17.8 billion, has a minority stake in 14 exploratory blocks (including the BM-S-9 Block in the pre-salt area) and in blocks already producing oil, such as Albacora Leste (operated by Petrobras).

By the end of 2010, Chinese NOCs had invested US$10.2 billion in Brazil, ramping up their access to resources from zero to 476 square miles in the Campos, Santos, and Espirito Santo basins. Fortuitous circumstances abetted this remarkable expansion, as did Chinese NOCs' willingness to relinquish control in several projects to more experienced companies. This accommodating attitude will help Chinese NOCs as Brazil's new oil regime comes into force. Unlike many US and other international oil companies (IOCs), Chinese NOCs lack the technology to explore the pre-salt finds and therefore do not oppose Petrobras, a leader in offshore production, being the sole operator. They also have easy access to the volumes of cash required to bring the pre-salt reservoirs to production, putting them in an advantageous position vis-à-vis most competitors; however, Chinese NOCs, like the IOCs, prefer equity oil and being able to determine the rate of production at their own fields. Unfortunately, under Brazil's new legal regime neither will be possible. Facing this environment, Chinese NOCs publicly declined to participate in the first licensing bid under the new regime.

In Brazil, a democratic country, the dramatic expansion of Chinese NOCs was at first welcomed by Brazilians; however, eventually several prominent figures accused Chinese foreign direct investment (FDI) of being qualitatively different from that of other countries because of the close State–state-owned enterprise (SOE) ties (Maciel and Nedal 2011). Also Brazil's companies have not been treated in China with the same openness Brazil has given to Chinese investors in China. CDB's insistence that Petrobras procure Chinese machinery and equipment was a source of concern, particularly because it resulted in Petrobras not being given access to the remaining US$3 billion of the 2009 loan, as well as a much publicized additional US$10 billion loan (Fick 2010; personal communication with Petrobras Representative, Rio de Janeiro September 1, 2009). Chinese NOCs holding substantial shares of Brazilian equity oil would also exacerbate the sector's growing dependency on China, which was Brazil's major oil export destination in 2010 (Table 10.2). Brazilians fear deeper economic dependence on commodity exports, especially when the Brazilian industry sector is losing ground to China, because it could reproduce the "core-periphery pattern" of decades past.

These deep concerns about China led the newly elected President Rousseff to abandon President Lula's accommodating policy towards Beijing. Instead, Rousseff used Brazil's leverage – China's need for a stable source of commodities – to balance their trade/investment relations. This new posture paid off, as evidenced by the billions of dollars in deals signed during her first visit to China in April 2011. Included in the 22 agreements were several manufacturing and infrastructure projects in Brazil, with the potential to generate thousands of jobs for Brazilians. Rousseff and Hu pledged to expand and diversify their countries' investments through company partnerships, especially in high-tech sectors, but also in mining and energy. To this effect Petrobras signed agreements with Sinochem and Sinopec to develop E&P technologies and exchange geological data (Menezes 2011). The pre-salt fields were not mentioned publicly, a sign that the relationship's focus had moved from what Brazil can offer China to what China can offer Brazil in exchange for access to its valuable commodities.

The 2011 agreements placated, at least temporarily, Brazil's harshest China critics, creating an environment favorable to investment by Chinese NOCs. In November 2011, Sinopec purchased a major, non-controlling stake (30 percent)

Table 10.2 The rising importance of oil and gas in China–Brazil's strategic agreements

Protocol for Cooperation on Energy and Mining (PCEM) (2009)	Joint Action Plan (2010–2014) (Art. No. 6)	10-Year Cooperation Plan (2013–2022) (Chapter No. 2)
Oil and gas is one of several areas of cooperation	Oil and gas is the number 1 priority in energy and mining	Oil and gas has dedicated sub-section
Focus: • Law/regulations • Advanced and efficient technologies • E&P in China/Brazil • Joint E&P in third countries • Engineering and services • Heavy oil refining	Same focus as PCEM plus strong commitment to: • Deepen Brazilian companies' participation in E&P in China and vice versa • Boost purchase of equipment • Increase investments in oil and gas supply chain	Same focus as PCEM and Joint Action Plan plus: • Development of value-added activities • Cooperation in petrochemicals and storage and transport of oil and gas

Source: author.

in a foreign oil company – Galp (Portugal), which had existing operations in the pre-salt. The Sinopec–Galp transaction, valued at US$6.4 billion, closed by the end of 2012 after securing the necessary approvals from relevant authorities (Ma 2011; Goncalves 2012). In January 2012, Sinochem announced that it would fund the Brazilian unit of Perenco (France) in exchange for acquiring a 10 percent stake in five offshore blocks in the Espirito Santo basin, subject to approval from Brazil's ANP (*Bloomberg* 2012).

In February 2012, China Deputy Prime Minister Wang Qishan attended the second meeting of COSBAN. During 2012, several high-level exchanges occurred between China and Brazil, including the visit of Premier Wen Jiabao, at which time he and President Rousseff agreed to upgrade bilateral ties to a comprehensive strategic partnership and set up a comprehensive strategic dialogue mechanism.

In this positive context, CNPC and CNOOC pre-registered for the auction for the exploration of Brazil's Libra pre-salt field. Two US companies, Chevron and ExxonMobil, which had shown initial interest, failed to pre-register for the Libra auction. In October 2013, an international consortium led by Petrobras won the bid. Brazil's NOC, which has a 40 percent stake in the consortium, will partner with Royal Dutch Shell and Total (each with 20 percent stakes), as well as CNPC and CNOOC (10 percent each), to develop the pre-salt oil field, which is estimated to contain between 8 and 12 billion barrels of oil. The winning consortium was awarded a 35-year production sharing agreement in exchange for the payment of a high signing bonus of US$6.8 billion to the Brazilian government and the equivalent of 41.7 percent of its output to the state once production is under way (OGJ editors 2013).

In November 2013, Brazil Vice-President Michel Temer visited China to participate in the third meeting of COSBAN. At this meeting both countries agreed to intensify cooperation in agriculture, energy, and infrastructure, reviving these sub-commissions in COSBAN. Vice-President Temer was quoted saying that,

> the welcomed participation of Chinese companies in the consortium that will explore the Libra field, in the Bay of Santos' pre-salt, opens important new prospects for our joint action in the oil and gas sector. I support Chinese participation in these projects that will only further strengthen mutual understanding and trust and generate shared gains.
>
> (Sarres 2013)

In addition, Vice-President Temer signed a 10-year Cooperation Plan (2013–2022) with China, which contains a dedicated sub-section to bilateral collaboration in oil and gas, testament to the growing importance of this sector in the succession of high-level agreements signed between China and Brazil since 2009 (see Table 10.2).

President Xi Jinping's first official visit to Brazil in July 2014, marking the 40th Anniversary of diplomatic relations, boosted China–Brazil collaboration in oil and gas. The joint statement issued by Presidents Rousseff and Xi during the visit touted the great potential for cooperation in energy and mining, which was underscored by the CNOOC and CNPC participation in the consortium to explore Libra. Despite this, none of the 32 agreements signed during this visit focused specifically on oil and gas cooperation (Itamaraty 2014).

Still, Brazilian oil only represents 2.2 percent of China's total oil imports, while China has quickly risen to become the second most important destination for Brazilian oil, even briefly replacing the US as top destination in 2010. More significantly, Brazil became a major destination for Chinese NOCs' investments and was responsible for a growing share of their total production. Chinese NOCs have also become top producers in Brazil, ahead of companies with years of experience in the country, such as Chevron. Thus, as of May 2014, among the top 10 oil-producing companies in Brazil, Chinese firms generate 3.5 percent of their total output, holding at least a 30 percent share in four of them (see Table 10.3).

Brazil's new oil reserves: a boon or bane for Sino–American–Brazilian relations?

The previous section indicates that Brazil's massive oil discoveries elicited distinct responses from the US and China. For example, China made larger and more frequent disbursements than the US, largely through multiple billion-dollar acquisitions by its NOCs, and yet both engaged in "tied aid": the Chinese government's loans to Petrobras were linked to oil supply contracts and purchases of PRC goods and services, while the US Ex–Im Bank credit guarantees were tied to purchases of US goods and services.

Table 10.3 Top 10 oil production by concession holder, May 2014

Concession holder	Oil (bbl/d)	Gas (Mm³/d)	Total (boe/d)
Petrobras	1,882,702	69,220	2,318,096
BG Brasil	61,736	2,555	77,808
Statoil Brasil[1]	38,566	60	47,500
Shell Brasil	29,051	589	42,269
Repsol Sinopec	31,415	943	34,986
Sinochem Petroleo	101	40	31,667
Parnaiba Gas	13,716	3,690	23,310
Petrogal Brasil[2]	266	660	17,866
Queiroz Galvao	15,446	2,625	16,780
Chevron Frade	16,009	164	16,478

Source: author using ANP/SDP/SIGEP data.

Notes:

1 In 2010, Sinopec acquired 40% of Statoil's Peregrino field, so the Chinese NOC owns part of this production.
2 In 2010, Sinopec acquired 30% of Petrogal Brasil, so the Chinese NOC owns approximately 30% of this production.

US experts debated the necessity and cost-effectiveness of Chinese NOCs' decision to seek security of supply through long-term contracts and their effort to take advantage of the economic downturn and lower asset values to step up their global acquisitions. They speculated whether such efforts will "lock up" supplies and deny them to Western competitors (Ellis 2010). Chinese experts, for their part, resented how Americans mistrust what they themselves consider to be a pragmatic and opportunistic strategy to secure access to resources. China, they feel, is filling the gap left by the US's underinvestment, while Chinese NOCs' funds will speed up the explorations of the pre-salt fields, which will benefit all oil importers, including the US (personal communication with Sinopec International Representative, Rio de Janeiro June 10, 2010).

China and the US had similar bureaucratic and symbolic responses to Brazil's new circumstances. Both introduced new bilateral agreements, dialogues, or MOUs to promote cooperation in oil and gas. The US and Brazil began a new presidential-level Strategic Energy Dialogue (2011) and the US–Brazil Commercial Trade Dialogue (2012). China and Brazil, on their side, established a new, ministerial level MOU on Oil, Equipment and Financing (2009) and the Protocol for Cooperation on Energy and Mining (2009), and oil and gas assumed growing importance in the Joint Action Plan and the 10-year Cooperation Plan (2013–2022). They also revived the COSBAN's Mining and Energy Sub-commission. The only exception is in the focus of the intensified cooperation: the US focuses primarily on technology/innovation/best practices, while China gives preference to investments and industrial development. Both the US and China also responded with a series of high-level visits by President Obama (in 2011) and Vice-President

Biden (in 2014), and by President Hu Jintao (in 2009), Premier Wen Jiabao (in 2012), Vice-Premier Wang Qishan (in 2012), and President Xi Jinping (in 2014).

The earlier section discussing the US's and China's response to the oil finds also shows how Brazil's successive governments reacted to the US and China's response to Brazil's changing oil fortunes. Overall, there has been a convergence in Brazil's approach towards the US and China, from positive/negative rhetoric to a more pragmatic, results-oriented approach – the National Security administration incident notwithstanding. In the case of the US, initial negative rhetoric under Lula gave way to a deeper, if at times strained, engagement under Rousseff. In the case of China, we see the reverse: an initial focus on "South–South" brotherhood gravitated towards a deeper, more realistic and perhaps more critical engagement. In return, Brazil received access to US funding and technology, while demanding that China buy Brazilian manufactured products, increase and diversify its investments, and expand access to the Chinese market. This evolution has helped Brazil secure funds and technical expertise needed for its own national development, while upgrading and intensifying Brazil's ties with both the US and China.

As for Sino–US competition within Brazil, no serious controversy involving oil or gas emerged. Even faced with perhaps one of the world's most promising hydrocarbon frontiers, Washington and Beijing know that they have a worldwide, multilevel relationship, which both sides are unwilling to jeopardize. More importantly, China's growing presence in Brazil's hydrocarbons sector has been irrelevant for US access to Brazilian resources. The recent decline in US oil imports from Brazil arises from the shale/tight oil and gas boom in the US, not from Chinese competition. Moreover, US companies have not been shut out of bidding processes by unfair competition from Chinese NOCs; instead, they opted to avoid all participation in such processes because of the high geological risk, high costs, and disadvantageous Brazilian government terms.

China, in turn, has not been affected by the US's presence in Brazil's oil and gas sector. Chinese oil imports from Brazil have grown unimpeded over the years and surpassed those from the US since 2010. But although oil imports from Brazil represent around 2 percent of China's total oil imports (similar to the US's imports from Brazil), the growing presence of Chinese NOCs in Brazil – and their active contribution to the exploitation of the massive Libra oil field and other costly pre-salt oil and gas prospects – could see that share rise.

Conclusion

In an increasingly unstable world where demand for oil continues to rise, Brazil's stable, strong institutions and predictable investment environment make it extremely attractive to all oil investors, even after the government dramatically increased its control over its resources.

Brazil's government is counting on these international investors to help develop its pre-salt reserves and new promising discoveries in the Amazon. Production from these reserves is expected to account for 22 percent of Brazil's gross domestic product by 2020, bringing as much as US$42 billion into the government's coffers (*O Globo* 2011). Emboldened by this prospect and the continued expansion of its economy, Brazil is becoming an increasingly assertive political and economic power in the world.

Dealing with a newly powerful regional neighbor may not be easy for the US, and although bilateral relations are improving, many points of contention and distrust exist. A sustained constructive relationship needs significant effort and attention by Brasilia and Washington. Both governments must better understand each other's priorities, interests, and positions on important regional and global issues, including energy security. They must strengthen communication channels to keep leaders abreast of decisions and actions. Most important, however, is to identify areas of mutual interest – such as successfully developing shale oil and gas prospects – which can become anchors for this progressively complex relationship.

Brazil's newfound wealth has also affected Sino–Brazilian relations. The jump in FDI has deepened them, even as Brazil's new resource discoveries underscore the "core-periphery" attributes of bilateral ties. Brazilian exports of oil to China have also increased dramatically (Table 10.1). As a result, Brazil started demanding more balance with its biggest trade partner and investor. China's response of showering President Rousseff with billions of dollars in pledged investments in successive high-level exchanges bodes well for Sino–Brazilian relations. Continued success, however, depends on the willingness of both countries to address thorny issues while ensuring that promised agreements are fulfilled.

Finally, Brazil's oil wealth has had no impact on Sino–US relations, particularly because Brazil is a minimal source of oil for both countries, the size of its investment are limited, and the development of alternative sources of oil and gas in the US has made Brazilian reserves less relevant for the US.

References

Banco Central, 2011. Investimento Estrangeiro Direto (Direct Foreign Investment [author's translation]). Available online at: www.bcb.gov.br/rex/ied/port/ingressos/htms/index3. asp?idpai=INVEDIR (accessed July 21, 2014).

Bloomberg, 2012. Sinochem to buy stakes in Brazil deep water blocks from Perenco. *Bloomberg,* January 6.

Cable 08BRASILIA1325, 2008. Scenesetter for CEO forum – Oct 9/10 – Secretary Guiterrez and AP Price. October 7. Available online at: https://wikileaks.org/cable/2008/10/08BRASILIA1325.html (accessed June 2, 2013).

Cable 08BRASILIA1356, 2008. Scenesetter for October 16–17 visit of Under Secretary of State for Political Affairs William J. Burns. October 10. Available online at: https://wikileaks.org/cable/2008/10/08BRASILIA1356.html (accessed June 2, 2013).

Cable 08RIODEJANEIRO91, 2008. Brazil's chief petroleum regulator inadvertently confirms rumors of mega oil find in Santos Basin. April 15. Available online at: https://wikileaks.org/cable/2008/04/08RIODEJANEIRO91.html (accessed June 3, 2013).

Cable 08RIODEJANEIRO165, 2008. Ambassador Sobel meets with key energy entities in Rio. June 8. Available online at: https://wikileaks.org/cable/2008/06/08RIODEJANEIRO165.html (accessed June 2, 2013).

Cable 09BRASILIA671, 2009. Brazil scenesetter: Codel Thompson May 27–28. May 27. Available online at: https://wikileaks.org/cable/2009/05/09BRASILIA671.html (accessed June 3, 2013).

Cable 09BRASILIA1041, 2009. Brazil's thinking on Colombia–US Defense Cooperation Agreement. August 20. Available online at: https://wikileaks.org/cable/2009/08/09BRASILIA1041.html (accessed June 3, 2013).

Cable 09BRASILIA1113, 2009. August 4–5 visit of US National Security Advisor to Brazil. September 4. Available online at: https://wikileaks.org/plusd/cables/09BRASIL-IA1113_a.html (accessed June 3, 2013).

Cable 09BRASILIA1411, 2009. Brazil: scenesetter for the December 13–14 visit of WHA Assistant Secretary Arturo Valenzuela. December 10. Available online at: https://wikileaks.org/cable/2009/12/09BRASILIA1411.html (accessed June 3, 2013).

Cable 09RIODEJANEIRO288, 2009. Rio's oil players react to speculation on pre-salt regulations. August 27. Available online at: https://wikileaks.org/cable/2009/08/09RI-ODEJANEIRO288.html (accessed June 3, 2013).

Cable 09RIODEJANEIRO369, 2009. Can the oil industry beat back the pre-salt law? December 2. Available online at: https://wikileaks.org/cable/2009/12/09RIODEJA-NEIRO369.html (accessed June 3, 2013).

Chatwynd, Gareth, 2010. Sinopec leads Chinese investment charge in Brazil. *Upstream.* May 28.

China–Brazil Business Council, 2014a. Bilateral Trade Data, 2010. Available online at: www.cebc.org.br/pt-br/dados-e-estatisticas/comercio-bilateral/balanca-comercial?y=2010 (accessed June 5, 2013).

China–Brazil Business Council, 2014b. Brazil's exports to China (first trimester of 2014). Available online at: www.cebc.org.br/pt-br/dados-e-estatisticas/comercio-bilateral/pauta-de-exportacoes (accessed June 4, 2013).

Colitt, Raymond, 2010. China Sinopec signs Brazil Oil Deal with Petrobras. *Reuters,* April 15.

Dow Jones Newswires, 2010. Brazil Petrobras, China Sinopec to sign gas pipeline deal. *Dow Jones Newswires,* April 27.

Downs, Erica, 2011. *Inside China, Inc: China Development Bank's cross-border energy deals.* John L. Thornton China Center Monograph Series, No. 3. Washington, DC: Brookings.

Ellis, R. Evan, 2010. What China will fight for in Latin America in the coming generation. *Security and Defense Studies Review.* Fall–Winter: 111–20.

Exame, 2012. Visita de Dilma aos EUA tera seminario e forum de executivos. *Exame,* March 2.

Export–Import Bank of the United States, 2008. Facts about the proposed Ex–Im Bank loans for Petrobras. Brazilian Offshore Oil Exploration and Development. Available online at: www.exim.gov/newsandevents/Facts-About-Ex-Im-Bank-Loans-To-Support-Petrobras.cfm (accessed June 2, 2013).

Export–Import Bank of the United States, 2010. Ex–Im Bank Signs Framework Agreement with Brazilian Development Bank BNDES, July 1. Available online at: www.exim.gov/newsandevents/releases/2010/ex-im-bank-signs-framework-agreement-with-brazilian-development-bank-bndes.cfm (accessed June 2, 2013).

Fick, Jeff, 2009. Chevron: new oil law reduces opportunities in Brazil. *Dow Jones Newswires,* October 8.

Fick, Jeff, 2010. Petrobras in talks for additional $10 billion loan with China. *Dow Jones Newswires,* March 29.

Fiori, Mylena, 2009. China e Brasil condenarão medidas protecionistas de países ricos em reunião do G20. *Agencia Brasil,* February 19.

Goncalves, Ana Maria, 2012. "Negocio entre Galp e Sinopec Fechado ate ao Final de Marco', *Diario Economico,* March 6.

ICIS Chemical Business, 2006. Petrobras and Sinopec to Build Gas Pipeline in Brazil. *ICIS Chemical Business,* April 24.

International Business Times, 2011. Obama has broad agenda in Latin America trip, Brazil key. February 19.

Itamaraty, 2004. MoU by and between Petróleo Brasileiro SA – Petrobras, China Petrochemical Corporation – Sinopec, the Export–Import Bank of China – China Exim Bank, Banco Nacional de Desenvolvimento Económico e Social – BNDES. Beijing, 6 September.

Itamaraty, 2014. Declaracao Conjunta entre Brasil e China por Ocasiao da Visita de Estado do Presidente Xi Jinping. Note No. 158, July 17.

Johnson, Miles and Andrew Ward, 2010. Sinochem buys stake in Brazil Oil Field. *Financial Times,* May 21.

Kong, Bo, 2010. *China's international petroleum policy.* New York, NY: Greenwood Publishing Group.

Langevin, Mark, 2010. Brazil's Big oil play: how this nation is charting national energy security. *Journal of Energy Security,* September 29. Available online at: www.ensec.org/index.php?option=com_content&view=article&id=264:brazils-big-oil-play-how-this-nation-is-charting-national-energy-security&catid=110:energysecuritycontent&Itemid=366 (accessed June 2, 2013).

Laroi, Vibeke and Meera Bhatia, 2010. Statoil sells 40 percent stake in Brazil field to Sinochem. *Bloomberg,* May 22.

Lores, Raul Juste, 2014. We want to rebuild trust with Brazil, says US Vice-President Joe Biden. A Folha, June 16. Available online at: www1.folha.uol.com.br/internacional/en/sports/worldcup/2014/06/1471088-we-want-to-rebuild-trust-with-brazil-says-us-vice-president-joe-biden.shtml (accessed July 20, 2014).

Ma, Wayne, 2011. Sinopec's Galp deal deepens China–Brazil energy ties. *Dow Jones.* November 11.

Maciel, Rodrigo Tavares and Dani K. Nedal, 2011. China and Brazil: two trajectories of a strategic partnership. *In:* Adrian H. Hearn and José Luis León-Manríquez, eds. *China engages Latin America: tracing the trajectory.* Boulder, CO: Lynne Rienner, pp. 235–56.

Meckler, Laura, 2011. Obama faces policy obstacles in Latin America. *Wall Street Journal,* March 23.

Menezes, Bruno, 2011. Brazilian president signs 20 trade deals on first trip to China. *Epoch,* April 19.

MercoPress, 2012. Rousseff expected at the White House but with no "state visit" privilege. *MercoPress,* March 19.

O Globo, 2009. Lula Vai a China Para Impulsionar Comercio entre os Paises. *O Globo,* May 15.

O Globo, 2011. Petroleo Garantira aos Cofres Publicos R$66 bilhoes. *O Globo,* January 17.

OGJ editors, 2013. Libra presalt rights awarded offshore Brazil. *Oil & Gas Journal,* October 21.

Oppenheimer, Andres, 2011. Obama's trip may lead to US–Brazil honeymoon. *Miami Herald,* January 27.

Pamplona, Nicola, 2010. Chineses Vao Explorar Petroleo no Brasil. *Estadao,* April 13.

Petrobras, 2006. Gasene Project Financing. December 14.

Petrobras, 2007. Petrobras to Begin Construction of the Third Gasene. December 27. Available online at: www2.petrobras.com.br/ri/spic/bco_arq/GaseneIng.pdf (accessed June 2, 2013).

Romero, Simon and Jackie Calmes, 2012. Brazil and US accentuate the positive. *New York Times,* April 9.

Sarres, Carolina, 2013. Brasil e China Intensificam Parceria com Visita de Michel Temer. Agencia Brasil. November 6. Available online at: www.ebc.com.br/noticias/brasil/2013/11/brasil-e-china-intensificam-parceria-com-visita-de-michel-temer (accessed July 20, 2014).

SinoCast China Business Daily News, 2008. Sinopec to Build Pipeline for Brazil. *SinoCast China Business Daily News,* January 23.

Sinopec Brasil, 2011. Landmarks of Gasene Project. Available online at: www.sinopecbrasil.com.br/pt/ (accessed June 3, 2013).

The Economist, 2007. Energy-hungry China looks to Latin America. *The Economist,* April 10.

US State Department, 2011. Remarks with Brazilian Foreign Minister Antonio Patriota after their meeting. February 23. Available online at: www.state.gov/secretary/20092013clinton/rm/2011/02/157029.htm (accessed June 4, 2013).

White House, 2011a. The United States and Brazil: the fact sheets. March 19. Available online at: www.whitehouse.gov/the-press-office/2011/03/19/united-states-and-brazil-fact-sheets (accessed June 4, 2013).

White House, 2011b. Fact sheet: US–Brazil strategic energy dialogue. March 19. Available online at: www.whitehouse.gov/sites/default/files/uploads/Brazil_Strategic_Energy_Partnership.pdf (accessed June 4, 2014).

Xinhua, 2009. Brazil's Lula on China visit secures 13 deals. May 20. Available online at: http://en.people.cn/90001/90783/91300/6661358.html (accessed June 2, 2013).

Yapp, Robin, 2011. Lybia: Barack Obama hails Brazil as example to follow. *The Telegraph,* March 20.

Part IV

America's 'pariahs' and China's energy supply

11 The US factor in Sino–Iranian energy relations

John W. Garver

China's balancing act

During the first decades of the twenty-first century, China expanded access to Iran's oil and gas resources even while US–Iranian confrontation was mounting *and* while Beijing faced mounting US pressure to draw away from energy cooperation with Iran. Beijing has contradictory policy objectives: on the one hand, China sought access to Iran's rich energy resources and on the other hand, it sought to maintain amicable relations with the US. As tensions between Washington and Tehran escalated in the 2000s, Washington repeatedly sought Chinese support for various coercive efforts against the Islamic Republic of Iran (IRI). An unequivocal Chinese rejection of those US demands could seriously injure Sino–US comity and the favorable macroclimate underpinning China's development effort; however, if China abandoned Iran in the face of US pressure, China's relations with Iran would suffer, including China's efforts to gain a foothold in Iran's energy sector and its aspirations for a long-term strategic partnership with Iran, perhaps the strongest state in the Persian Gulf. Beijing has found ways to satisfy first Washington and then Tehran, attempting to advance its conflicting interests with both capitals. Neither Washington nor Tehran has been fully satisfied with Beijing's equivocal approach, but Beijing's linked efforts to befriend Tehran while cooperating with Washington has not produced any significant tension in the Sino–US relations. Conversely, most Iranian leaders have absorbed the hard lesson that Beijing will not confront the US on Tehran's behalf, even if they value the support that Beijing is prepared to give.

Beijing used disassociation from US policies pressuring Iran, along with elements of support for Iran in the face of US pressure, to gain access to Iranian energy resources. By refusing to follow US wishes on sanctions against Iran, Beijing positioned Chinese firms to move into Iranian energy projects when Western firms withdrew from Iran in accord with US policy. Moreover, by giving the IRI a degree of support against US pressure, Beijing encapsulated the Sino–Iranian energy relation in security cooperation that was meaningful to Tehran. Beijing hoped that by encapsulating the energy supply relation in a cocoon of political cooperation, China's foreign energy supply would become less vulnerable to various shocks. This strategy was a core aspect of China's energy security strategy

(Downs, 2000). Yet Beijing avoided going so far in supporting Iran as to anger Washington. Too much Chinese support for Tehran against US pressure would be dangerous; it might easily convince Washington that China was, after all, a strategic rival – as some in the US argued – again spoiling the favorable macro-climate for China's development drive. Beijing carefully weighed the intensity of US demands along with probable injuries to both Sino–US and Sino–Iranian relations, and then acceded to or rejected particular US requests. In effect, Beijing balanced between Tehran and Washington, trying to safeguard its interests with both.

Beijing's Persian Gulf dilemma intensified in the 2000s as US-led international concern over Iran's nuclear programs mounted. The international debate over Iran's nuclear programs entered a new phase in 2002 when confirmed intelligence indicated the existence of two previously undisclosed nuclear facilities. Growing hard-liner control over Iran's diplomacy during the second half of the 2000s further rendered Tehran's diplomacy more confrontational, a fact that influenced US policy and Beijing's calculation of the gains and losses associated with alignment with either Tehran or Washington. The Iranian government under President Mohammad Khatami (in office from August 1997 to August 2005), pursued a relatively conciliatory policy – cooperating with the International Atomic Energy Agency (IAEA), signing an Additional Protocol providing for expanded inspections, suspending enrichment pending final status negotiations, and embracing negotiations with the Europeans. In October 2003, Tehran disclosed a wide array of secret nuclear activities over the previous 18 years, including several activities involving China. The revelations involving China further raised the stakes for China's reputation as a responsible power (Ansari 2006; Garver 2006). By 2003 the US was pushing to refer the Iran nuclear issue to the UN Security Council, the body charged with dealing with threats to international peace and security. The conservative take-over of Iran's parliament in February 2004, followed by the June 2005 election of Mahmoud Ahmadinejad as president, led to a far tougher, more confrontational approach by Iran. Ahmadinejad and his followers believed that an attempt to negotiate agreements with Western countries was futile; instead, Iran should confront the West and refuse to compromise. This Iranian shift to a hard-line, confrontational approach increased the costs to China of supporting Tehran against Washington.

China fills the vacuum created by US pressure

European firms were Iran's primary energy partners until 2007. Chinese firms played a minor role, greatly exceeded even by those of South Korea and Malaysia (see Table 11.1).[1] Until 2010, European firms continued to play a major role, as did those of South Korea and Malaysia. In that year, unilateral US sanctions (i.e. outside the UN framework and based on US law and Executive Orders) began to persuade many of these firms to decline further investment in Iran. During 2009, Chinese firms entered into eight new energy deals, many of which had been abandoned by Western firms under fear of US sanctions. China had become Iran's major energy partner.

Table 11.1 Foreign investment in Iran's energy sector, 1997–2010 (US$ billions)

	1997¹	1999	2000	2001	2002	2004	2006	2007	2008	2009	2010
Europe and Canada	2	0.1	2.37	1.23	2.73	0.027		10.25	15.4	8.93	0.23
South Korea and Malaysia	2				1.6				1.76	10.7	1
India, Brazil and Belarus					0.027	0.034			3	7.07	3.5
Russia									2	2	
Iran's neighbors²									0.35	3.8	1.922
China				0.09	0.08		0.05	21.75		15.6	4.843

Sources: Katzman, Kenneth, n.d. *The Iran Sanctions Act, CRS Report for Congress, RS 20871*. Global Business in Iran Database. Available online at: www.irantracker.org/global-business-in-iran (accessed 14 December 2013). Iran Taskforce of Pittsburgh, n.d. *Companies investing in Iran's oil and natural gas sector since 1996*. Available online at: www.irantaskforce.org (accessed 14 December 2013).

Notes:

1 Years are missing if no FDI reported for that year.
2 Armenia, Azerbaijan, Turkey, Qatar and UAE.

The premise for China's filling the vacuum created by the exit and growing disinterest of other nation's firms was China's relative policy independence from the US. Beijing was less willing than the European countries and Japan to follow US policy advice on Iran or to bow before US unilateral actions penalizing non-US firms for involvement in Iran's energy sector. Chinese firms were willing to move into Iran's energy sector as European and Japanese firms exited. Beijing's greater independence from Washington allowed it to penetrate Iran's energy sector, while Beijing's relative refusal to bow to US pressure made China attractive to Tehran. In sum, despite the relative technological backwardness of China's petroleum technology, within a few years China was able to seize the opportunity presented by the withdrawal of Western and Japanese oil firms from Iran, persuade economic nationalist Iranian officials to grant commercially attractive terms to Chinese firms, and establish Chinese majors in a leading position in a country with vast, unexploited energy resources. Of course, the flight of European and East Asian oil majors from Iranian projects left Iran with few choices other than Chinese firms.

But no sooner had Chinese firms filled the vacuum than they began hesitating about moving forward with their various deals – many of which were only Memorandum of Understanding (MOU). China National Offshore Oil Corporation (CNOOC) cancelled the US$16 billion deal initialed in May 2007 just before signing the contract. In mid-2010, China National Petroleum Company (CNPC) reportedly halted work on the South Azadegan project that was agreed to only the previous August. CNPC also delayed drilling at the South Pars gas field, which was agreed to in March 2009. These moves were in line with a mid-2010 instruction from China's government to slow down implementation of the recently concluded deals in Iran (Chen 2010; Lin 2010). China's investment fell off precipitously in 2010.

China's go-slow and pullback approach to energy cooperation with Iran was related to Sino–US bargaining. Two factors, both tied to the US, were in play. First, the US was implementing more comprehensive and stringent sanctions against Iran, and Chinese oil firms had subsidiaries listed on US stock exchanges and thus vulnerable under a new US law. The US market was simply bigger and more lucrative than Iran's, and Chinese firms, like those of Europe, followed their economic interests. Second, negotiations between Beijing and Washington over Iran were underway, with Washington proposing increased Chinese access to US-energy suppliers and energy suppliers of US-allies (especially Saudi Arabia and the United Arab Emirates [UAE]) in exchange for China's drawback from Iranian energy projects. Beijing carefully adjusted ties with Washington and Tehran in pursuit of its interests with both.

US lobbying for greater Chinese support on Iran's nuclear programs

Beijing and Washington bargained continually throughout the 2000s over China's ties with Iran. Washington put strong pressure on Beijing to cooperate more with US policies intended to pressure Iran to verify to the international community the non-military nature of its nuclear programs. China did not reject cooperation with the US in this regard, but Chinese cooperation came only after strenuous US diplomatic effort and was configured so that it did not overly constrain China's burgeoning economic relation – including but not limited to energy cooperation – with Iran. China cooperated with the US enough to keep the US happy, but not enough to undermine China's economic cooperation with Iran.

Washington made cooperation on the Iran nuclear issue a key litmus test of the sincerity of China's many professions of a desire for partnership with the US. To cite one prominent example, a September 2005 speech by US Deputy Secretary of State Robert Zoellick called for China to become a "responsible stakeholder" in the international system in cooperation with the US (Zoellick 2005: A6). Zoellick highlighted the Iran nuclear issue as one requiring increased US–People's Republic of China (PRC) cooperation: "China's actions on Iran's nuclear program will reveal the seriousness of China's commitment to non-proliferation" (Zoellick 2005: A6). Only a week before Zoellick's speech, President George Bush lobbied hard with President Hu Jintao for China not to block referral of the Iran nuclear issue by the IAEA to the UN Security Council, an action that was then imminent (Sanger 2005). Hu gave a non-committal reply, but in the end China did not block IAEA transmission of the issue to the UN Security Council. When President Hu visited the White House in April 2006, President Bush declared, "As stakeholders in the international system, our two nations share many strategic interests." The first of those issues listed by Bush was "the nuclear ambitions of Iran." In his reply, Hu declared, "We are ready to continue to work with the US side...[on] the Iranian nuclear issue through diplomatic negotiations to uphold the international non-proliferation regime and safeguard global peace and security" (US Department of State 2006).

Washington believed that China's expanding energy cooperation with Iran was antithetical to being a "responsible stakeholder." When Sinopec signed its landmark US$2 billion Yadavaran contract in December 2007, a US State Department spokesman protested: "Major new deals with Iran, particularly ones like these involving investment in oil and gas, really undermine international efforts to pressure the Iranians to comply with obligations already in place under UN Security Council resolutions" (Fars News Agency 2007). Beijing was probably confident that moving forward with the Yadavaran project would not antagonize Washington too badly because, by December 2007, China had already supported two sets of UN Security Council sanctions resolutions and the previous month had seen the release of the US National Intelligence Estimate declaring that Iran had suspended nuclear weapons programs.

The Obama Administration continued the Bush era drive for greater Chinese cooperation on the Iran nuclear issue. In September 2009, Deputy Secretary of State James Steinberg enunciated a vision of the US–PRC relationship very similar to Zoellick's, but using the phrase "strategic reassurance" in lieu of "responsible stakeholder" (Steinberg 2009). According to Steinberg, strategic reassurance was a sort of tacit bargain. The US and its allies "must make clear that we are prepared to welcome China's 'arrival'...as a prosperous and successful power" (Steinberg 2009). Steinberg did not mention access to resources, but that was implicit in the US "welcoming" of a "prosperous and successful" China. Beijing, for its part, "must reassure the rest of the world that its...growing role will not come at the expense and security and well-being of others." Iran was fourth – after the global economic crisis, climate change, and North Korea – on Steinberg's list of areas in which expanded Sino–US cooperation was needed. "It will be important for us to demonstrate [to Congress and the US public]," Steinberg said, "the possibility of cooperation in dealing with Iran's nuclear programs through the P5+1 [The five permanent Security Council members plus Germany]." The month after Steinberg's speech, Assistant Secretary of State Kurt Campbell traveled to Beijing to lay the ground for P5+1 talks scheduled for Geneva later that month. "If we are to make real progress on sending a consolidated message to Iran, we are going to need the support of China," Campbell said. "We're going to need to see more cooperation and coordination between the US and China if we are going to be effective in Iran" (Agence France-Presse [AFP] 2009).

President Obama made his first-ever visit to China in November 2009 with a high-ranking objective being securing greater Chinese support for a push for a new round of sanctions against Iran. During several hours of talks with Chinese leaders, Obama pushed without success for China to commit to supporting a fourth traunch of UN sanctions (Cooper 2009: A1). Continuing differences were apparent in the joint press statement (*The Washington Post* 17 November 2009). Secretary of State Hillary Clinton continued to press Beijing during a meeting with

Foreign Minister Yang Jiechi in late January 2010. Clinton publicly aired several reasons why Beijing should support new sanctions on Iran. "We understand that right now that [further sanctions] is something that seems counter-productive to you [to] sanction a country from which you get so much of the natural resources your growing economy needs. But think about the longer term implications," Clinton urged (Landler 2010: A5). A nuclear-armed Iran could set off an arms race in the Persian Gulf, Clinton asserted, or because Israel viewed a nuclear Iran as an existential threat, would provoke an Israeli military strike against Iran. Other countries – European countries, Russia, Middle East countries – were moving towards further sanctions and "China will be under a lot of pressure to recognize the destabilizing effect that a nuclear armed Iran would have" in the Persian Gulf "from which they receive a significant percentage of their oil supply" (Landler 2010: A5). If China failed to support new sanctions against Iran, it would face economic insecurity and diplomatic isolation, Clinton said. Treasury Department specialists, whom Clinton had brought along as part of her team, explained to Chinese officials how Iran's government was using Asian banks to circumvent restrictions on its transactions in Western banks.

Parallel to Clinton's discussions with Yang Jiechi, US representatives lobbied Saudi Arabia and the UAE to increase oil output to meet China's oil requirements in the event that sanctions led to a cut in Iranian oil exports to Iran. A senior Obama Administration official told the media that Yang Jiechi had listened to US arguments, but then reiterated Beijing's preference to stick with negotiations. Sanctions, Yang said, would imperil efforts at a negotiated settlement of the problem. Yang took the same position at a security conference in Munich early the next month (Ministry of Foreign Affairs 2010a).

Washington rallied friendly countries concerned with Iran's nuclear programs to urge China to give greater support to the US-led effort – Israel, Saudi Arabia, the UK, and France were all greatly concerned about possible IRI possession of nuclear weapons. As international pressure on China intensified in 2010, so too did the costs to China of aligning too closely with Iran. China's standing with the Arab countries and Israel might suffer, and China's reputation as a responsible stakeholder who did not pursue its own interests at the cost of other countries security could suffer seriously. Moreover, if China failed to support the US push, there would be no reason for Washington to use its influence with Saudi Arabia and the UAE to meet any oil shortfalls that might result from a crisis in the Gulf, and, returning to the triangular focus of this chapter, the costs of China's non-co-operation in terms Sino–US relations might be especially grievous. Americans might conclude that China was not a reliable strategic partner, but was one who would put its own narrow economic interests above efforts of the US and the international community to maintain peace and security.

These potential injuries to China must have been discussed by Chinese leaders after Yang Jiechi reported on his European discussions. A shift in Chinese policy soon became apparent. When Deputy Secretary of State James Steinberg and National Security Council Senior Asian Director Jeffrey Bader visited China in March 2010, President Obama followed up with a two-hour phone conversation with President Hu, it was clear that Chinese policy had shifted back toward cooperation with the US. Beijing was now willing to accept a new set of Security Council sanctions against Iran (Jacobs and Landler 2010: A10). President Hu also agreed to attend a nuclear security conference in Washington in April. US officials had feared that Hu would avoid the conference in order to express Beijing's anger over recent US moves toward Taiwan and the Dalai Lama. Following the Steinberg–Bader visit, China's UN representative participated in two conference calls with other P5+1 representatives to work out modalities for beginning negotiations over the content of a fourth traunch of UN Security Council sanctions (Batson 2010).

While Beijing finally acceded to Washington's demands for a fourth round of UN sanctions, Beijing tried to minimize damage to Sino–Iranian relations. Dai Bingguo invited the Secretary of Iran's Supreme National Security Council and chief nuclear negotiator, Saeed Jalili, to visit Beijing to discuss Iran's nuclear issue (Buckley 2010). In his public comments, Dai did not mention sanctions – although it is almost certain that he explained the logic of Beijing's recent shift in private. Publicly, Dai stressed that diplomacy and peaceful talks were the most effective way to solve the Iran nuclear issue. "Peacefully solving the Iranian nuclear issue through diplomacy and talks is in the interests of the parties involved and regional peace and stability," a PRC Ministry of Foreign Affairs spokesman said of Jalili's visit (PRC Ministry of Foreign Affairs 2010: 11). The spokesman reiterated China's opposition to Iranian possession of nuclear weapons, but paired that with a declaration of acceptance of Iran's "right to peacefully develop nuclear energy." Dai and Jalili discussed ways of expanding Sino–Iranian cooperation. In plain speech, Beijing's message to Iran was that forthcoming sanctions would not substantially effect PRC–IRI economic cooperation, and that China would not endorse use of military force against Iran's nuclear programs. Of course, China's opposition to the US use of force on earlier occasions in the Persian Gulf (in 1987 during the "tanker war," against Iraq in 1990 and in 2003) had been inconsequential.

Iran was unhappy with China's balancing approach. In June 2010, shortly after China supported UN Security Council Resolution 1929, President Ahmadinejad visited Beijing in association with the Shanghai World Expo. Shortly before Ahmadinejad arrived in Beijing, the head of Iran's nuclear program, Ali Akbar Salehi, slammed China's weak support at the UN. Speaking to the Iranian media, Salehi said: "There was a time when China branded the US as a paper tiger. I wonder what we can call China for agreeing to this resolution." Beijing had "double standards" – supporting North Korea even though it has abandoned the Nuclear Non-Proliferation Treaty (NPT), and sanctioning Iran even though it adhered to the NPT (AFP 2010). Neither Tehran nor Washington was completely satisfied with Beijing's balanced approach, but both were minimally satisfied.

Divergent Chinese and US perspectives

The US position in discussions with China over Iran was that China and the US had common interests in preventing proliferation of nuclear weapons and preventing an arms race that could destabilize the Persian Gulf. China and the US should, therefore, cooperate on the basis of those common interests. From the US perspective, this seemed very persuasive, almost common sense. Chinese representatives, on the other hand, framed the issue in terms of China doing the US a favor by helping it deal with essentially US problems (Steinberg 2010). This was a good bargaining tactic, but it also reflected underlying Chinese views.

Beijing viewed the Iranian situation very differently to the US. Although China upheld the non-proliferation regime, it had no fundamental conflicts of interest with the IRI, such as those between Tehran and the US. Since the early 1980s, China had had quite cordial and mutually beneficial relations with IRI. Chinese thought US arrogance and ignorance was largely to blame for the poor state of Iranian–US relations. While the US obsessed with the IRI's repression of dissent and lack of democracy, Beijing viewed Iran's internal governance a matter for Iran to determine and of no concern to any other country. As Yang Jiechi said in a speech on Iran at Chatam House in London: "The people of the world have the right to determine their own fates and access their own needs" (*Al-Quds al-Arabi* 2007). In Chinese eyes, the US preoccupation with Iran's internal governance was simply hegemonistic behavior. Washington's refusal to meet with Iranian representatives, believing such meetings conferred an aura of US approval on the regime, also struck Chinese representatives as great power arrogance. Beijing sympathized with Iranian efforts to resist US hegemonistic interference, and suspected that the US's real objective in Iran was regime change – overthrow of the IRI. Following the overthrow of Saddam Hussein in Iraq, the US aspired to achieve a similar result in Iran. Many Chinese suspected that this was the ultimate objective behind Washington's drive for sanctions.

Beijing believed that US policy was itself largely to blame for deepening instability in the Persian Gulf because Washington's rapid resort to sanctions created anger and polarization. Sanctions would not cause their target to back down – at least not if the target was a proud, politically mobilized and strong country like Iran. Sanctions would instead lead to counter-moves and increased tension, and could complicate resolving the issue by negotiations. China also saw the US inclination to use military force as a deep source of insecurity and instability in the Persian Gulf. The removal of Saddam Hussein, for example, had produced anarchy and terrorism with wide ranging consequences. If Iran was enhancing its military preparedness, the US's deep hostility and inability to accommodate the IRI was the major reason for that military buildup. In Chinese eyes, if the US and/or Israel struck militarily at Iran, the result could be region-wide instability. Such a war could cut off the flow of oil from the Gulf to China. In discussions between Chinese and Iranian representatives, agreement on negative assessments of US Gulf policy helped establish a sense of commonality, *ganqing* in Chinese (Brady 2003). When Iranian Deputy Foreign Minister Mehdi Safarai met with his Chinese counterpart in mid-2007, Iran's press agency reported that "China, a permanent member of the United Nations Security Council, has always stressed the need for a peaceful resolution of the nuclear dispute and repeatedly expressed its opposition to America's unilateral policies and the sanctions imposed against Iran" (Press TV 2007).

China's balancing at the United Nations

Beijing did not sell lightly its cooperation with the US. As one European diplomat noted regarding the US–PRC negotiations in early 2010 over the fourth round of UN Security Council sanctions, by playing hard to get China put itself at the heart of the negotiating process (Landler 2010: A5). From late 2002 when the world learned about Iran's extensive, secret nuclear activities through to June 2010, first the IAEA Board of Governors and then the UN Security Council debated the Iran nuclear issue. Four UN Security Council key resolutions were eventually passed with China's support. Throughout this process, Washington and Tehran both sought Beijing's support, while Beijing balanced between them trying to protect its interests with both. In negotiations over each resolution, China watered down and delayed sanctions levied ultimately against Iran. Yet in each case, when continued obstruction risked identifying China with the IRI too closely, thereby endangering China's ties with Washington, Beijing eventually agreed to sanctions. But although cooperating with the US in its push for UN Security Council sanctions, Beijing helped Tehran by delaying action by many months, securing the deletion of tough rhetoric, weakening sanctions, and most importantly ensuring that sanctions did not interfere with Iran's production and export of oil and gas.

In November 2003, nearly a year after the first revelations about Iran's extensive secret nuclear activities, the IAEA determined that Iran had violated its NPT obligations by failing to report a wide range of nuclear activities. IAEA rules required a rapid report of such findings to the UN Security Council, but it was only in February 2006, after a delay of 26 months, that the IAEA finally reported the issue. Apparently from November 2003 to January 2006, Beijing resisted referral of the Iran nuclear issue from the IAEA to the UN Security Council. Chinese opposition was a major reason for this two-year delay in IAEA referral of the issue to the UN Security Council. Once the IAEA's report was in the UN Security Council, that body first passed a Presidential Statement demanding that Iran cease all uranium enrichment and plutonium reprocessing, declaring such activities a "threat to international peace" under Article 41 (providing for imposition of sanctions) of Chapter VII of the UN Charter, and threatening sanctions if Iran did not comply. When Iran did not comply, over a period of four years the UN Security Council passed four resolutions with steadily, if only marginally, more severe sanctions. China ultimately voted yes to the IAEA report of the issue to the UN Security Council, to a UN Security Council declaration threatening sanctions, and also to all four sets of UN Security Council sanctions resolutions.

Although it is a simplification to term China's various stances in these debates "pro-US" and "pro-IRI," doing so is perhaps justified with the caveat that this merely means elements of China's policy mix supported one side or the other in the UN Security Council debates. Table 11.2 disaggregates China's policy stances during the 2006–10 UN Security Council debates into these two clusters.

During the 2003–10 UN Security Council debates, China declared its support for the nuclear NPT regime and its willingness to cooperate with the US in supporting that regime. China had joined the NPT through a process starting in 1984 (when it

Table 11.2 Balancing of pro-US and pro-IRI elements of China's UN position, 2002–2014

Pro-US	Pro-IRI
Oppose IRI possession of nuclear weapons	Accept at face value IRI professions of non-military intentions
Endorse NPT and demand IRI uphold its obligations under the NPT	Support IRI "right to peaceful use" under NPT
Support transmission from IAEA to UN Security Council	Reject as illegitimate intelligence gathered by "national means"
Approve four UN Security Council resolutions imposing sanctions on IRI	Insist "report" rather than "refer" under Article VII
Support UN Security Council demand that IRI cease enrichment	Oppose, delay and water-down sanctions
Agree accept reference to Article VII of UN Charter	Reject as illegitimate all extra-UN sanctions
Press IRI to be flexible and genuinely seek negotiated solution	Secure elimination of some threatening language
	Insist on negotiated solution and oppose threat or use of military force

joined the IAEA) and culminating in 1992 when it acceded to the NPT. US leaders saw China's gradual embrace of the NPT as a major gain for US policy and for the international community generally (Medeiros 2007). They also saw nuclear non-proliferation as a major area of common US and Chinese interest, explicitly identifying and stressing it during the major effort to "re-normalize" US–PRC relations in 1996–97. Had China failed to cooperate with the US on non-proliferation, it could have undermined a major US hope for strategic partnership with China. Although supporting the principle that Iran did not possess nuclear weapons, Beijing, following Iran's lead, married that principle to Iran's "right" under the NPT to peaceful use of nuclear energy, a "right" that Tehran insisted included both full uranium enrichment and plutonium reprocessing. Early on, the US was quite skeptical about Iran's profession of non-military purposes. The European countries increasingly shared that skepticism after the failure of their own effort from late 2004 to early 2006 to hammer out an agreement with Tehran. China, however, was still willing to take Tehran at its word, although it also seems to have become increasingly skeptical of Tehran's intentions during the 2000s.[2]

Beijing repeatedly stated its clear opposition to Iranian possession or development of nuclear weapons. A *Renmin ribao* article of January 2006 laid out Beijing's position:

> The crux of the Iranian nuclear issue lies in the nature of Iran's nuclear program, that is, whether it is for peaceful use or for developing nuclear weapons. As a signatory to the Nuclear Non-proliferation Treaty...Iran enjoys the

right to the peaceful use of nuclear energy, but it absolutely cannot become a nuclear state possessing nuclear weapons. This is the international community's consensus...The reason why Iran's resumption of uranium enrichment activity has intensified the international community's suspicions is because highly enriched uranium can be used to make nuclear bombs.

(Renmin ribao 2006)

In discussions with President Mahmoud Ahmadinejad in Tehran in November 2007, Foreign Minister Yang Jiechi linked China's support for Iran's "right to peaceful use" to its non-development of nuclear weapons. According to Yang, China "appreciates...Iran's repeated declaration of no intention to develop nuclear weapons....China stands for the maintenance of the international non-proliferation system" (Iranian Foreign Ministry 2007). Despite warning that China could not support Iran's acquisition of nuclear weapons, Beijing was generally accepting of Tehran's disclaimers that it was not seeking nuclear weapons. For example, Zhang Yan, PRC representative to the IAEA, said during a September 2004 debate that the uranium enrichment facility at Natanz was not itself evidence of nuclear weapons intent – enriched uranium could either generate electricity or make bombs. Enrichment per se was not banned by the NPT, Zhang said (*Xinhua* 2004). Beijing also insisted that only the IAEA, and not the US government and its intelligence agencies, could make authoritative, internationally binding determinations regarding the military nature of Iran's nuclear programs. This also supported Tehran.

After Iran rejected a set of "incentives" offered by the EU-3 with US support, in September 2005 the IAEA Board of Governors adopted a resolution that was very critical of Iranian actions. That resolution declared that "Absence of confidence that Iran's nuclear programs is exclusively for peaceful purposes have given rise to questions that are within the competence of the Security Council as the organ bearing the main responsibility for maintenance of international peace and security" (IAEA 2005: 28). China *abstained* from voting on the resolution. Beijing tried, although not too firmly, to keep the issue in the IAEA and away from the UN Security Council. A Ministry of Foreign Affairs spokesperson explained China's view that the resolution did not reflect fairly the progress made by the IAEA in its work with Iran and, more critically, "would likely take the issue beyond the IAEA framework" (*Xinhua* 2005). According to the spokesperson, "there is still room for the solution of the Iran nuclear issue *within the IAEA framework*" (emphasis added).

In January 2006, the US and the EU renewed the push to transfer the issue to the UN Security Council. Again China and Russia objected. Beijing opposed transfer of the issue from the IAEA to the UN Security Council because it correctly saw such a transfer as a step toward economic sanctions and perhaps ultimately the use of military force against Iran. As China's representative Wang Guangya said following the IAEA report (not referral) of the issue to the UN Security Council: "All we want is to work for a diplomatic [solution]...invoking Chapter 7 will [make things] more complicated and...lead events in a direction that is uncertain... We all know what Chapter 7 is....this would be the beginning of a series of resolutions" (DemocraticUnderground 2006). In fact, Beijing's suspicions proved correct. Once the issue was "reported" to the UN Security Council, the US and the EU began pushing for sanctions.

China's efforts during the weeks prior to the June 9, 2010 promulgation of UN Security Council Resolution 1929 offer a good example of Beijing's dilution of sanctions. China was able to limit to two the number of Iranian shipping and air cargo firms targeted, to limit banks targeted only to those linked to trafficking in nuclear contraband, and limit a travel ban to only a few individuals rather than to the entire Revolutionary Guard, its entities, and supporters as the US had initially proposed. Most importantly, Beijing secured dropping the bar to investment in Iran's energy sector (Lauria and Solomon 2010: A11). China's opposition was a major factor canceling out Washington's calls for tougher sanctions. According to the *New York Times*, "Some of the toughest proposals were barely even discussed as the US sought support from China...Along with the Russians, the Chinese blocked any measure that would stop the flow of oil from Iranian ports, or gasoline into the country" (Sanger and Landler 2010: A1). According to one anonymous official, Washington tried but failed to resolve differences with Beijing over China's investment in Iran's energy sector. Beijing also opposed the application of sanctions outside the UN framework. Responding to a question about the US imposition in October 2007 of unilateral sanctions against Iran, a Ministry of Foreign Affairs spokesman said:

> China does not approve of easily resorting to the use of sanctions in international relations. Dialogue and negotiations are the best way to resolve the Iranian nuclear issue. Imposing new sanctions against Iran while the international community and Iran endeavor to resolve the issue through dialogue will only complicate the issue.
>
> (PRC Ministry of Foreign Affairs 2007)

Following adoption of Resolution 1929, China condemned the US and European imposition of additional financial sanctions outside the UN framework. "China has noticed the unilateral sanctions announced by the US and others over Iran," a Ministry of Foreign Affairs spokesman said in early July. "China believes that the [recent Security Council] resolution should be earnestly, accurately and fully implemented, instead of being arbitrarily interpreted and expanded (Ministry of Foreign Affairs 2010b).

China's support for Iran against US pressure

Aside from support in UN debates, China supported the IRI diplomatically and in the areas of regime and national security. As one Iranian commentator said about a mid-2007 fifth meeting of the PRC–IRI Political and Economic Commission, "China's growing influence in the world affairs [sic] has become a counter balance to the hegemonic tendencies of the US" (Press TV [Tehran] 2007).

The most mundane but also the most important form of Chinese support was robust diplomatic interaction. According to China's annual diplomatic almanac, there were six high-level (vice-ministerial and above) Chinese and Iranian official exchanges in 2003, eleven in 204, fourteen in 2005, ten in 2006, seventeen in 2007, twelve in 2008, eight in 2009, seven in 2010, and nine in 2011 (*Zhongguo waijiao* 2003–2012). The breadth of these high-level exchanges is also notable: transportation, agriculture, environmental protection, ship building, training of diplomats, information technology, labor and social security, internal security, and military industry. The nuclear issue was a frequent topic of discussion during these interactions, with China's position paralleling its stands in UN debates. In a situation where other world powers had few interactions with the IRI, China's willingness to treat the IRI as a respected major regional power and say a few words in support of Iran in its debates with the Western countries constituted important political support. Iran had other friends – Syria, Hezbollah, North Korea, and Venezuela – but none with the global stature and influence of China. The range of Sino–Iranian exchanges made clear, not least to the Iranian people, that Iran had capable partners willing and able to step in for Iran's increasingly hesitant Western ex-partners. Iran, with China's assistance, was not isolated.

Admittance of the IRI as an "observer" of the Shanghai Cooperation Organization (SCO) in June 2005 was another manifestation of Chinese support. The SCO became a mechanism for (nearly) annual meetings between President Ahmadinejad and top Chinese leaders. Ahmadinejad attended the SCO summit in Shanghai in June 2006 where he addressed the summit and held talks with President Hu Jintao. Ahmadinejad's presence at the 2006 SCO summit constituted a rebuff of US efforts to isolate Iran (Savadore 2006: A6). Ahmadinejad attended the 2007 SCO summit in Kyrgyzstan and exchanged views thoroughly on the Iran nuclear issue with President Hu and Yang Jiechi. In 2008, Ahmadinejad visited Beijing again, this time in association with the Beijing Olympic Games. In talks with Ahmadinejad, President Hu restated China's standard support for Iran's "peaceful nuclear programs" and noted the "strong economic complimentarily and great potential for cooperation" between China and Iran (*Xinhua* 2008). In 2009, Ahmadinejad again met President Hu at an SCO summit where President Hu told Ahmadinejad that the two countries should "further strengthen the mutually beneficial cooperation" in various areas (PRC Ministry of Foreign Affairs 2007).

Sino–Iranian cooperation included internal and national security areas. In 2003, the commander of the IRI Interior Ministry's Law Enforcement Forces visited China. In 2004, the deputy director of China's Committee on Science, Technology, and Industry for National Defense (COSTIND) visited Iran. COSTIND is the heart of China's military–industrial complex and oversees China's military modernization drive. During the August 2005 visit by the commander of China's Nanjing Military Region, Lieutenant General Zhu Wenquan, the chief of joint staffs of the Iranian military pushed for the establishment of a joint technical committee to expand bilateral cooperation in military training and research (*Mehr News Agency* 2005a,b). The Chinese response to this Iranian proposal was equivocal (General Zhu "welcomed" the Iranian proposal) but subsequent Chinese diplomatic almanacs do not report establishment of a joint technical military committee. Other more innocuous appearing mechanisms existed for sensitive Sino–Iranian cooperation. In October 2005, Iran joined the China-led Asia Pacific

Table 11.3 Foreign weapons sales to Iran, 2002–2012 (US$ millions)

	2002	2003	2004	2005	2006	2007	2008	2009	2010	2011	2012	Total
Russia	92	85	15	15	409	321	15	15	41	33	15	1056
China	80	83	79	42	54	47	47	47	62	62	44	647
North Korea	116	114	27									257

Source: Arms Transfer Database, Stockholm International Peace Research Project. Available online at: www.sipri.org

Space Cooperation Organization (APSCO) designed to facilitate cooperation in space and satellite technologies (Fisher 2006). China helped Iran develop ballistic missiles capable of launching satellites (Iranian Student News Agency 2005). Reports by the US intelligence community stated that there had been continuous assistance by "Chinese entities" to Iran's ballistic missile programs.

China continued to serve as Iran's second ranking arms supplier as tension over the Iranian nuclear issue mounted and as US officials periodically stated that "all options remained on the table," a euphemism for a possible military strike if Iran refused to come to terms. According to the Arms Transfer Database of the Stockholm International Peace Research Institute (SIPRI) Project, presented in Table 11.3, China supplied US$647 million worth of arms to Iran during 2002–12, ranking only behind Russia in this regard. Iran was the fourth ranking recipient of Chinese munitions during that decade, behind Pakistan, Myanmar, and Bangladesh (SIPRI 2005–9).

China's munitions sales to Iran during the 2002–12 period centered on anti-ship and anti-aircraft missiles, including hundreds of anti-ship missiles for Fast Attack Craft supplied by China in the 1990s, helicopter-launched anti-ship missiles copied from an Italian design, and over a thousand portable surface-to-air missiles (SIPRI 1995–2009). Iranian transfer of some of its Chinese-origin weapons to third parties, such as Hezbollah and Hamas, has been a source of friction in Tehran–Beijing relations. Such transfer endangered China's ties with Arab states and Israel, and with the US. The prospect of China-origin weapons being used to attack US warplanes and ships in the event of a US and/or Israeli strike against Iran's nuclear facilities raised the potential injury to Sino–American relations even higher.

Resolution 1929 of June 2010 provided that "all States shall" prevent the sale or transfer to Iran of seven categories of heavy weapons, including "missiles or missile systems." This would presumably mean the end or at least the substantial reduction of Chinese (and Russian) arms sales to Iran. Significantly, however, a Chinese assisted factory for producing a variant of a Chinese anti-ship missile opened in Iran in March, just before the passage of Resolution 1929 (Brown 2010). This sequence parallels the 1980s when China finally capitulated to sustained US pressure and stopped selling Silkworm anti-ship missiles to Iran, only to set up a factory in Iran to produce that and even more advanced types of anti-ship missile.

Between the start of 2003 and July 2010 there was a steady flow of strategically sensitive technologies and materials from China to Iran in contravention of US law. During that period the US levied sanctions 60 times against "Chinese entities" for such transfers (Dubowitz and Grossman 2010). Chinese military and nuclear materials also reached Iran by covert means. A number of Chinese firms, including some of China's largest and politically well-connected firms, supplied Iran with a wide array of industrial equipment and materials (Godsey 2009).

Friendship under condition of US pressure

The US has played an important role in China's resource diplomacy with Iran. There has been sharp triangularization of the China–Iran resource relationship. UN and US sanctions during the 2000s that targeted European and East Asian firms, involvement in Iran's energy sector pushed most of those firms out of Iran and/or made them reluctant to participate in the major new energy development schemes offered by Iran. This created opportunities for Chinese oil majors – and these firms seized the opportunities. US-mobilized economic and diplomatic pressure on Iran caused Tehran to seek support to offset that pressure, and the fact that China was willing and able to render various types of support made China an attractive energy partner for Iran. Beijing was able to ensure that multilateral sanctions carrying UN Security Council imprimatur did not interfere with foreign investment in Iran's energy sector. Beijing also refused to go along with *unilateral* US legislation and executive orders barring investment in Iranian energy or transfer of sensitive dual-use technologies and materials to Iran, although individual Chinese firms weighted their economic interests in the US against their Iranian interests.

Chinese and Iranian leaders shared many similar negative views of US policy in the Gulf region. The US, they agreed, was prone to interfere in the internal affairs of non-Western countries with traditions and values different from those of the US. The US arrogantly presumed to serve as international judge and policeman regarding matters such as which countries could and could not use nuclear technology. The US was too ready to resort to economic sanctions and the use or threatened use of force against other countries, making the US itself a major cause of regional insecurity and instability. The US objective in doing all this was to achieve hegemony over the Persian Gulf and its rich energy resources. Chinese representatives generally preferred to keep private expression of these views, but the convergence of views was indicated by occasional bland statements that the two sides had common views on regional and international issues. China and Iran see one another as potential and important partners in an emerging world order in which the US role will be much reduced to the benefit of both China and Iran, and indeed the entire non-Western world.

But China's entente with Iran poses great potential danger to China in the form of possible US hostility. Precisely because so much US energy and attention was focused on the Persian Gulf, and precisely because controlling that region's resources were so important to Washington's global strategy (at least are that strategy was constructed by Chinese leaders), openly thwarting US policies in that region was dangerous. If US leaders concluded that China was encouraging and supporting Iran against the US, or that China was enabling Iran's nuclear and missile activities (which China had promised in 1997 to terminate), US policy toward China might harden, possibly spoiling the Sino–US comity that has underpinned the favorable macroclimate for China's post-1978 development drive. It might also tarnish China's reputation as a responsible rising power upholding, rather than seeking to undermine international regimes such as the NPT. To prevent this, China needed to cooperate with the US to some degree on the Iran nuclear issue, most prominently by endorsing four sets of UN Security Council sanctions. US hostility to China's "rise" could create great difficulties and conceivably even abort China's development drive. Those difficulties, along with associated reverses and instability, could endanger the survival of the Chinese Communist Party (CCP) regime.

Supporting the IRI against the US improved China's position vis-à-vis the US in several ways beyond helping China obtain the oil it needs:

- A strong Iran confronting the US keeps US diplomatic attention and military power tied down in the Persian Gulf, diverting both away from the Western Pacific where China's core interests lie.
- It exhausts US strength and will, and making it less likely that the US will challenge China's rise.
- A Sino–Iranian entente hedges against the remote but distinct possibility of a US–PRC war in which China would need major oil producers willing to supply China over US objections.
- It demonstrates to Iranian leaders that China is a capable and trustworthy partner willing to support the policy objectives set by Iran's government.

By building a relationship of cooperation and trust with Iran, China is courting a significant regional power that will look with favor on China rise to a position of greater eminence in Asia and the world. China's current support for the IRI against US pressure can be seen as a long-term investment in a Sino–Persian entente that is to be one basis of a stronger Chinese position in Asia.

On the other hand, and at the opposite end of China's policy calculus, resolution or at least amelioration of a US–IRI conflict would ease China's difficulties, allowing it to expand energy cooperation with Iran rapidly and without "triangular" concerns about spoiling Sino–US ties. Since the mid-2000s, voices within China's Foreign Ministry have advocated an effort to use China's good offices to mediate the US–IRI conflict (Garver 2007). In line with this thinking, during the negotiation of the interim nuclear deal in November 2013 China's UN representative played a crucial role in bridging differences between the US and Iran. "When the two parties came across irresolvable problems, they would come to China which would 'lubricate' the negotiations and put things back on track," according to ex-PRC ambassador to Tehran Hua Liming (Wan 2013). Foreign Minister Wang Yi welcomed the November 2013 interim agreement; China as an important member of the dialogue always advocated peaceful solution to the

Iran nuclear issue through negotiation and kept close contact with all parties and had played its role and shouldered its responsibility for the conclusion of the deal, Wang said (*Xinhua* 2013).

It is doubtful that these Ministry of Foreign Affairs efforts to ease US–Iran tension would be welcomed by hardliners in China's foreign policy elite. The more jaundiced hardliner view sees continued enmeshment of the US in confrontation with Iran in the Gulf as constraining Washington's efforts to contain China by pivoting to Asia. From this perspective, Ministry of Foreign Affairs efforts to ease Washington's quagmire in the Gulf would be muddleheaded.

Notes

1 It must be stressed that these are very rough approximations. Deals that were canceled after being agreed upon were not counted. Thus Japan does not appear. Deals that press articles indicated as merely "under discussion" were also not included. On the other hand, there was a fair amount of agreement among the sources indicated in Table 11.1.
2 During interviews at Beijing think tanks in 2007, several Chinese analysts responded to direct questions by saying they thought Iran aspired to nuclear arms – although this was because of US threats. It may be that Beijing's 1997 discontinuation of cooperation with Iran's nuclear program was due to injury to China's reputation if it was associated with Iran nuclear effort when it, sooner or later, "went nuclear." Regarding China's disengagement from Iran's nuclear programs, see Garver 2006: 219–26.

References

Agence France-Presse (AFP), 2009. Campbell says US needs Chinese support on Iran nuclear issue. *AFP,* 14 October. World News Connection (accessed 5 January 2010).

Agence France-Presse (AFP), 2010. Ahmadinejad starts China trip. *AFP,* 11 June. Available online at: www.LexisNexis.com (accessed 11 March 2011).

Al-Quds al-Arabi, 2007. Chinese Foreign Minister outlines position in Iran at UK lecture. *Al-Quds al-Arabi,* 6 December. World News Connection (accessed 8 January 2008).

Ansari, Ali M., 2006. *Confronting Iran.* New York, NY: Perseus Books.

Batson, Andrew, 2010. China visit suggests thaw over Iran, Yuan. *Wall Street Journal,* 2 April, p. A1.

Brady, Anne Marie, 2003. *Making the foreign serve China; managing foreigners in the People's Republic of China.* Lantham, MD: Rowman and Littlefield.

Brown, Peter, 2010. US mum over China's links to Iran. *Asia Times Online,* 22 May. Available online at: www.atimes.com (accessed 7 March 2011).

Buckley, Chris, 2010. Iran envoy goes to China as sanctions talk rises. *Reuters,* 31 March. Available online at: www.reuters.com (accessed 7 March 2011).

Chen, Aizhu, 2010. China slows Iran oil work as US energy ties warm. *Iran Focus,* 28 October. Available online at: www.iranfocus.com (accessed 15 March 2011).

Cooper, Helene, 2009. China holds firm on major issues in Obama's visit. *New York Times,* 17 November, p. A1.

DemocraticUnderground, 2006. *Chinese envoy opposes resorting to Chapter 7 resolution on Iran.* 28 April. Available online at: www.democraticunderground.com (accessed 16 January 2007).

Downs, Erica S., 2000. *China's quest for energy security.* Santa Monica, CA: Rand.

Dubowitz, Mark and Grossman, Laura, 2010. *Iran's Chinese energy partners: companies eligible for investigation under US sanctions law.* Washington, DC: Foundation for Defense of Democracies, September. Available online at: www.defenddemocracy.org (accessed 21 March 2011).

Fars News Agency, 2007. China defends commercial oil deal with Iran. *Fars News Agency,* 12 December. World News Connection (accessed 8 January 2008).

Fisher, Richard, 2006. China's alliance with Iran grows contrary to US hopes. Washington, DC: International Assessment and Strategy Center, 20 May. Available online at: www.strategycenter.net (accessed 17 March 2007).

Garver, John W., 2006. *China and Iran, ancient partners in a post-imperial world.* Seattle, WA: University of Washington Press.

Garver, John W., 2007. Can China mediate US–Iran relations?. *China Currents,* 6 (1). Available online at: www.chinacenter.net/can-china-mediate-us-iran-relations/ (accessed 21 January 2008).

Godsey, Matthew, 2009. Chinese companies evade US trade ban. Wisconsin Project on Nuclear Arms Control, 15 December. Available online at: www.wisconsinproject.org/pubs/ reports (accessed 18 March 2010).

International Atomic Energy Agency (IAEA), 2005. Implementation of Safeguards Agreement in the Islamic Republic of Iran. GOV/2005/77. Available online at: www.iaea.org (accessed 21 March 2010).

Iran Foreign Ministry, 2007. Iranian President Mahmoud Ahmadinejad met at the Presidential Palace. 13 November. Available online at: www.ministryofforeignaffairs.ir.gov.ir (accessed 21 February 2009).

Iranian Student News Agency, 2005. Aerospace official says Iran building satellite-carrying missiles. 15 March. *Iranian Student News Agency.* Available online at: http://dialog.com (accessed 20 November 2005).

Jacobs, Andrew and Landler, Mark, 2010. Strains easing, Chinese leader plans US visit. *New York Times,* 2 April, p. A10.

Landler, Mark, 2010. Clinton presses China to support penalties for Iran. *New York Times,* 30 January, p. A5.

Lauria, Joe and Solomon, Jay, 2010. UN slaps Iran with new curbs. *Wall Street Journal,* 10 June, p. A11.

Lin, Justin, 2010. Chinese investment in Iran: one step forward and two steps backward. *East Asia Forum,* 3 November. Available online at: www.eastasiaforum.org (accessed 4 June 2011).

Medeiros, Evan S., 2007. *Reluctant restraint: the evolution of China's nonproliferation policies and practices, 1980–2004.* Stanford, CA: Stanford University Press.

Mehr News Agency, 2005a. Iranian, Chinese armed forces to form Joint Technical Commission. *Mehr News Agency,* 20 August.

Mehr News Agency, 2005b. Iranian, Chinese military officials hold more talks. *Mehr News Agency,* 20 August. World News Connection (accessed 18 March 2007).

Ministry of Foreign Affairs, 2010a. Press conference: Foreign Minister Yang Jiechi on 5 February.

Ministry of Foreign Affairs, 2010b. Ministry of Foreign Affairs press conference on 6 July.

PRC Ministry of Foreign Affairs, 2007. *Foreign Ministry spokesperson Liu Juanchao's remarks on US sanctions against Iran.* 26 October. Available online at: www.fmprc.gov.cn (accessed 25 February 2008).

PRC Ministry of Foreign Affairs, 2010. *Foreign Ministry Spokesperson Qin Gang's Regular Press Conference.* 6 July. Available online at: www.fmprc.gov.cn (accessed 27 April 2011).

Press TV (Tehran), 2007. Iran's deputy FM arrives in China. *Press TV (Tehran),* 30 July. Available online at: www.presstv.ir.com (accessed 17 March 2008).

Renmin ribao, 2006. Door to negotiations has not been closed. *Renmin ribao,* 19 January. World News Connection (accessed 15 March 2007).

Sanger, David, 2005. Bush puts Iraq, China and Iran on agenda. *New York Times,* 14 September, p. A6.

Sanger, David and Landler, Mark, 2010. US strikes deal on new penalties by UN for Iran. *New York Times,* 19 May, p. A1.

Savadore, Bill, 2006. Tehran seeks allies through energy cooperation. *South China Morning Post,* 16 June, p. A6.

Steinberg, James, 2009. *America's vision for a China partnership.* Speech at the Center for a New American Security on 29 September. Available online at: www.realclearworld.com.

Steinberg, James, 2010. Comments at forum on Sino–US global cooperation. Brookings Institution, 20 May.

Stockholm International Peace Research Institute (SIPRI), 1995–2009. *Transfer of major conventional weapons, sorted by supplier, China to Iran, 1995–2009.* Available online at: www.sipri.org (accessed 11 March 2011).

Stockholm International Peace Research Institute (SIPRI), 2005–2009. *Trend Indicator Value of arms exports from China, 2005–2009."* Solna: Stockholm International Peace Research Institute.

The Washington Post, 2009. President Obama delivers joint press statement with President Hu Jintao of China. *The Washington Post,* 17 November. Available online at: www.washiingtonpost.com (accessed 25 March 2015).

US Department of State, 2006. *Remarks by President Bush and President Hu of the People's Republic of China in arrival ceremony.* 20 April. Available online at: www.state.gov (accessed 21 February 2008).

Wan, Adrian, 2013. China plays key broker role in Iran nuclear deal. *South China Morning Post,* 25 November. Available online at: www.scmp.com (accessed 5 December 2013).

Xinhua, 2004. RRC [sic] Envoy to UN Zhang Yan urges peaceful resolution of Iranian nuclear issue. *Xinhua,* 19 September. World News Connection (accessed 21 March 2009).

Xinhua, 2005. PRC FM spokesman: Iran Nuclear issue should be resolved through negotiations. *Xinhua,* 27 September.

Xinhua, 2008. Hu Jintao, Ahmadineahad [sic] discuss bilateral ties, nuclear issue. *Xinhua,* 6 September. World News Connection (accessed 21 March 2009).

Xinhua, 2013. China's FM hails agreement on Iran's nuclear issue. *Xinhua,* 24 November. Available online at: http://news.xinhuanet.com (accessed 5 December 2013).

Zhongguo waijiao (China's Diplomacy), 2003–2012. Annual volumes. Beijing: Shijie zhishi chubanshe.

Zoellick, Robert, 2005. *Wither China: from membership to responsibility?* Speech to the National Committee on US–China Relations, 21 September. Available online at: www.ncuscr.org (accessed 15 January 2006).

12 The case of Darfur

Diplomacy under influence of Sino–US resource rivalry

Sonja Regler[1]

Introduction

For many years, both the US and China depended heavily upon the import of oil to keep their economies going. As a rapidly growing state, China will need to expand its outward activities, possibly colliding with the spheres of interest of other states. As occurred in the early 20th century (Choucri and North, 1975: 1), China may become embroiled in international conflicts. US and Chinese companies are already jockeying for position in the volatile oil fields of the Middle East, Central Asia, the Americas and Sub-Saharan Africa. As Zweig suggests in Chapter 1, it seems reasonable to assume that tightening oil supplies will increase Sino–US competition over access to resources (Hatemi and Wedeman, 2007: 109).

This chapter will focus on the conflicting interests of the US and China in Sudan. The main statement of this chapter is the triangularity: as China deepened its seemingly bilateral relations with Sudan, it found itself in a triangular relationship that included the US. The US faces a similar triangular relationship because China is the largest external investor in Sudan. This triangularity becomes a challenge because of the conflicting interests. Beyond moral values, political, strategic, and economic interests in Sudan have increased the frictions between the US and China (Shichor, 2010: 61). Alden (2007: 118) observes that,

> the shadow boxing over the Darfur issue between Beijing and Washington, while rooted in the local dynamic of the Sudan crisis and the ongoing ethnic cleansing, none the less carries with it dimensions of global rivalry over influence and access to vital energy resources.

What he sees as "(t)he spectacle of international disputes, albeit diplomatic, over events in Sudan is the clearest and most recent expression of the potential for conflicts over oil, human rights and sanctions" (Alden, 2007: 6).

However, this chapter will demonstrate that although the Darfur crisis initially placed China and the US in a confrontational situation over how to deal with the crisis, both countries learned that they could not advance their own interests outside of triangle, finally achieving limited cooperation in line with their own interests. The issues behind both the confrontational and cooperative behavior show that, first and foremost, the perceived resource interests of the US and China and the mistrust between these two countries complicated a settlement of the Darfur crisis. Although US sanctions prevented US companies from competing directly with China in Sudan, both countries saw the situation in the Sudan through the prism of oil – and the oil-related real or perceived interests of the other country. In this respect, Darfur does not directly constitute a case of "resource diplomacy," but rather a case of "diplomatic interaction under the influence of resource rivalry." As a case study, the Darfur crisis illuminates the source of many other civil wars in Africa: open or hidden conflict among the great powers and competition over resources impact directly on policies adopted by the international community regarding countries in conflict. Since human security depends on cooperative solutions, energy interests easily stand in the way of human security. The question of what it takes to make China and the US move beyond confrontation and seek cooperation is crucial for vulnerable populations suffering from wartime atrocities in African countries.

The Darfur crisis in the light of Chinese and US energy interests

The Darfur crisis that started in February 2003 was heavily influenced by Sudan's North–South war. This North–South conflict initially constituted a war over power, gravitating around cultural identity and self-determination, between Southern rebels and the Northern government in Khartoum. The second phase of the North–South war from 1983 until 2005 was significantly fueled by the fact that Khartoum undertook steps to start oil exploitation. The Southern rebel group, the Sudan People's Liberation Army (SPLM), attacked oil fields right from the beginning of the second period of the war (Behrends, 2008: 45), arguing that if the Sudanese government was able to continue oil exploration, its position in the civil war would be strengthened (Ali, 2006: 72). Controlling the region's oil wealth became a primary goal for both sides (Human Rights First, 2008: 33). As peace negotiations between Khartoum and the SPLM resumed in 2002 (right before the outbreak of the Darfur crisis in 2003), much of the negotiations focused on wealth sharing, and wealth in Sudan, strictly speaking, is oil (Interview with Abdelgadir Saleh, June 19, 2008).

The Darfurian insurgency in 2003 resembled the rebellion of the South in many ways. Both major rebel groups – the Justice and Equality Movement (JEM) and the Sudan Liberation Army (SLA) – focused on alleviating economic and political marginalization (Daly, 2007: 275–6, 280). Oil played a role as well because oil was explicitly referenced in the Darfur Peace Agreement (UN Peacemaker, 2005a: Art. 117, 118, 161). The amount of oil discovered in Darfur is fairly small and mainly in Nyala, the capital of South Darfur state, as well as in Northern Darfur state (Sanders, 2007). However, there is an ideological dimension to oil exploitation: the inability to exploit oil resources on their own behalf became part of the rebel's identity as an oppressed people (Behrends, 2008: 40). Marginalized populations, who consider the oil as theirs, have no leverage over its extraction or the distribution of oil income, nor can they assert their demands through democratic means. The only leverage they have is fighting, which melds the oil issue into wider ones of self-determination and identity.[2]

Oil motivated both sides. Although in Darfur, the initial impetus for the Sudanese government to fight the rebels was fear that the country's periphery was asking for the same things that the South was demanding – a share of power and wealth. The later issue became the clearing of land for future oil exploitation (Interview with Andrew Natsios, May 18, 2010). Since oil exploitation exacerbates existing conflicts in a country with a fragile ethnic balance and major inequalities, external actors bear at least indirect responsibility for resource-related conflict in the host country. US companies are therefore just as involved as the Chinese; in fact, until 1986 the major foreign oil company in Sudan was Chevron, which then withdrew its assets due to the unstable security situation and increasing pressure from human rights groups (Patey, 2006: 13–15).

After the US sanctioned Sudan as a state sponsor of terrorism, and China became a net oil importer in 1993 (Alden, 2007: 11–12), Chinese companies positioned themselves to become the biggest stakeholders in Sudan's oil sector. Chinese companies therefore clearly challenged Washington's interests because their investments made US sanctions ineffective. But US companies are also indirectly involved. Even though they pulled out of Sudan in 1986, they remain connected to the Darfur crisis through their business in neighboring Chad.

In 2003, a consortium composed of US-based Exxon Mobil, and Chevron, plus Malaysian-owned Petronas, began oil exploration in Chad. The US is the largest purchaser of Chadian oil, accounting for over 90 percent of Chad's oil exports, and this even after Chinese companies entered the extractive industries in Chad in 2006. Chad is the sixth leading African exporter to the US, making it of considerable economic importance to the US (Hansen, 2011); however, oil exploitation in Chad is causing similar problems over the distribution of oil income, as in Sudan. According to Behrends, (2008: 41) "In Chad oil (…) was, however, a likely trigger for the rebellion against Chadian President Idriss Déby, which originated in his closest and political kin circles." The rebellion, however, also indirectly involved Darfurian oil because Déby's power base is mainly composed of his kin from the Zaghawa tribe, one of the three Darfurian tribes rebelling against Khartoum. These kin expected him to support the Darfurian rebels against the Sudanese government, but efforts by Déby to remain neutral eroded his power base further, increasing instability in Chad. Moreover, Chadian rebels trying to overthrow him operated from Darfur. Déby, who was helped into power by Sudanese leaders, was therefore put in a considerable quandary by the Darfur crisis (Behrends, 2008: 48).

In the beginning, neither the US nor China were willing to address the Darfur crisis in the international community (Interview with Günther Pleuger, August 4, 2009). After Khartoum had committed itself to the War against Terror in the wake of 9/11, the US reengaged in the region and brokered the Comprehensive Peace Agreement (CPA) between Khartoum and the Southern rebels. Even as the Darfur crisis peaked in 2003 and early 2004, the US remained focused on North–South peace negotiations. Beyond US interest in the Sudanese oil sector, this focus was also motivated by religious and humanitarian concerns – the North–South war had gone on for decades, and the Southern Sudanese were mostly Christians. Only in April 2004, the 10th anniversary of the Rwandan genocide, did the US administration turn its attention to Darfur and start pushing the envelope; however, initial US initiatives were consistently blocked by China.

Struggling over Darfur: Sino–US standoff in 2004–2006

Between July 2004 and the end of October 2007, the UN Security Council voted on 20 resolutions concerning Darfur, most drafted by the US. Among those resolutions, China abstained six times in resolutions voted upon from 2004 to 2006, foremost on Resolutions 1556, 1564, and 1591 that concerned economic sanctions; Resolution 1593, referring the situation in Darfur to the International Criminal Court (ICC); and Resolution 1706, extending UNMIS (the UN peacekeeping mission for the North South conflict) to Darfur. Resolution 1672 only included individual sanctions and therefore is of no importance in this analysis. The motivations for abstention in the vote differ for each case.

Resolution 1556, passed on 30 July 2004, was deeply affected by Sino–US disagreements (Zygar, 2004). Submitted by the US, the resolution demanded both the disarmament of the Janjaweed Militia, which committed crimes against civilians in Darfur and that they are brought to justice. The resolution also placed an immediate weapons embargo on "all non-governmental entities and individuals, including the Janjaweed" operating in Darfur, but excluding the Sudanese government forces (Nabati, 2004). China had opposed restrictions affecting the regular Sudanese armed forces in the drafting process of Resolution 1556 (Holslag, 2007: 8), and China were keen on preserving Khartoum's control over the country. In addition, the resolution had originally provided for "sanctions" against Sudan in the event of non-compliance (i.e. in case Khartoum failed to disarm and prosecute the Janjaweed). Including China, seven of the 15 UN Security Council members were reluctant to endorse any explicit threat of sanctions (Nabati, 2004). When the US lobbied vigorously to retain those sanctions, China threatened to veto the resolution. The US finally agreed to delete the word "sanctions" from the draft (Zygar, 2004), and it was substituted with a reference to "further actions, including measures" as provided for in chapter 41 of the UN Charter. The softer language, however, had limited consequences: under Chapter 41 of the UN Charter, the UN Security Council may decide measures including complete or partial interruption of economic relations and of means of transport and communication, as well as the severance of diplomatic relations. Chapter 41 basically has the same effect as "sanctions," except it does not include military intervention. In the voting, China and Pakistan still abstained (Nabati, 2004); however, the softening of the language enabled both countries to compromise over their differing policies, while basic disagreements over how to address the Darfur crisis persisted. In particular, some aspects of Resolution 1556 may not have convinced every member of the UN Security Council, for example there was reason to doubt that Khartoum was actually able to disarm the Janjaweed militia (Interview with Andrew Natsios, May 18, 2010).

The Chinese, however, generally suspected that ideological and economic motives were behind any attempt to sanction Sudan. In Chinese eyes, pressuring Khartoum through UN resolutions shifted the power in favor of the rebels whom they considered Western allies. In this interpretation, Western countries sought regime change in order to access Sudanese oil, which at the time was largely exploited by Chinese companies. Chinese experts also suspected that Western countries supported Darfurian rebels by delivering weapons through NGOs, which indirectly breached the weapons embargo formulated in Resolution 1556 (Interview with Miaofa Wu, October 29, 2008).

Chinese mistrust of the US originates largely due to the war against Iraq. Even though UN sanctions prevented China from exporting oil from Iraq, the Chinese had struck a deal in 1997 under the expectation that the sanctions against Iraq would be lifted. After the toppling of the Iraqi regime, the interim authority "froze" contracts signed by Saddam Hussein's government, sparking intense Chinese oil activity elsewhere because Iraqi oil had been a major part of China's long-term energy plan (Hatemi and Wedeman, 2007: 106). All in all, the Chinese, as well as many other members of the international community, experienced the war against Iraq as a complete breach of trust. According to Bronner (2003), "the justification for war against Iraq was not simply based on 'mistaken' interpretations, or 'false data', but on sheer mendacity. ...US policy was propelled by thoughts of an Iraqi nation 'swimming in oil'". Darfur was therefore suspected to be a repetition of Iraq, insofar as the US would use human rights as a pretext to intervene in China's economic interests.

The international community's mistrust became even clearer when on September 9, 2004, the US declared the Darfur crisis a case of "genocide". US human rights groups had been consistently claiming that Darfur was an act of genocide, so this declaration seemed aimed at alleviating the domestic pressure on the US administration (Mamdani, 2009: 25) The immediate international response to the declaration was to question its motives because many countries suspected that the US would use the declaration to justify a military invasion, driven by oil interests, of yet another Muslim country and achieve regime change – as with Iraq (Stedjan and Thomas-Jensen, 2010: 168). The moral outrage over Darfur seemed a way to divert international attention away from the disaster in Iraq and to polish the damaged image of the US (Interview with Chinese Senior Scholar, November 28, 2008).

A genocide designation would also have consequences for Sudan's sovereignty. Genocide is an international crime under customary international law, and parties to the Convention can punish it wherever it occurs. Any party may refer the matter to the ICC, and Sudan need not be a party to the Genocide Convention for this to happen (Nabati, 2004). The genocide designation, however, remained a unilateral measure by the US from which other countries refrained. Although it is indeed debatable whether Darfur constitutes a case of genocide, the fact remains that unbearable war crimes took place there.

Resolution 1564 and Resolution 1591 were largely repetitions of Resolution 1556. Resolution 1564, passed on September 18, 2004, generated new dissonance between China and the US. The draft once more raised the threat of future sanctions on Sudan's oil sector should Khartoum fail to take appropriate actions regarding Darfur. In his explanatory remarks after the vote, the Chinese ambassador, Wang Guangya, said that China's position remained unchanged on sanctions (Permanent Mission of the People's Republic of China to the UN, 2004). Six months later, the US hoped to break the deadlock on sanctions in the UN Security Council with Resolution 1591 (Lederer, 2005). The US had been pushing a single Resolution after mid-February 2004 and finally decided to break it into three parts, setting aside the divisive issue of punishing atrocities and continuously working on sanctions and a peacekeeping mission by the UN. The draft resolution had contained a renewed threat of an oil embargo against Khartoum in the event of continued non-compliance, but the revised version dropped the oil embargo, leaving only a partial arms embargo, a travel ban, and an asset's freeze against perpetrators of atrocities (Leopold, 2005). After China had watered down all resolutions concerning sanctions (i.e. Resolutions 1556, 1564, and 1591), it was accused of protecting Khartoum from action by the international community out of its own oil interests.

However, China may not have been motivated entirely by oil interests. The prevailing wisdom in the West regarding the Chinese investments in Sudan is that they are part of a highly coordinated quest for oil and gas assets in which the companies are mere puppets of the state (Downs, 2007: 48). Indeed, the Chinese government has a direct stake in the financial performance of China National Petroleum Corporation (CNPC), Sinopec, and China National Offshore Oil Corporation (CNOOC). The Chinese government also receives income tax in addition to the shares in the profit of the companies (Congress of the United States, 2006: 12). Beijing, therefore, certainly wants these companies to be competitive internationally.

The details of government ownership are complex (Congress of the United States, 2006: 11). The economic reform, including restructuring state-owned enterprises (SOEs), was just under way in China when the Darfur crisis broke out, and the State-owned Assets Supervision and Administration Commission (SASAC), which manages all major SOEs, was only founded in 2003. Until that time, Chinese national oil companies (NOCs) should be considered state-invested but not state-run. Even though the CEOs were political appointees, officials in the Energy Bureau of the National Development and Reform Commission (NDRC) – the country's main economic decision-making body – were generally powerless in the face of pressure from state-owned energy firms (International Crisis Group [ICG], 2008: 6) At that time, the Chinese government had been unable to build up independent expertise and remained largely dependent on self-serving advice from NOCs (ICG, 2008: 5), meaning that China's energy agenda was largely driven by the corporate interests of China's energy companies rather than by the interests of the state (Downs, 2006: 16). However, although China's energy policy was in a transition phase at the outbreak of the Darfur crisis, economic interests, albeit important in Beijing's foreign policy decision making, still do not sufficiently explain China's position in Darfur (Huang, 2007: 828).

Sudan needs China much more than China needs Sudan, with a clear asymmetry in the power dynamic between two. Although the Sudanese are highly dependent on China for economic and political reasons, China can easily diversify its oil imports away from Sudanese oil (Interview with State Department Officer, June 3, 2010). Sudan's oil exports to China oscillated somewhere between 65 and 82 percent of its total annual output, which constitutes 5–7 percent of China's total oil imports (Shinn, 2009: 88). This 5–7 percent, in turn, represents less than 1 percent of China's total energy consumption.

There is, however, a strategic component in Sudan's energy supply for China because Sudan is one of the major countries where Chinese NOCs possess equity shares (ICG, 2008: 6). The importance of "equity oil" – where the Chinese company allegedly has control over the disposition of its share of production – stems from the belief that if a crisis leads to a shortage of energy in world markets, the Chinese NOCs could be pressed into service by the government (ICG, 2008: 4). Chinese leaders also generally distrust international oil markets (Herberg, 2007) and fear that the most influential players in the market, particularly the US, could one day deny China access to the energy source (ICG, 2008: 3). The fact that Chinese investments in Sudan are partly equity oil might therefore serve as a "last insurance" to the Chinese leadership.

Moreover, 2003–2005, when the Darfur crisis peaked, were challenging years for China in terms of energy security. In 2004, China had become the second largest consumer of petroleum worldwide, just behind the US. In that year, global demand in petroleum had increased by 3.3 percent, more than twice the annual rate of increase over the previous decade. Nearly 30 percent of that additional demand originated in China, where demand for petroleum grew at an annual rate of 15 percent (Congress of the United States, 2006: 5). At the same time, growth of petroleum production in China nearly stopped in 2004 (Congress of the United States, 2006: 5–6), triggering serious energy shortages in China in 2003 and 2004 (ICG, 2008: 6). The percentage of imported oil increased from 36 to 45 percent in 2003–2004 (Ziegler, 2006: 6), even as prices on the international market soared, rising from US$30/barrel in 2003, doubling to US$60/barrel by August 2005, and peaking at US$147.30/barrel in July 2008 (Cabalza and Manalo, 2010: 35).

Nevertheless, the Chinese government did not impose a quota on Chinese oil companies regarding the amount of equity oil they shipped to China. The majority of Sudanese equity oil went to local or international markets, not to China (Jiang and Sinton, 2010: 17; ICG, 2008: 3), and, in some cases, was carried out by marketing subsidiaries located outside the headquarters of the NOCs (Jiang and Sinton, 2010: 7). While Chinese-controlled oil production in Sudan increased over 2004–2005, Sudan's share of Chinese oil imports actually decreased, indicating that Chinese NOCs sold it elsewhere – maybe in swapping contracts. In fact, by 2006 almost 39 percent of brut exports from Sudan went to Japan and only 31 percent went to China. Still, in 2007, Sudanese exports to China came back to 6.5 percent of Chinese total oil imports (Zweig, Chapter 1 this volume; Shichor, 2010: 55–6).

In January 2005, while the international energy market remained volatile, the signing of the Comprehensive Peace Agreement (CPA) between Khartoum and the Southern rebels actually consolidated the "last insurance" China had in Sudan's equity oil: the CPA included the protection of existing oil contracts from renegotiation (UN Peacemaker, 2005b). This point certainly made China even less dependent on any good understanding with Khartoum. Nevertheless, China's opposition to sanctions remained steadfast – it vetoed Resolution 1591 in March 2005 – suggesting once again that China's motivation to refuse sanctions was not based solely on its energy needs but more on a principled policy.

Two days later, however, China's vote on Resolution 1593 on March 31 proved China's independence from Sudan and the threshold of its limited willingness to use its veto to shield Khartoum from demands of the international community. This resolution referred Darfur to the ICC, an action that would clearly compromise the sovereignty of Sudan in order to protect civilians from war crimes. Khartoum had repeatedly asked Beijing to veto the resolution and was deeply angered when Beijing abstained. Following the vote, Sudanese presidential advisor, Nafie Ali Nafie, addressed a crowd and criticized China publicly for not blocking the resolution (Ali, 2007). The Chinese ambassador to the UN, in turn, explained that China definitely did not favor the referral to the ICC without the consent of Khartoum but, on the other hand, "believed that the perpetrators must be brought to justice" (UN Security Council, 2005).

The source of China's opposition to economic sanctions and other measures proposed by the US is more complex than simple oil interests. China sees itself as a leader among developing countries and wants to speak on their behalf. Moreover, China also suspects that US actions are targeting China. As one Chinese scholar phrased it, the US regards China as a potential economic threat and will use all possible means to pressure China (Interview Chinese Senior Scholar, November 28, 2008). Many Sudanese locals also believe that the goal of US sanctions is to weaken Khartoum with the aim of eliminating Chinese influence in Africa (Shichor, 2010: 71). US observers do realize that China might view any interference in Sudan's oil sector – a blockade, a voluntary embargo, or denying port access to ships transporting Sudanese oil – as an aggressive act against them (Swilla, 2004). To the extent that China both needs Sudan's oil supplies and considers US a realistic threat, China's UN voting behavior should be seen as part of its efforts to shield itself from undesired US influence on its oil investments. For the most part, however, China votes on political principles, reflecting its image as an international actor.

Sino–US rapprochement in the Darfur crisis 2006–2008

Realizing that it was caught in a stagnant triangular relationship, the US shifted its approach towards China in the second term of George W. Bush's presidency. In September 2005, then Deputy Secretary of State Robert B. Zoellick gave a ground breaking speech on Sino–US relations, laying out future cooperation between the US and China (Zoellick, 2005). Zoellick said that,

> China's economic growth is driving its thirst for energy. In response, China is acting as if it can somehow 'lock up' energy supplies around the world...

[Instead], China should work with the US and others to develop diverse sources of energy. We should encourage the opening of oil and gas production in more places around the world.

He admitted to a gulf in perceptions on both sides regarding the other country and admonished China to recognize how its actions are perceived by others. He referred to Sudan by saying that,

China should take more than oil from Sudan, it should take some responsibility for resolving Sudan's human crisis. It could work with the US, the UN, and others to support the African Union's peacekeeping mission, to provide humanitarian relief to Darfur, and to promote a solution to (Sudan's conflicts).

Zoellick's goal was to transform the Sino–US relationship and enhance cooperation within a larger framework where the parties recognize a shared interest in sustaining political, economic, and security systems that provide mutual benefits, thereby encouraging China to become a "responsible stakeholder" in the international system (Zoellick, 2005).

US diplomacy regarding energy security became increasingly proactive towards China. When President Hu Jintao visited Washington in April 2006, part of the discussion related to access to oil. In March 2006, the revised National Security Strategy was published, which warned that the Chinese might try to "lock up" global supplies instead of opening them to the world. Chinese and US negotiators were debating a joint study of both nations' energy needs as a way to ward off conflict in coming decades when China's expanding needs for imported energy could collide with the needs of the US, Europe, and Japan. Talks on the presidential level also concerned Iran, where China's sensitivity about its own oil supplies was one reason President Bush had rejected any sanctions against the Iranian oil sector. By this means, the US was trying to get the Chinese to come around on Iran and the nuclear issue (Sander, 2006). Thus began a series of talks on Iran, North Korea, and Sudan that President Bush kept on the US–Chinese agenda in subsequent phone calls and bilateral meetings (Kleine-Ahlbrandt and Small, 2008). These talks were accompanied with regular visits to Beijing of US special envoys on Sudan and North Korea. The increasing frequency of communication also permitted Beijing to take a greater part in joint decision-making (Kleine-Ahlbrandt and Small, 2008).

Intensified communication between the US and China, as well as the changing economic and political situation on the ground, made Sino–US cooperation on Darfur possible because Beijing finally felt that the US was taking its concerns seriously. As for the situation on the ground, many factors made China aware that a more proactive approach on Darfur was in its own interests. In the summer and fall of 2006, the economic environment in which Chinese companies operated changed considerably after Chad broke ties with Taiwan in August 2006 and renewed links with China. Chinese companies entering the Chadian oil business needed a stable business environment. Chad's president also communicated his country's need for stability to Beijing. Suddenly Chinese companies were involved in three of the four countries that bordered Darfur and northern Sudan, a factor Beijing certainly considered as it calibrated its policy on Darfur (Shinn, 2009: 88). In addition, the African Union (AU) had denied the aspirations of Sudan's President, Omar al-Bashir, to become AU president, first in 2005 and again in 2006. At the Forum of Chinese–African Cooperation (FOCAC), held in Beijing in November 2006, Chinese President Hu Jintao expressed China's concern about Darfur and made critical remarks (Shinn, 2009: 91). Thereafter, Beijing aligned its policies with the views of most African and Arab states (Shinn, 2009: 94).

Given the changing political and economic environment, the US's push for a UN peacekeeping mission became a feasible option for China. The UN peace-keeping mission in Darfur had been a priority for the US administration since the early summer of 2005 and in January 2006. When the AU requested the replace-ment of its exhausted troops, the US proposed "bluehatting' the AU troops in Darfur (de Waal, 2007: 1042).

For China, contributing to UN peacekeeping missions had become a corner-stone of its efforts to contribute positively to global peace and security, to establish a harmonious image beyond its borders, to reassure neighbors about its peaceful intentions, and to softly balance US and Western influence, while gradually estab-lishing China's acceptance as a great power (Gill and Huang, 2009: 4). Neverthe-less, Beijing cautiously abstained in the vote on August 31, 2006 on Resolution 1706, which extended the North–South peacekeeping mission, UNMIS, to Darfur, claiming that Khartoum's consent should be included in the resolution otherwise the resolution may be interpreted as permitting troops without Khartoum's prior consent. Following the vote, Beijing was in an awkward situation: it needed Khar-toum's support to be consistent in its stance, but Khartoum continued to reject UN troops in Darfur. Thereafter, and according to a plan worked out by Kofi Annan, the mission became a hybrid one, combining UN and AU troops in order to gain Khartoum's acceptance. At this point, Beijing began actively lobbying Khartoum to accept the hybrid mission (Holslag, 2007: 6).

The US enhanced its coordination with China on Darfur. The US special envoy to Darfur, Andrew Natsios, who had previously served as an adminis-trator of United States Agency for International Development (USAID), flew to Beijing in January 2007, right before President Hu Jintao's trip to Sudan and several other African countries. During his meetings with Chinese counter-parts, Natsios clarified that the US administration did not favor regime change; rather it wanted to avoid the dissolution of Sudan as had happened in Somalia. The main goal of the US was to protect civilians from violence and find a diplomatic solution, which would insert Darfur into the Comprehensive Peace Agreement (Interview with Andrew Natsios, May 18, 2010). As one of his Chi-nese interlocutors recalled, this was the first time he had actually heard a US official say that the US did not favor regime change (Interview with Weizhong Xu, December 3, 2008). Until then, China had seen US policy as confusing and inconsistent, due partly to the fact that different actors in the US held very different opinions.[3]

Natsios believed that if he could gain the trust of his Chinese counterparts, they would accept his information on atrocities committed in Darfur (Interview with Andrew Natsios, May 18, 2010). Natsios also avoided confrontational meetings with his Chinese counterparts because cooperation with China would give the US more leverage in dealing with Khartoum. According to Natsios, "what would make the Sudanese government much more nervous was if they thought we were cooperating with the Chinese....That would put more pressure on them than anything the Chinese would do overtly" (Interview with Andrew Natsios, May 18, 2010). The "China-card", as Natsios called it, would be more effective than any confrontational statements (Interview with Andrew Natsios, May 18, 2010).[4] After meetings with Chinese officials, Natsios felt assured that Washington and Beijing were largely working in concert on Darfur (*iol news,* 2007). As Natsios recalled, the Chinese had realized that their interests were more in line with those of the US than they had previously believed. They had also misunderstood US policy because US advocacy groups had distorted them. In addition, China had also built links in South Sudan (Interview with Andrew Natsios, May 18, 2010).[5] The prevailing wisdom in the US that China changed its Darfur policy due to pressure from civil society groups is therefore not supported by Natsios' account.

In late 2007, when Khartoum still had not accepted the peacekeeping mission, President Hu Jintao publicly criticized Sudan, something China had never done before. His public statements were considered very "forward leaning" in US eyes (Interview with Andrew Natsios, May 18, 2010). In 2007, therefore, China went a step further as it sought to negotiate the terms of the troop deployment and assisted in the creation of a workable road map to achieve tangible progress (Holslag, 2007: 8). In US policy-making circles, military intervention was once more discussed in case Khartoum would not accept the UN hybrid mission (Rice *et al.,* 2006). China certainly preferred to have the UN hybrid mission brought into place under its oversight rather than any kind of US military intervention. The Chinese therefore strongly advised Khartoum to accept Resolution 1706 (Interview with Andrew Natsios, May 18, 2010).

Beijing agreed to operate as a mediator between the US and Khartoum through which the US pressured Khartoum (van der Meulen and van der Putten, 2009: 41). Downs also recalled private conversations with Chinese foreign policy officials who maintained that the US, which usually plays the "bad cop," needed China to assume the role of a "good cop" for progress to happen in negotiations with Sudan (Downs, 2007: 81). Essential for this cooperation was the realization that the interests of the US and China were much more in line than both countries had originally thought, and that they could only safeguard their interests if they worked in concert, ever aware of the triangularity of the relationship.

After 2008, once the UN peacekeeping mission was finally established on the ground, both great powers continued to cooperate over Darfur. Although Natsios' direct successor, Rich Williamson, did not intensify the cooperation with China, his successor, Scott Gration, did. When Gration visited Beijing in May 2009, both countries agreed to work together in regards to Darfur. Gration subsequently stated that China and the US had similar goals and that both countries agreed to integrate their humanitarian activities in the region (Shinn, 2009: 94).

The importance of China's role in the region increased even more after South Sudan's separation from Sudan in 2011. After the country declared independence on July 9, 2011, China sent a message of recognition, saying that the founding of South Sudan was the common aspiration of the South Sudanese people and their own decision (*China Daily*, July 9, 2011).

After achieving formal independence, South Sudan took 75 percent of Sudan's oil fields. It now has the third-largest oil reserves in sub-Saharan Africa, but still relies on Sudan for the pipelines to get its crude to market. A dispute between Juba and Khartoum over pipeline fees led to a 14-month shutdown beginning in early 2012. According to the US Energy Information Administration, crude from South Sudan made up only 1 percent of China's total crude imports in 2012. The year before, South Sudan's crude had accounted for 5 percent of Chinese oil imports (Fortin, 2014). China consumed about 10.7 million barrels per day in 2013, which makes South Sudan's contribution – or lack thereof – a drop in the bucket (Fortin, 2014). On the other hand, China was importing about 77 percent of South Sudan's crude before the conflict began (Fortin, 2014), and therefore was the most important economic partner for South Sudan before the crisis. Although China's economic motivation for intervening in the crisis was not great, Beijing also wanted to be seen as a responsible player by the international community for doing their share to build peace alongside other international actors, particularly the US (Walker, 2012).

Internal conflict in South Sudan added to the crisis between Sudan and South Sudan that started as a political dispute between the new country's president and vice-president. Fighting largely ran along ethnic lines of Dinka (Kiir) versus Nuer (Machar). At least 10,000 people were killed and more than 1.3 million displaced since the start of fighting. China aligned with other actors in the region and planned to send a battalion of troops to join the peacekeeping mission. This would be the first time China had contributed a full infantry battalion of about 850 troops to a UN peacekeeping mission (*South China Morning Post*, June 10, 2014). Building on experience gained in the Darfur crisis, China continues its engagement of an international diplomatic actor in crisis management, establishing itself as one of the main international interlocutors for conflict resolution in the region. Without China, the US will not be able to tackle future regional conflicts.

Conclusion

This chapter has examined the process by which the US and China moved beyond confrontation over Darfur to work in concert on resolving the Darfur crisis. The concrete interests of the Chinese companies played a minor role in the decision making by Beijing. Instead, China's *misperception* of Washington's goals, believing that it wanted to replace the Sudanese government with a friendlier one and thereby getting access to Sudanese oil, played a much more significant role.

Protecting itself and Khartoum against "imperialist" policies was a primary motivation for Beijing's policy choices.

The US special envoy Andrew Natsios shared information and reassured his Chinese interlocutors that the US was not looking for a pretext to achieve regime-change in Sudan, but actually had a serious interest in tackling the Darfur crisis. This knowledge made it easier for China to develop proactive approaches to solving the Darfur crisis. After the US had considerably reinforced communication with China, both countries finally adopted a common policy. China was willing to take the lead in diplomacy for the implementation of the hybrid UN peacekeeping mission and actively sought consensus for the benefit of both countries' interests in regional stability – which in turn fosters energy security.

Throughout this process, both countries learned that the triangular relationship is inevitable. Both countries had to deal with the nature of the other country's relationship to Sudan and both countries depended on each other for a feasible solution to the Darfur crisis. Due to economic sanctions, the US lacked any non-military leverage over Sudan. With its troops deployed in Afghanistan and Iraq, the US administration lacked the will to intervene militarily in Darfur. Cooperating with China, which had enormous economic and political leverage (as a P5 member of the UN Security Council), was the most effective way for the US to make its demands clear to the Sudanese government.

All in all, Darfur is not a case of resource diplomacy but a case of diplomacy under the influence of resource conflict – locally and internationally. If both China and the US are to avoid future conflict and manage humanitarian crises efficiently, they must accept the triangular relationship as a given and resolve their mutual misperceptions and deep-seeded mistrust effectively. In this way they can satisfy their energy needs on the African continent and find constructive ways to help the African continent leave poverty and instability behind.

The Darfur crisis has been a pivotal experience for China through which is has developed its position as a mediator between warring parties. China's intervention proved valuable to US efforts in its search for a solution to the Darfur crisis, an experience that China used in the subsequent conflicts between Sudan and South Sudan and South Sudan's internal conflict.

Notes

1 This chapter is based on research conducted for a PhD thesis entitled "Struggle over Darfur: the development of Chinese security politics in the Darfur crisis in interaction with the United States," completed at Free University of Berlin in September 2012.

2 Darfur is comparable to the movement for self-determination in Aceh, Indonesia, where oil and gas resources were a major catalyst of conflict (Wennmann and Krause, 2009: 5). Likewise, the Niger Delta region, Nigeria's oil belt, has been the site of an ethnic and regional struggle for self-determination since 1998 (Osaghae *et al.,* 2007: 1).

3 The political elite in the US was divided into two groups on the Darfur crisis. On one side were political "realists," such as the State Department, the CIA and the Defense Intelligence Agency. On the other side were rights advocates, such as the pro-Garang Lobby on Capitol Hill and the USAID. John Garang was the leader of the Southern rebels who was murdered in 2005 (Prunier 2008: 138–40).

4 The "China-card" in US politics refers to the question debated during the Carter and the Reagan administrations over whether or not US policy should use improved relations with China as a source of leverage against the Soviet Union.
5 The idea that both governments were more in agreement than the public is also supported by Xu Weizhong, who is a renowned specialist on Sudan, although he might have conceived this idea from Natsios during their meeting in Beijing.

References

Alden, Chris, 2007. *China in Africa.* London/New York: Zed Books.
Ali, Abdalla, 2006. *The Sudanese-Chinese Relations Before and After Oil.* Khartoum: Sudan Currency Printing Press.
Ali, Wasil, 2007. The threat of the ICC to the Sudanese regime. *Sudan Tribune,* 14 January. Available online at: www.sudantribune.com/spip.php?page=imprimable&id_article= 19753 (accessed 4 May 2015).
Behrends, Andrea, 2008. Fighting for oil when there is no oil yet: the Darfur–Chad border. *Focaal–European Journal of Anthropology,* 52: 39–56.
Bronner, Stephen Eric, 2003. American landscape: lies, fears and the distortion of democracy. *Logos,* 2.3 (Summer). Available online at: www.logosjournal.com/bronner.htm (accessed 3 May 2015).
Cabalza, Chester B. and Manalo, Rosario G., 2010. *The Diplomatic Dimension of National Security.* National Defense College of the Philippines, Quezon City.
China Daily, 2011. China recognizes independence of South Sudan. *China Daily,* 9 July. Available online at: www.chinadaily.com.cn/china/2011-07/09/content_12869896.htm (accessed 3 May 2015).
Choucri, Nazli and North, Robert C., 1975. *Nations in Conflict. National Growth and International Violence.* Cambridge, MA: Massachusetts Institute of Technology.
Congress of the United States, Congressional Budget Office (CBO), 2006. *China's growing demand for oil and its impact on US petroleum markets.* Washington, DC: Congress of the United States. Available online at: www.cbo.gov/sites/default/files/cbofiles/ftp-docs/71xx/doc7128/04-07-chinaoil.pdf (accessed 3 May 2015).
Daly, M. W., 2007. *Darfur's Sorrow – A History of Destruction and Genocide.* New York, NY: Cambridge University Press.
de Waal, Alex, 2007. Darfur and the failure of the responsibility to protect. *International Affairs,* 83 (6): 1039–54.
Downs, Erica S., 2006. *China.* The Brookings Foreign Policy Studies Energy Security Series, (December): 1–67. Available online at: www.brookings.edu/~/media/Files/rc/reports/2006/12china/12china.pdf (accessed 3 May 2015).
Downs, Erica S. 2007. The fact and fiction of Sino–African energy relations. *China Security,* 3 (3) (Summer): 42–68. Available online at: www.brookings.edu/~/media/research/files/articles/2007/6/summer-china-downs/downs20070913.pdf (accessed 3 May 2015).
Fortin, Jacey, 2014. China in the middle: South Sudan's biggest oil importer learns to wield its clout. *International Business Times,* April 9. Available online at: www.ibtimes.com/china-middle-south-sudans-biggest-oil-importer-learns-wield-its-clout-1568133 (accessed 3 May 2015).
Gill, Bates and Chin-Hao Huang, 2009. *China's expanding peacekeeping role: it's significance and its policy implications.* Stockholm International Peace Research Institute (SIPRI), Stockholm.

Hansen, Ketil Fred, 2011. *Chad's relations with Libya, Sudan, France and the US.* Norwegian Peacebuilding Resource Centre, April 15. Available online at: www.peacebuilding. no/layout/set/print/Regions/Africa/Publications/Chad-s-relations-with-Libya-Sudan-France-and-the-US (accessed 3 May 2015).

Hatemi, Peter and Wedeman, Andrew, 2007. Oil and conflict in Sino–American relations. *China Security,* 3 (3): 95–118.

Herberg, Mikkal, E., 2007. China's energy consumption and opportunities for US–China cooperation to address the effects of China's energy use. Testimony before the US–China Economic and Security Review Commission. June 14. Available online at: www.uscc. gov/Hearings/hearing-china%E2%80%99s-energy-consumption-and-opportunities-us-china-cooperation-address-effects (accessed 3 May 2015).

Holslag, Jonathan, 2007. China's diplomatic victory in Darfur. *Brussels Institute of Contemporary China Studies (BICCS) Asia Paper,* 2 (4): 1–12. Available online at: www.vub.ac.be/biccs/documents/Holslag,%20Jonathan%20%282007%29,%20China%27s%20Diplomatic%20Victory%20in%20Darfur,%20Asia%20Paper%202%20%284%29,%20BICCS,%20Brussels%5B1%5D..doc.pdf (accessed 3 May 2015).

Huang, Chin-Hao, 2007. US–China relations and Darfur. *Fordham International Law Journal,* 31 (4): 827–42.

Human Rights First, 2008. *Investing in Tragedy – China's Money, Arms and Politics in Sudan.* New York; Washington: Human Rights First. Available online at: www. humanrightsfirst.org/wp-content/uploads/pdf/080311-cah-investing-in-tragedy-report. pdf (accessed 3 May 2015).

International Crisis Group (ICG), 2008. China's thirst for oil. Asia Report No. 153. June 9. Available online at: www.crisisgroup.org/~/media/files/asia/north-east-asia/153_china_s_thirst_for_oil.ashx (accessed 3 May 2015).

iol news, 2007. China assumes positive role in Darfur. *iol news,* January 12. Available online at: www.iol.co.za/news/africa/china-assumes-positive-role-in-darfur-1.310714 (accessed 3 May 2015).

Jiang, Julie and Sinton, Jonathan, 2010. Overseas investments by Chinese national oil companies. Assessing the drivers and impacts. Information paper, Standing Group for Global Energy Dialogue. International Energy Agency. September 25. Available online at: www.iea.org/publications/freepublications/publication/overseas_china.pdf (accessed 3 May 2015).

Kleine-Ahlbrandt, Stephanie and Small, Andrew, 2008. China's new dictatorship diplomacy: is Beijing parting with pariahs? *Foreign Affairs,* 87 (1).

Lederer, Edith M., 2005. United States propagates UN sanctions against Sudan. *The Associated Press,* March 23.

Leopold, Evelyn, 2005. Annan calls emergency Sudan session of UN council. *ReliefWeb,* March 7. Available online at: http://reliefweb.int/report/sudan/annan-calls-emergency-sudan-session-un-council (accessed 3 May 2015).

Mamdani, Mahmood, 2009. *Saviors and survivors. Darfur, politics, and the war on terror.* New York, NY: Pantheon.

Nabati, Mikael, 2004. The UN responds to the crisis in Darfur: Security Council Resolution 1556. The American Society of International Law (ASIL). Available online at: www.asil.org/insights/volume/8/issue/18/un-responds-crisis-darfur-security-council-resolution-1556 (accessed 3 May 2015).

Osaghae, Eghosa, Ikelegbe, Augustine, Olarinmoye, Omobolaji, and Okhonmina, Steven, 2007. *Youth militias, self-determination and resource control struggles in the Niger-Delta region of Nigeria.* Leiden, Netherlands: Leiden University, African Studies Centre.

Patey, Luke A., 2006. *A complex reality: the strategic behaviour of multinational oil corporations and the new wars in Sudan.* Copenhagen: Danish Institute for International Studies. Available online at: www.isn.ethz.ch/Digital-Library/Publications/ Detail/?id=19131&lng=en (accessed 3 May 2015).

Permanent Mission of the People's Republic of China to the UN, 2004. Explanatory Remarks by Chinese Permanent Representative Mr. Wang Guangya at Security Council on Sudan Darfur Draft Resolution. September 18. Available online at: www.china-un. org/eng/xw/t158034.htm (accessed 3 May 2015).

Prunier, Gérard, 2008. *Darfur. A 21st Century Genocide.* Ithaca, NY: Cornell University Press.

Rice, Susan E., Lake, Anthony and Payne, Donald M., 2006. We saved Europeans, why not Africans? *The Washington Post,* October 2. Available online at: www.washingtonpost. com/wp-dyn/content/article/2006/10/01/AR2006100100871.html (accessed 3 May 2015).

Sander, David E., 2006. China's oil needs are high on US agenda. *New York Times,* April 19.

Sanders, Edmund, 2007. Darfur oil seen as a curse now, blessing later. *Sudan Tribune,* March 3. Available online at: www.sudantribune.com/spip.php?article20564 (accessed 3 May 2015).

Shichor, Yitzhak, 2010. Influence sans ingérence: les relations de la Chine avec le Soudan. *Les Temps Modernes,* 657 (Janvier–Mars): 54–72.

Shinn, David H., 2009. China and the conflict in Darfur. *Brown Journal of World Affairs,* 16 (1): 85–100.

South China Morning Post, 2014. China adopts a more hands-on approach to the conflict in South Sudan. *South China Morning Post,* June 10. Available online at: www.scmp. com/news/china/article/1528859/china-adopts-more-hands-approach-conflict-south-sudan (accessed 3 May 2015).

Stedjan, Scott and Thomas-Jensen, Colin, 2010. The United States. *In:* David R. Black, R. David and Paul D., Williams, eds. *The International Politics of Mass Atrocities. The Case of Darfur.* New York, NY: Routledge, pp. 157–75.

Swilla, Nelly, 2004. *The threat of international sanctions on Sudan's oil sector. How feasible? What likely impacts?* Center for Strategic and International Studies. Africa Notes 24, December. Available online at: http://csis.org/files/media/csis/pubs/anotes_0412.pdf (accessed 3 May 2015).

United Nations Peacemaker, 2005a. Darfur Peace Agreement. Available online at: http:// peacemaker.un.org/node/535 (accessed 3 May 2015).

United Nations Peacemaker, 2005b. Section 5 of the Protocol on Wealth Sharing. Available online at: http://peacemaker.un.org/sites/peacemaker.un.org/files/SD_060000_The%20 Comprehensive%20Peace%20Agreement.pdf (accessed 3 May 2015).

United Nations Security Council, 2005. Security Council refers situation in Darfur, Sudan, to prosecutor of International Criminal Court. Press Release SC/8351, March 31. Available online at: www.iilj.org/courses/documents/SecurityCouncilResolution1593.pdf (accessed 3 May 2015).

van der Meulen, Emma and van der Putten, Frans-Paul, 2009. *Great powers and international conflict management. European and Chinese involvement in the Darfur and Iran Crises.* Netherlands Institute of International Relations Clingendael (January): 1–46. Available online at: www.clingendael.nl/publications/2009/20090225_cscp_sec_ paper_great_powers.pdf (accessed 3 May 2015).

Walker, Beth, 2012. China's uncomfortable diplomacy keeps South Sudan's oil flowing. *chinadialogue.* Available online at: www.chinadialogue.net/article/show/single/en/5378-China-s-uncomfortable-diplomacy-keeps-South-Sudan-s-oil-flowing (accessed 3 May 2015).

Wennmann, Achim and Krause, Jana, 2009. *Managing the economic dimensions of peace processes: resource wealth, autonomy, and peace in Aceh.* Geneva: Centre on Conflict, Development and Peacebuilding (CCDP).

Ziegler, Charles E., 2006. The energy factor in China's foreign policy. *Journal of Chinese Political Science,* 11 (1): 1–23.

Zoellick, Robert B., 2005. Whither China: from membership to responsibility? Remarks to National Committee on the United States and China Relations. New York City, September 21. Available online at: http://2001-2009.state.gov/s/d/former/zoellick/rem/53682.htm (accessed 3 May 2015).

Zygar, Mikhail, 2004. Behind the UN Security Council Resolution: Chinese, Russian and Indian oil interests in the Sudan. *GlobalResearch,* September 19. Available online at: www.globalresearch.ca/behind-the-un-security-council-resolution-chinese-russian-and-indian-oil-interests-in-the-sudan/612 (accessed 3 May 2015).

13 Resource diplomacy under hegemony

The peculiar case of Venezuela under the Bolivarian Revolution

Cynthia Watson and David Zweig

One of the most vexing problems for the US administrations of George W. Bush and Barack Obama has been the government in Caracas, Venezuela. A vital source for petroleum for Washington, the northern republic of South America has been an increasingly volatile nation since the end of the Cold War because of gross political ineptitude on the part of the two traditional governing parties. At least as important has been the proclaimed desire by the previous and current Venezuelan presidents to shift Venezuela's international relations away from their historic Washington-first position.

Venezuela's increasing links to the People's Republic of China (PRC) have worried some in the US, although it is merely a manifestation of an increasingly developed and energy-hungry China protecting its national interests. Those concerned about the evolving relationship point to the broadening nature of the ties beyond energy to military, economic, and overall strategic cooperation. President Chávez Frías's actions, leading to the fairly common designation of 'strategic partnership' with the PRC, have made Venezuela a notable asset for China, while making many in Washington view the country as a 'pariah'. His successor, President Maduro Moros's behaviour has shown the difficulties that the energy relationship can produce.

Chávez Frías' election in 1998 marked a new era for Venezuela and coincided with changes in the nation's relations with Washington and Beijing. This paper explores those changes within the context of the triangular relationship among the US, China and Venezuela. The hypothesis for the chapter is that the Venezuelan government's desire to engage in anti-US behaviour is the driving factor in the relationship, rather than competition between China and the US for hegemony. The Venezuelan case appears unlikely to become a 'source of conflict' in the Sino–US relationship because the driving factor in this triangle is the Venezuelan government's own desire to enhance its Bolivarian character, rather than Beijing's desire to pull Caracas away from the US or to leverage Venezuela in its own challenge to the US for regional dominance. In fact, although willing to accept Venezuelan oil, and is amenable to enhanced ties, China remains cautious in its dealings so as to avoid any confrontation with the US over this Resource Rich State (RRS).

Venezuela in the old days: 1900–1998

Located in north-eastern South America, Venezuela was a relatively remote back-water of the Spanish empire, and was part of the newly independent República de Gran Colombia until it became independent in 1830. In the first three-quarters of a century of independence, the somewhat isolated republic suffered a series of *caudillos* – strong-willed individuals seeking power through the gun and personal persuasion – rather than the ballot box. Although petroleum's discovery along the Colombian border at the beginning of the twentieth century brought the country greater international attention, the authoritarianism of the nineteenth century survived through the dictatorship of General Marcos Pérez Jiménez. His rule ended in 1958 with the implementation of the National Front government, leading to the emergence of two powerful political parties in the Republic.

The 40 years of democracy in the second half of the twentieth century coincided with Venezuela's rise as a prominent petroleum-producing state and the dramatic influx of revenue into the country. Many analysts credit Venezuela as the driving founding state for the Organization of Petroleum Exporting Countries (OPEC) in the 1960s. The National Front governments of the Acción Democrática (AD) and Partido Social Cristiano de Venezuela (COPEI) proved superb at retaining power, albeit through alternating their tenures, whilst showing almost kleptocratic behaviour about national assets from petroleum revenues. While the petroleum revenues flowed in, domestic expenditures for infrastructure and social development stalled, and despite the state having a relatively small population by Latin American standards, the vast wealth did not increase the standard of living. In 1980, for example, Venezuela had a population of roughly 12 million people, but close to a third were Colombians who had fled across the Orinoco River boundary to find a better standard of living and to flee the violence plaguing Colombia over the prior 30 years.

In the 1980s, often referred to as the 'Lost Decade' for the region, Venezuela had a staggering foreign external debt in the range of US$50–60 billion (Sachs *et al.*, 1988). For a good portion of the decade, Venezuela's petroleum industry produced ever-growing quantities of petroleum for export, with an overwhelming portion going to the US. In the eyes of the US, politics in Venezuela and the relationship with Washington appeared firmly anti-communist, pro-free growth and a model of how democracies could be 'grown' after decades of harsh, ineffectual one-man rule. The fact that the Venezuelan democracy had beaten back a Fidel Castro-inspired insurgency in the 1960s was icing on the cake, leading many scholars and international affairs analysts to see Venezuela as the quintessential (US-driven) success in Latin America. One of the authors had a conversation met with a prominent Latin American scholar in 1995 who had more than 35-years experience studying the region. In relaying his analysis when he returned to the US after a trip to Caracas, the scholar voiced shock at the deteriorating political conditions there: 'But we thought they were the model! We thought they had solved the problems of democracy' (personal communication, 1995). Many scholars watching Venezuela closely had begun to see this disconnect growing slowly but inexorably, although no one really thought democracy was actually on the ropes.

In 1989, under President Carlos Andrés Pérez, energy-wealthy Venezuela had to appeal to the International Monetary Fund (IMF) for assistance, which allowed the IMF to insist that Caracas embrace the economic orthodoxy requiring the government to reduce subsidies for the range of goods it supported. The responding February 1989 *'caracazo'*, a widespread violent incident in Caracas, led to the deaths of 200 citizens and fuelled public anger that a state with the wealth of Venezuela could be forced to act upon IMF requirements. This massive social unrest, distrust – and ultimately impeachment and house arrest – of President Carlos Andrés Pérez, opened the door to new actors on a political scene that had been dominated for more than a generation by a small number of wealthy men from the two dominant parties. The petroleum curse, so common to states in the Middle East and Africa where oil proved destabilizing rather than helpful to regimes, was an obvious problem for Venezuela. The expectations that the resources of the subsoil would benefit the average citizen were proving untrue. The resources seemed only to be helping a small elite bent on retaining power, which ignored the rest of the country.

In 1992, the long docile Venezuelan military launched two *golpes de estado*, or armed overthrows, against the civilian, democratically elected governments. Army officers, led by Lt Col Hugo Chávez Frías, instigated the first attempt in February with the argument that the military had to defend the nation against a political system that would place the interests of the global economic system over those of the Venezuelan citizenry, as the 1989 austerity program illustrated. These coup-makers were taking on the traditional Latin American perspective that they needed to protect the *patria* against foreign exploitation. Chávez Frías went to prison for two years, but assumed great national prominence upon his release because of his nationalist argument that the leadership of the international economic system undermined the Venezuelan people. The outside demands contributed to Venezuelans seeing their standard of living decrease despite almost a century of petroleum wealth coming out of their soil. He advocated 'gradually eliminat[ing] the savage system of income distribution that exists in Venezuela, that progressively eliminates the great difference that exists between a minority that has everything and a huge majority that has virtually nothing except hope' (McCoy and Neuman, 2001: 80).

In 1998, Chávez Frías ran what initially appeared an unlikely ultra-nationalist campaign for presidency but won in December that year. In campaigning, Chávez Frías portrayed himself as a champion of the people, an adversary for those (somewhat unspecified) who sought to exploit Venezuela, and an architect of a new political, legal, social and economic path for the Republic. Soon after assuming the sash of power, he adopted the motto of instituting a 'Boliviarian Revolution', named after the independence leader, Simón Bolívar, credited with liberating northern Latin America from Spain. Chávez Frías's appeal to the downtrodden, who were subjects in his words of outside imperialist aggression and perpetrated by those seeking to steal these subsoil resources, provided him with sufficient votes to overturn the tired COPEI–AD monopoly on politics in Venezuela. Under both colonial Spanish and then independent Latin American law, the resources under the ground are subsurface assets that the government holds on behalf of the population, not the assets of an individual landowner who may claim to own a parcel above ground. The collective rights of the *pueblo* trump those of any individual landowner or profit-making entity.

Upon assuming office, Chávez Frías vowed to help the disadvantaged, the exploited and the masses of Venezuelans without a firm plan but with much fiery rhetoric and petroleum resources. As one reporter noted, Chávez Frías 'had nonetheless struck a chord in a country whose political elite seemed to have lost its way and become a self-perpetuating clique, mired in corruption' (Gunson, 2006: 58), rather than having a clear cut set of policy changes to implement. Further, '*Chavismo* took root in response to the undeniable fact that the increasingly impoverished mass of the Venezuelan population was excluded from any meaningful participation in the political, social, and economic life of the country' (Gunson, 2006: 60).

The new regime

President Chávez Frías did not initiate diplomatic relations with China – that had occurred when Venezuela transferred its embassy from Taipei to Beijing in 1974, at a time when most states in the region followed suit. The ties between the two states were not particularly important, however, because only three state visits had transpired by the turn of the twenty-first century (Ellis, 2009).

The cool relations had two sources. First, between 1978 and 2000, Beijing focused almost entirely on domestic developments following the end of the Cultural Revolution, concentrating on exploitation of rapidly increasing foreign investment through the Pearl River Delta and other special economic zones to accelerate economic growth. Also, in this initial period of reform, China sought technology from advanced states, having little interest in developing countries, such as Venezuela. Second, the two traditional parties in Venezuela vowed nationalist proclamations about protecting petroleum assets. Thus, as the 28 December 2010 obituary for Carlos Andrés Pérez in the *Financial Times* noted, he had used strong anti-foreign rhetoric during his late 1980s presidency, but had also turned to the outside assistance in 1989 that led to the Venezuelan riots over ending subsidies (Rathbone, 2010). The truth, however, was that the ties between Caracas and Washington were firm. Washington took this relationship for granted, seeing it as successfully guaranteeing access to cheap petroleum from a 'long-standing democratic ally' – Venezuela. In fact, few US analysts questioned the strength of Venezuela's democratic institutions. Similarly, governments in Caracas saw no reason to look elsewhere for petroleum sales.

Chávez Frías had a different goal – he was interested in eradicating the long-standing 'dependence' on sales to the US. He wanted to aggressively and deliberately find alternate partners who, in his eyes, would not cheapen Venezuela's position as a sovereign state, as his predecessors had allowed in the relationship with Washington. China ultimately became the most logical state in his plans for several reasons.

Almost immediately upon taking the oath of office, Chávez Frías reached out to several controversial leaders, including Saddam Hussein, and sought partnerships with countries that might antagonize the US, such as China. Chávez Frías travelled to China in the autumn of 1999. More noteworthy was that Chinese President Jiang Zemin, who was on a trip through Latin America, kept Caracas on the itinerary, even as the E-P3 incident reached resolution in April 2001, illustrating the growing importance of Venezuela in China's decision-making. In his December 2004 visit to China, Chávez Frías expressed his willingness to work with China in a provocative way that fed US concerns. Addressing a meeting of Chinese businessmen in Beijing, Chávez said:

> We have been producing and exporting oil for more than 100 years, but these have been 100 years of domination by the United States. Now we are free, and place this oil at the disposal of the great Chinese fatherland.
>
> (Cited in Forero, 2005)

The increasing ties went beyond presidential visits as Defense Minister Chi Haotian visited Venezuela in August 2001 promoting both friendly relations and military cooperation between the two states.

China appealed to the Venezuelan leader because it was a guaranteed source of demand for petroleum, making it a counter-weight to the US. China was also a relative novice in this region, rather than a tired partner over the past centuries, as was the case with the US, the UK and Spain. Chávez Frías also hoped that China would actively assist him in his conflict with the US. In August 2006, on his weekly radio talk show 'Alo Presidente', President Chávez announced that 'China will help us', thereby welcoming China's role in Latin America and its rise in making the world more multi-polar.

Energy diplomacy: pulling closer to China

Natural resource exploitation remains the key to Venezuela's economy, with the petroleum sector providing 90 per cent of the export earnings and 30 per cent of the gross domestic product in 2009. While the Venezuelan government privatized much of the industry in the 1970s, Chávez Frías renationalized most of it under a 2001 Hydrocarbons Law. The only exceptions to state control were joint ventures for 'extra-heavy crude oil production' (US Energy Information Administration [EIA], 2014). This type of petroleum is harder to refine, but Venezuela has long believed it has a much greater reserve than most of the rest of the world in this form of oil, giving it a particular value for joint ventures in the future. The state that has become a favourite partner for these joint ventures is the PRC.

Venezuela has been producing petroleum since World War I, much of it going to the US because of the its refining capabilities for light crude, which was where the original reserves were exported to. With the advanced exploration equipment of the past decade, Venezuela's heavy crude reserves in the Orinoco area have begun to change the nation's profile in petroleum. Venezuela is now considered to have the second largest reserve of petroleum in the world, largely as a result of the more recently discovered extra heavy crude. The overwhelming majority of the exports, however, still go to the US.

Latin America's disappointment with the George W. Bush administration was deep and profound – although it had begun with enthusiasm that this new president, a Texan with knowledge of the bordering Latin American states, would appreciate the region. Two events in the early 2000s, however, generated fundamental distrust of the Bush administration's intentions in Latin America. First, Washington rejected an economic bailout when Buenos Aires confronted a politico-economic meltdown only six months after the Secretary of Treasury, Paul O'Neill, had noted publicly that Argentina needed to take reform seriously. The resulting inability to meet its national responsibilities threw millions of Argentines – a population living in a country that a hundred years earlier had been the most optimistic and best endowed nation in the region – into chaos because lines of panicked depositors could not get their assets from the banks. Second, Argentina went through five governments in ten days as the Republic closed out 2001, with much blame aimed at Washington for its heartless policies. This crisis occurred almost at the same time that the Bush government began to invest and encourage the rest of the world to provide enormous assistance to the new Karzai-led Afghanistan in the months shortly following the 9/11 attacks. As attention moved sharply to the Islamic world, Latin America saw its prominence as a partner region for the US disappear.

In general, Latin Americans strongly opposed the 2003 decision to invade Iraq because the southern states have traditionally been bulwarks in opposing violation of any state's sovereignty. The region's experiences, particularly in the nineteenth and twentieth centuries at the hands of perceived imperial interventionists in London and Washington, make defending national sovereignty one of the strongest and most viscerally held beliefs in the region.

The Bush administration's impact on Venezuelan, however, lingered longer than with any other state in the hemisphere, primarily because of Chávez Frías's impassioned views of the bilateral relationship and his overwhelming desire to rebalance his country's ties with Washington. The problems between Washington and the region's capitals took its most sustained hit in Caracas, bottoming out in April 2002 when Chávez Frías and a substantial portion of Venezuelans believed that Washington was behind a *golpe de estado* (coup d'état) against the president. Some of the Venezuelan military were encouraged by a chaotic scene of street protests and ousted Chávez Frías. They named business leader Pedro Carmona, who was not in the constitutional chain of leadership, as temporary president. Chávez Frías survived the initial house arrest and ultimately forced those armed forces loyal to him to reject the successor, who had temporarily claimed to be President of the Republic. Chávez Frías and his supporters assumed the Bush administration had orchestrated the event, making Washington the target of massive criticism throughout the region. Venezuela promoted this episode as another example of Washington talking about democracy even although it undermined democracy by putting regimes into positions of power that supported its hegemonic aspirations.

More relevant, suspicions of US instigation of an anti-Chávez Frías coup in April 2002 cast a spotlight on Washington's 'imperialist' actions because this obvious undermining of Venezuela's legitimate rights occurred as the largely unilateral US invasion of Iraq became an unavoidable topic throughout that year. Latin America has long held the same views as China regarding the need to keep outsiders from dictating the future of a society and not allow them to humiliate weak states.

The criticism of imperialistic politics following the invasion of Iraq roughly coincided with a significant increase in Chinese involvement in Venezuela. Chinese inroads into Latin America had gradually increased in the 1990s as the number of military and government visits increased after the middle of the first decade of the twenty-first century (*China Daily*, 2009). Trade and economic inter-action accompanied these official interactions as China reached out to the international community more actively.

More importantly for this analysis, China's need for minerals and energy to perpetuate the double-digit economic expansion, characteristic of contemporary China since the late 1970s, expanded dramatically. Press accounts worldwide have stressed China's desire to lock up petroleum resources around the globe, focusing on Venezuela as a state where it could apply 'resource diplomacy'. China's increase in petroleum correlates nicely to the improved ties with Caracas.

The increasingly vitriolic relationship between Chávez Frías and the White House, especially during the Bush administration, is noteworthy because some critics have accused Beijing of pressuring the Venezuelans to export oil to China instead of Washington. Although this chapter argues that Venezuelan leaders adopted that position on their own, the petroleum from Venezuela does present an oddity in China's quest for unfettered access to resources. Venezuelan heavy crude is not compatible with current PRC refining technology, and the petroleum connections being made could therefore be even more important in the future.

The current Venezuelan government sees this idea as a highly desirable position. Opening the door to long-term links to China will provide a guaranteed alternative market for petroleum that deliberately excludes the US, thus making a statement about the US's inability to dominate Venezuela (and Latin America). China has a series of high profile agreements with Venezuela for petroleum ventures, but only a single major joint venture in the lucrative Orinoco Belt. That project involves a bilateral agreement between China National Petroleum Company (CNPC), with a 40 per cent share, and Venezuelan national petroleum company, PDVSA. It was to begin in 2012 in the Junin-4 oil field, which should produce 400,000 barrels per day.

Existing and growing ties

Official government and military visits occur, along with other high profile interactions, such as petroleum-backed development lending through the China Development Bank. These loans have continued apace throughout the decade, leading to a much broader relationship than a few years ago, although the starting point was low. As Downs (2011) notes, Venezuela is a regime with an avowed position of opposing the US, and yet the decision-making in Beijing is not unalloyed strategic opposition to US interests. As is true with many trends in China's expanding global involvement, the effects and the motivations are more mixed than singularly positive or negative for China or for the US.

The problems that Venezuelan presidents face, however, hinder their efforts to deepen ties with the PRC. The growing discontent with Chávez Frías, and now with his successor, may give Beijing pause in its desire to treat Venezuela more closely than other more stable partners in Latin America, most notably Brazil and Chile. Although the levels of resources sold to China by these two states remains limited, longer-term ties appear sound in a way that is not true with Venezuela (see Chapter 10, this volume).

Even more dramatically, Chávez Frías's bout with cancer raised doubts about his ability to make his intended shift in energy relations from Washington to Beijing permanent. His frequent absences from Venezuela during the last year of his presidency, along with the admitted need for follow-up medical visits, led many in the Republic to scramble to re-evaluate, if not lead, the political system. More importantly, however, was his government's inability – along with an even more obvious failure by his successor over the first twelve months of his rule – to deliver meaningful improvements in the standard of living for the majority of Venezuelans, leading to a growing sense of crisis in the political landscape of the nation (Toothaker, 2011).

The implications of China–Venezuela–US relations over resources

China is using Venezuela as a tool against US hegemony, but the driver of this triangular relationship is Venezuela, whose leaders need the proof of growing ties to China as leverage to maintain their successive regimes in the face of mounting impatience with their domestic failures. The US remains the primary purchaser of Venezuelan petroleum, accounting for 924,000 barrels per day of crude oil (EIA, 2011), but these levels have fallen to the lowest level since 1972 (Blas, 2012). The decline in exports also relates to the deteriorating conditions in Venezuela's petroleum industry, as well as because of a shift from exports to the US to those aimed at the Middle Kingdom. China's petroleum from Venezuela has not grown as substantially as the rhetoric of the bilateral relationship – or, to put it another way, the massive Chinese takeover of Caracas feared by some quarters in Washington simply has not occurred. The petroleum associated with the energy-based loans that China provided Venezuela with between 2010 and 2014 was roughly half the size of China's energy-based loans to Russia, for example, even though Venezuela is the greatest recipient of these loans (Downs, 2011).

Counter-factual evidence that China is not the driver in this relationship is provided by China's relationship (or lack thereof) with Venezuela's neighbour, Colombia. Colombia's history as a nation-state is quite different from Venezuela's, although both states have been known broadly as 'democratic' states since 1958. Unlike Venezuela's greater tendency towards single man rule, Colombia has a long tradition of elected central governments, despite other conditions in the Republic.

What must be explained is why Beijing has shown curiously little interest in Colombia's petroleum, including a massive field discovered in the 1990s, precisely during the same period that the Chávez Frías government sought to increase bilateral ties with China. One explanation could be that Colombia's petroleum fields and pipelines lie largely in an area where the government has had limited sovereignty and where the smaller of two long-standing guerrilla groups, the *Ejército de Liberación Nacional* (ELN), was heavily engaged. Open sources indicate little, if any effort, on Beijing's part to cultivate ties with the ELN or to try to build notably better petroleum relations with Colombia.

This oddity may indicate that Beijing respects Washington's firm commitment to Bogotá through multi-billion dollars provided under Plan Colombia (1999–2007) or to President Álvaro Uribe Vélez (2002–2010), who was known as President George W. Bush's closest ally in the region. Perhaps Beijing preferred to wait to cultivate better resource ties with the Juan Manuel Santos, who assumed the Colombian presidential sash in August 2010.

Alternatively, Beijing may have dismissed Colombia as simply too close to Washington's orbit, while Caracas offers an obvious choice of a state seeking to distance itself from the US. If true, the key aspect of that condition is driven by Venezuela's own initiative to move away from Washington, rather than China's unilateral power play to alter the current balance in the international system. Even after Santos moved Colombia to a more centrist position, Beijing is still keeping its distance, particularly when compared to a number of other governments in the region, such as Bolivia, Ecuador and Venezuela where the leaders build their political base on opposing Washington and embracing a completely new partner – China.

Many analysts have argued that China's long-term petroleum relationship with Venezuela has exacerbated the problems between Washington and Caracas, highlighting the US's deteriorating position in the region (Ellis, 2010). Still, the PRC's ties with Venezuela do not appear on a trajectory towards conflict with Washington, although problems between the US and the president of Venezuela seem to grow daily.

For example, after Chávez's 2004 visit to Beijing, Congress called on the Congressional Research Service to study the impact on the US of a cut-off of Venezuelan oil. Since then, the US has imposed an arms ban on Venezuela, prohibiting the sale of US defence articles and the licensing of services. On 2 March 2006, the US House of Representatives International Relations Committee Subcommittee on Western Hemisphere held a hearing entitled 'Energy Security in the Western Hemisphere'. At the hearing, Dan Burton, Chairman of House Subcommittee asserted that the US must 'always look at Latin America in relation to the Monroe Doctrine, we have concerns: Chávez, Castro, Ortega, Morales in Bolivia, and their connections with communist China' (Hawksley, 2006). Some at the hearing said China's efforts in the region meant that the US would face growing competition from energy-hungry nations and could not take Western hemisphere energy for granted.

At the beginning of 2011, Chávez Frías stated that he would not accept the Obama nominee for ambassador, Larry Palmer, was seen as only further evidence of Caracas seeking to push Washington away. President Maduro Moros has continued pushing US diplomats away by expelling several during the street protests in February/March 2014 and he continues to blame foreign intervention, with indirect but clear allusions to Washington, with fostering the economic and political problems plaguing his regime.

The effects of the Sino–Venezuelan axis

Chávez Frías sought to strengthen relations with China to reduce the traditional reliance of ties with the US. China began to increase ties as it sought to diversify its sources of petroleum imports, although the density of Venezuela's petroleum was not useful for China immediately. The Chinese recognized that building ties with Caracas would be beneficial once it could reconcile the orimulsion (heavy) product from South America, which is quite different from the product refined for Chinese use. The recognition of this longer-term goal of gaining sustained access to one of the world's second largest reserve bodies of petroleum illustrated Chinese motives to think several steps ahead for the future.

Chávez Frías sought much greater involvement by Chinese firms, as he proclaimed publicly often. Sometimes those proclamations were made whilst he was visiting China or hosting reciprocal delegations from the Middle Kingdom to Venezuela, while other comments came from his weekly radio broadcasts to promote the Bolivarian cause at home. When Chávez Frías announced his desire in 2007 to raise China's participation in the Venezuelan petroleum industry to the top position, Beijing responded favourably, leading to obvious fulfilment of the Chinese declaration made the prior year of a 'strategic partnership' between the two states.

Along with the declared elevation of bilateral ties, Chinese leaders welcomed Chávez Frías visiting the Middle Kingdom, which he did several times during his tenure as president. Additionally, military-to-military ties broadened somewhat; China welcomed Venezuela's military for training and education because the ties with Washington had frayed considerably. Venezuela also began buying some military platforms from China, such as the sale of K-8 fighters in 2008 (Ellis, 2009).

For the US, however, Venezuela has remained a major source of petroleum throughout the period of Chinese involvement (2000–present), although the number of annual barrel exports to the US has decreased to roughly 730,000 barrels per month from the 2008 levels of more than 880,000 barrels (*Latin American Herald Tribune*, 2014). The exports remain significant because the US receives more Venezuelan petroleum than any other state, even though China has been a growing part of Venezuela's plans for a decade.

Venezuela's greater ties with China resulted from the China loaning Caracas money with the intention of firming up future relations. At least one long-standing analyst, R. Evan Ellis, opines that the Venezuelan situation in 2012 necessitated the US$40 billion in loans that Caracas received from the China Development Bank (Ellis, 2013). China's loans, important in the absence of such support from traditional US ties, became all the more notable as Venezuela's economy continued sliding.

Post-Chávez Frías

Chávez Frías first sought treatment for cancer during 2011 in Cuba, but despite assurances that the treatment had arrested the disease, he succumbed in March 2013. Chávez Frías had stood for re-election as president of the Republic in 2012 and won, but his health had never allowed an inauguration to take place, and he passed away on 5 March 2013 without having retaken the Oath of Office. Before passing away, he selected a relative unknown and political novice, Nicolás Maduro Moros, to serve on his 2012 ticket at vice president. Maduro Moros vowed, upon assuming office in 2013, to continue Chávez Frías's Bolivarian revolution and to strengthen ties with China, among other tasks. Many Chávez Frías critics believed Maduro Moros would pick up where the ailing Bolivarian left off without any troubles because Chávez Frías had been so public in anointing his successor. On the surface, this expectation was unrealistic for a number of reasons. Maduro Moros had little prior experience in politics, giving him little probability of navigating the political system that Chávez Frías left behind, which had many democratic trappings. Maduro Moros also lacked the political and personal charisma that characterized the former Venezuelan paratrooper, who proved able to take the national political 'temperature' when he came out of prison in the mid-1990s when the two traditional parties were falling further out of favour with the public. Finally, Maduro Moros could not fall back upon signature improvements in the standard of living for those *venezolanos* that Chávez Frías left behind after 15 years in power. Instead, the rhetoric that had long damaged the relationship with the US and declarations of an improved set of ties with China had proven a chimera for improving the standard of living. Most Venezuelans saw only empty promises and grew less patient as the weeks passed.

With Chávez Frías's passing, Nicolás Maduro Moros did assume the presidency in March 2013 as constitutionally mandated, but his experience – and that of the Bolivarian Revolution – continues to be an exceedingly rough one for the nation. Protests over a variety of programs and the inexorable slide into economic decay and chaos continues. Two years after assuming office, the new chief of state has proven unable to galvanize the popular support characterizing Chávez Frías's tenure in power. As of mid-March 2015, we see Maduro Moros becoming more authoritarian as he seeks to stay in office because his government is proving incapable of providing any semblance of growth or hope for the nation. At the same time, he is further splitting the country in two, with on half growing increasingly impatient with him and the other half blaming the *yanqi* power in Washington for a variety of woes, ranging from murdering Chávez Frías to deflating petroleum prices, which adversely affect Venezuela.

Curiously, China is hardly part of the discussion. All this chaos is occurring in Venezuela as China is increasingly consumed by its own domestic concerns (slowing economic growth, widespread corruption and environmental decay, amongst other issues) and with a host of tense relations in the Asian theatre. By all indications, China is taking a guarded and distant look at the deteriorating developments confronting Caracas, and does not appear on the verge of 'saving' the Bolivarian Revolution or its leadership.

Despite deepening Sino–Venezuela relations, Washington's concerns about China's activism in its backyard have subsided even further for several reasons. Most importantly, whilst Chávez Frías reduced petroleum exports to the US, the explosion of fracking technology in the high plains of the US has led to expectations in some quarters that the nation will soon be 'energy independent'. This dramatic change in the energy context curtails the US anxiety significantly in a Venezuela with increasing business ties to Beijing. The US's indifference to these deepening ties, however, belies the Chinese hypothesis that the US would use China's energy needs to undermine its rise. In fact, Venezuela is the best counter-factual to that assumption because the US clearly has the ability to complicate China's energy diplomacy in Venezuela. What we see, however, is a significant increase in the role of Venezuela in China's energy diversification, a process that seems acceptable to the US, despite Venezuela's proximity to the US.

US–Venezuelan problems currently have little to do with the Venezuelan–Chinese petroleum ties; in fact these ties are a result of the problems in the US–Venezuelan relationship. The triangularization is therefore not because of US–Chinese relations. The deterioration in US–Venezuelan relations is due to the Bolivarian Revolution and the explicit goal of Chávez Frías, now adopted by Maduro Moros to distance their country from the century-long petroleum links to the US.

Additionally, the very survival of Chávez Frías's Bolivarian Movement looks increasingly problematic. As mentioned earlier, the heir lacks the charisma to hold the movement together. The decline in the price of oil, which cuts Venezuela's foreign exchange reserves, also increases the frustration of a population whose standard of living improve, despite rhetoric to the contrary, has not improved one iota over the entire generation since Chávez Frías won his first presidential run in December 1998. Venezuelan leaders neither invested in the country's infrastructure, nor have they articulated any policy path to move beyond the commodity trap against which Chávez Frías railed through the final two decades of his life. In a rich irony, in early January 2014 the Venezuelan regime increased gasoline prices at home, demonstrating that despite its enormous energy supplies, the Venezuelan government could not finance its populous-based subsidies without raising the cost of petroleum for its citizens. This type of inability to maintain subsidies on food in 1989 was the source of much of the anger that originally brought Chávez Frías to power through the ballot in 1998.

More broadly, Venezuela's turmoil may only be starting because petroleum revenue is falling for the Republic. Latin American commodity exporters remain dependent on a market outside their control (China in this case), and the commodity boom that fuelled the welcomed economic growth in Latin America (among other places), petered out as China's growth subsided in 2013 and 2014. Additionally, Beijing continues to diversify its energy mix and the sources of that energy. In June 2014, China finalized a massive agreement on natural gas with Russia.

The historic 'monoculture export economy trap', although originally assumed in the twenty-first century to be an artefact of the past, appears to have been an under-appreciated effect of China's massive economic modernization between roughly 1978 and 2014. Rather than financially powering Venezuela or China vis-à-vis the US, today's falling petroleum prices are dragging down Venezuela's economy, leaving it few effective options with which to exercise its relations around the world, particularly with Washington. Chávez Frías's most cherished goal of enhancing the nation's muscular position through resource diplomacy is proving far harder and less likely than anticipated, regardless of which countries it engages. For China and the US, an energy provider such as Venezuela remains completely secondary to the competition and cooperation in their bilateral relationship.

References

Blas, Javier, 2012. Washington's reliance on Caracas for oil declines: US net imports of Venezuelan oil hit 30-year low. *The Financial Times,* 13 December.

China Daily, 2009. Chinese Leaders Make Success of Foreign Visits. *China Daily,* 23 February. Available online at: www.chinadaily.com.cn/china/2009-02/23/content_7503602_2.htm (accessed 29 March 2015).

Downs, Erica S., 2011. *Inside China, Inc: China Development Bank's cross-border energy deals.* John L. Thornton China Center Monograph Series, No. 3, March.

EIA, 2011. Existing and Planned Orinoco Belt Projects. Available online at: www.eia.gov/countries/cab.cfm?fips=VE (accessed 14 July 2011).

EIA, 2014. International Energy Outlook. Available online at: www.eia.gov/forecasts/ieo/pdf/0484%282014%29.pdf (accessed 20 May 2015).

Ellis, R. Evan, 2009. *China in Latin America: the whats and the wherefores.* Boulder, CO: Lynne Rienner.

Ellis, R. Evan, 2010. Chinese soft power in Latin America. *Joint Force Quarterly,* 60 (winter).

Forero, Juan, 2005. China's Oil diplomacy in Latin America. *New York Times.* 1 March. Available online at: www.nytimes.com/2005/03/01/business/worldbusiness/01oil.html?pagewanted=all&_r=0 (accessed 29 March 2015).

Gunson, Phil, 2006. Chávez's Venezuela. *Current History,* February: 58–63.

Hawksley, H., 2006. China's new Latin American revolution. *The Financial Times,* 4 April. Available online at: www.ft.com/intl/cms/s/2/ad142990-c407-11da-bc52-0000779e2340.html#axzz3W9WmofjA (accessed 2 April 2015).

Latin American Herald Tribune, 2014. Venezuela oil price falls. *Latin American Herald Tribune,* 26 January. Available online at: www.laht.com/article.asp?ArticleId=1357201&CategoryId=10717&utm_source=dlvr.it&utm_medium=twitter (accessed 29 March 2015).

McCoy, Jennifer and Neuman, Laura, 2001. Defining the 'Bolivarian Revolution': Hugo Chávez's Venezuela. *Current History,* February: 80–5.

Rathbone, J.P., 2010. Carlos Andrés Pérez: Self-styled leader of developing world dies. *Financial Times,* 27 December. Available online at: www.ft.com/intl/cms/s/0/dab85ba0-11e6-11e0-92d0-00144feabdc0.html#axzz3VmUKnGqf (accessed 29 March 2015).

Sachs, J.D., Solomon, A.M., Ogden, W.S., Wiesner, E. and McNamar, R.T., 1988. 'Developing Country Debt', *International Economic Cooperation,* Martin Feldstein, ed., Chicago, IL: University of Chicago Press, 233–320.

Toothaker, C., 2011. Chavez says cancer fight 'my longest walk'. *Associate Press,* 12 July. Available online at: www.cubastudygroup.org/index.cfm/newsroom?ContentRecord_id=b857618d-7fc5-4782-80c9-2e628d406cce&ContentType_id=8c81d17c-7ffe-48d6-81e7-cd93fe3120eb&Group_id=0b3ad3ec-d24e-4d2a-b425-a97ae7617c16&MonthDisplay=7&YearDisplay=2011 (accessed 29 March 2015).

14 Conclusion

China's 'energy anxiety'

David Zweig

Introduction

Over the next two decades, China's demand for energy will top the world, with one Chinese source estimating that China's energy consumption will rise by 2.23 per cent per year. China consumed 20 per cent of the world's energy in 2011, but by 2035 it will use up 26 per cent of world supply (Xu, 2014). Unfortunately, domestic sources will fail to meet China's total needs because supply will grow by only 1.97 per cent per year[1] (World Energy China Outlook [WECO], 2014). Ever-growing energy requirements place enormous pressure on China's leaders to ensure that the shortfall can be filled from abroad. Although China now imports more than 50 per cent of its oil, this figure may reach 80 per cent by 2030 (Kennedy, 2010). These efforts involve primarily expanding ties with Resource Rich States (RRSs) through enhanced trade and economic relations. But how welcoming is the world to China's increased energy demands? Answering that question has been the major task of this book.

Diversifying suppliers alone will not resolve China's energy security, which arises from numerous sources. The US's worldwide interests include influence in many RRSs, such as Saudi Arabia, Australia, Mexico and Canada, which as US allies, are potentially vulnerable to US pressures not to deal with China. As a late-comer to the energy import game, China must seek oil from many unstable governments, such as Venezuela, Sudan and Iran, whose supplies are unpredictable. Some of these unstable states are also US-declared 'pariah' states, and therefore China, in some cases, faces pressure from the US to eschew the energy in these states. Much of China's oil also comes from global 'hot spots' – Nigeria, Syria, Libya, Iraq and Algeria – and existing arrangements are easily disrupted by civil unrest and political crises.

Assuming China can purchase the oil, getting it back home is another issue – oil Sea Lanes of Communication (SLOCs) are rife with pirates and vulnerable to terrorism. The main force protecting these 'choke points' is the US Navy, leaving China dependent for its own energy security on the good will of its hegemonic competitor – a state it firmly believes wants to contain its rise. China, therefore, sees the US as committed to controlling the 16 global 'choke points' and accuses the US of using anti-terrorism as an excuse for putting troops into the Straits of

Malacca (Dou, 2008). It also sees the US airbases in Central Asia as part of a plan to disrupt China's pipeline supplies (Dou, 2008). Guaranteeing acceptable prices is also a problem. Over the past decade, prices have oscillated due to currency and supply problems. Finally, as part of its 'going out' strategy (*zou chu qu*), China finds a new imperative to deal with the issue of human security because its oil workers have become the target of kidnapping in restive areas, such as the Niger River Delta (Information Office of the State Council, 2013: 12). In response, Chinese firms sometimes hire local or Chinese militia to monitor pipelines, as was rumoured in Sudan.

China's current energy mix compounds its problem. According to Odgaard and Delman (2014), China's energy imports run directly counter to the specific resource's security risk. Although coal has the lowest security risk, due to its ready supply from stable countries that are often tangential to China (Australia, Mongolia, Vietnam, Indonesia, Russia and North Korea), as well as China's own enormous supplies, moving away from coal is a must because it kills Chinese citizens and destroys the ozone. On the other hand, oil, whose imports will continue to rise, is the riskiest energy source because much of it comes from unstable countries through vulnerable sea lanes and at prices (and supplies) that can fluctuate enormously.

China's leaders are also committed to enhance energy efficiency, limit individual and enterprise energy consumption, and restructure the energy mix by employing more natural gas, whose global supplies are more abundant than oil and whose effect on the global environment is far less deleterious. The Third Plenum of the 18th Central Committee in October 2013 revealed that China's leaders know that ameliorating pollution at home and cutting the emission of greenhouse gases into the atmosphere must jump to the top of the CCP's agenda. Still, Chinese people, firms and officials are accustomed to subsidized cheap energy and a development strategy that privileges production over clean air, water and soil, and an economy driven by an expanded automobile sector.

Despite these pressures, China's energy crisis may be smaller than anticipated. The 'neo-energy revolution' created by shale has turned the US from an energy consumer to supplier, relieving global demand that helps China. According to Crooks (2014), US production of liquid petroleum has surpassed its 1970 peak due to the nation's shale oil boom. Removing the US from world energy markets makes much of the oil produced in West Africa available to China (Chapter 8, this volume). The world's ability to extract more 'tight gas' and 'tight oil' also brings new supplies onto the market. Central Asia and Russia, China's new suppliers, can ship oil and gas to China by land, avoiding dangerous waterways. Finally, China's financial wealth, and that of its national oil companies, facilitates investment in energy projects around the globe.

Similarly, the threat to China's oil supplies posed by the US, whether to China's ability to purchase equity oil from US pariahs and allies, or its energy supply routes, may be overstated. Few cogent observers believe that the US could establish effective blockades at the straits of Hormuz or Malacca. The US, like China, wants stable prices and has established the International Energy Agency (IEA) to manage energy shocks. US companies and the US Department of Energy appear willing to share technology with China, particularly if it relates to shale.

The purpose of this chapter is first to reflect on the findings of the various case studies presented in this book in terms of China's ability to diversify its supply and the role of the US in China's efforts in each of our ten case studies. The chapter will also look at other risks to Chinese energy security, the strategies the Chinese are likely to adopt, and the potential impact of the US on those strategies. As Table 14.1 shows, the risks have their domestic and international roots and

Table 14.1 Risks to Chinese energy security, strategic responses, and role of the US

Nature of risk	Strategies	Role of the US
Price increases	Long-term contracts with fixed prices, but these are hard to get	Stabilizes prices through production and coordination through International Energy Agency (IEA)
Threat to delivery (supply lines)	National tanker fleet	Protects sea lines of communication (SLOCs) for all nations, including China
	Pipelines	US Government competes for Central Asian oil to flow west to Europe
Cutoff in oil supply	Strategic petroleum reserves (SPR)	Encourage Saudi Arabia to work with China
	Drilling in South China Sea	Support for ASEAN states and Japan against China
	Diversify suppliers	Pressure China not to buy from 'pariah' states Compete for Canadian oil
	Develop local shale oil and gas	US–China cooperation on shale
	Restructure energy mix away from oil	n/a
	Strengthen own national oil companies	US companies are major competitors but can cooperate with Chinese national oil companies (NOCs)
Pollution and public health	Clean coal	Transfer of technology
	Shift to non-carbon, especially gas, through imports or shale gas development, and nuclear	Work with Chinese companies on shale deposits in China Transfer shale technology to China
Cancelling of contracts	Improve global governance	Establish global energy norms
	Create new international institutions	Persuade allies or neutrals not to join Chinese created organization
War among states including oil producers	Use own tanker fleet, which it will have to protect with a fleet of ships or use the strategic reserves	US could be protagonist (versus Iran?) Could establish near shore or SLOC blockade in case of war with China (or China with Japan)

involve both traditional energy security, such as military threats to delivery, as well as public policy issues that respond to the demands of Chinese society and the global community, such as limiting local pollution and greenhouse gas emissions. My goal in this conclusion is to expand our assessment of the US's potential impact on China's energy strategy beyond the widely addressed strategic aspects, as well as incorporate potentially positive contributions the US has made (and can make) to China's energy security.

As stated at the outset of this volume, the US's negative impact on China's search for energy is a hypothesis. It is what my former professor, Bob Putnam, used to call an 'empirical question' that needs to be tested in the real world, rather than an assumption that we should take as given. Testing that hypothesis was the charge given to the contributors of this volume to keep in mind, whilst assessing China's interaction with the RRS that was the focus of their case study. The question was did the US or its firms influence the bilateral ties between China and its energy suppliers? The analysis in this chapter incorporates their findings, which in Table 14.1 primarily fall under the category of risk to China's oil supply. However, rather than simply restate their findings, which Hao and Zweig have done in their introduction, I present a more comprehensive analysis of the strategies China has adopted in response to its energy anxiety and risks (Tunsjo, 2013), and the US's impact on that strategy.[2] My goal is a more comprehensive assessment of the true impact of the US on China's energy security.

Public policy and domestic anxieties

Domestic pressures contribute to the Chinese state's 'energy anxiety'. Over the past decade, concerns about energy and the environment among the Chinese public have increased dramatically. In 2006, a survey of 800 college students at eight universities in three cities found that 25 per cent were 'deeply concerned', seeing it as a 'crisis'; 50 per cent were 'concerned'; and 22 per cent were 'neutral'. Almost no one was unconcerned, necessitating a powerful and meaningful response from China's leaders (Zweig and Ye, 2008). Today, a new balance must quickly emerge between government, industry and the public interest (Xu, 2014). North China is now regularly engulfed in a bubble of soot and dirty air that explains partly why life expectancy in South China exceeds North China by 5 years (Reilly, 2013).

For the leadership, a critical problem is how to address this environmental crisis, while still maintaining strong industrial growth. A steady flow of energy to maintain economic growth remains intrinsic to the Chinese Communist Party (CCP)'s hold on state power. Oil is linked to the 'social contract' between the CCP and China's new middle class – the CCP promises to deliver cars, air conditioning, new homes, a higher quality of life and the right to partake in the 'China dream', in return for which citizens agree not to challenge CCP rule.

But how to do so remains a problem. A shift to greater reliance on nuclear could meet with strong resistance from a far more environmentally aware populace that, to date, has stopped the establishment of chemical plants in at least five cities in South China, including Kunming, Ningbo and Xiamen, as well as coal-fired power plants in Haimen and Shantou in Guangdong Province (China.org.cn, 2011). Should the supply of energy slow, the CCP's hold on power could suffer.

Cleaning up China's air involves, first and foremost, a shift away from coal as a source of energy to less polluting energy sources, particularly gas and renewables. China's current energy mix – relying on coal for 68 per cent of its energy supply – is unsustainable. WECO, which proposes a new 'eco-friendly energy strategy' to replace the 'failed strategy which has pushed industrial production above all else' (WECO, 2014: 4), insists that coal must drop dramatically to 48 per cent within 20 years. The shift to gas imports, however, involves the massive development of new infrastructure across the country. In the past, the state placed too much emphasis on traditional energy security, whilst allowing industrialization to continue apace; however, the urbanization proposed at the Third Plenum in autumn 2013 will increase demand for energy exponentially as urban residents use far more energy than rural residents. The Third Plenum also sets its own guidelines for energy efficiency, air pollution and social stability.

One strategy is coal liquidation. With its huge reserves of coal, China has also been working hard on coal liquefaction, a technology that is led globally by South Africa (Patel, 2012). China has huge deposits of shale oil and gas, but the locations are problematic; water is a major problem for China; China's national oil companies (NOCs) are unwilling to invest in China, and, private firms, who spearheaded the technological breakthroughs in the US, have not become reliable project developers (Bradsher, 2014).

Chinese students are acutely aware of China's energy insecurity, a view reflected in the survey by Zweig and Ye (2008). Of the 800 students who responded to the survey, 57 per cent feared that energy imports would let other countries control China, while many worried that unless China did something about it, the country would become a dependent nation. Many feared that disagreements over energy could lead to military conflict. Fortunately, their choice of strategies for ameliorating these problems involved peaceful efforts, such as diversifying suppliers, enhancing energy efficiency and building a strategic petroleum reserve (SPR). Troubling for the Chinese government, however, is that these students had a very strong belief that citizens had a responsibility to participate in programs concerning energy – 27.7 per cent reported speaking to their friends about energy four times or more in the past year, while over 60 per cent did so twice or more. They also felt that, in general, the Chinese government was very slow in establishing an SPR.

Maintaining the stability of China's oil supply

China has introduced various strategies to resolve its energy anxiety, and the most common focus of analysis is the 'diversification of its suppliers'. These suppliers include US allies, such as Canada, Australia and Saudi Arabia, as well as states that in the eyes of the US (and at times the UN), are labelled as 'pariahs'. As of 2007, China imported 22 per cent of its oil from such states, (Iran, Sudan and Venezuela), but only 14.7 per cent as of 2013 (Chapter 1, this volume). According to the US Department of Defense, as of 2013, China imported oil from over 50 countries (US Department of Defense, 2014). To ensure supply, China has invested in equity oil and has bought oil fields in these countries, as well as the oil in the ground. To succeed in such investments, China must also develop the capacity of its NOCs, particularly its big three: China National Petroleum Company (CNPC), PetroChina and China National Offshore Oil Company (CNOOC). Some analysts fault this strategy, arguing that in a time of war, China may not be able to access the oil and the supplier may not be willing to deliver it, particularly not at prices that may have been predetermined by contract (personal communication, Phillip Andrews-Speed, November 1, 2014). In other cases, observers assert that China overpaid for the oil concession. The Chinese government, as a latecomer to the energy game and in possession of trillions of US dollars, may support the firms who pay high prices to get their foot into the door of the energy rich country. Also, according to Chen (2008), China's oil diplomacy has been driven not only by the government's strategic concerns, but also by the NOCs' strong commercial motives to expand business abroad and their management's personal incentives.

China's banks promote overseas energy purchases. The State Development Bank (SDB) has poured billions into RRS in the form of long-term, low-interest loans as a quid pro quo for access to energy resources (Downs, 2011). The Export–Import Bank supports similar efforts, albeit on a smaller scale. Also, when China's leaders travel abroad, they support the NOC's search for oil by committing to build roads, schools, harbours and other infrastructure projects needed for modern development in these RRS (Economy and Levi, 2014). Although a common strategy even among developed countries, Chinese leaders bring along CEOs of major Chinese corporations to facilitate their access to ministers and other government leaders.

China is well known for its 'infrastructure for oil programs' inherited from Japan (Brautigam, 2009). As of 2013, 36 per cent of China's oil supplies came through 'oil for credit' programs, which gives China a leg up on many other consuming countries (Odgaard and Delman, 2014). As we saw in our case studies of Africa, this was a popular policy in the eyes of the former president of Nigeria (Chapter 8, this volume) and is often referred to as the 'Angolan model' (Chapter 7, this volume).

Relying on authoritarian regimes generates energy anxiety. Authoritarian regimes that are resource-rich tend to be more unstable and less transparent than other types of regimes. According to Friedman (2008), as the price of oil goes up, so does the state's power; but lower prices breed instability because authoritarian states suddenly lack the funds needed for maintaining tight security. In general, many RRS are also relatively unstable: note the cases of Libya, Iran, Algeria and Syria. In 2013, civil war made it impossible for Sudan to export any oil at all, wiping out a major Chinese supplier. In Venezuela, political instability, a weak economic environment and a lack of skilled workers has dramatically weakened Venezuela's ability to export oil (Chapter 13, this volume). Venezuela's energy production[3] peaked in 2005 when it produced 233,734 kilotons of oil equivalents, dropping 18 per cent in the following 5 years to 192,708 kilotons of oil by 2010.

Still, Venezuela's share of Chinese oil imports has risen from 1.5 per cent in 2007 to over 6 per cent by 2013.

China's energy strategy differs from that of the West's in that China is willing to work closely with NOCs in the RRS. Although Western international oil companies (IOCs) have always preferred to work on their own, China has adopted an alternative strategy that has proven successful. More than pragmatism may be at work here. As a developing country that felt exploited by colonial powers, China and its leaders want to recognize the sovereignty of the RRS and the right of those countries to control their resources. Over 70 per cent of the world's energy is owned by NOCs, and China must therefore link up with them to get access to old and new fields.

As a result, China accepts an equity share of less than 50 per cent in its energy deals (as in the case of Venezuela), which the majors have always eschewed; however, such strategies do not always succeed, particularly in 'neutral' countries that are not dependent on China. For example, a CNPC deal to buy Verenex, a company with interests in Libya, was pre-empted by Libya's state oil company, while a CNOOC/Sinopec bid to buy Marathon's stake in a large Angolan offshore oilfield, Jubilee, was pre-empted by Sonangol, Angola's national oil company (Faucon and Swartz, 2009).

The Russian Federation has proven to be a very important partner in China's effort at diversification. Home to 5.5 per cent of existing oil and 16.8 per cent of existing natural gas reserves (BP, 2014), Russia is an attractive partner for China from the perspective of the 'power transition model' (Tammen, 2008) due to its relatively hostile relations with the US. China should be seeking new allies that can balance the military preponderance of the reigning hegemon. Russia has supplied China military weaponry for many years, and China and Russia have carried out several rounds of joint military training exercises, and yet Russia is not the easiest of partners. It played China off against Japan for over 2 years, refusing to commit where the Angarsk–Nakhodka pipeline would end in the hopes of improving its negotiation leverage vis-à-vis both countries. The Chinese are not naïve about Russia. In its 2014 report on World Energy, the Energy Group at CASS recognized that 'there exist [clear] interest differential and even conflicts of preferences between the two powers' (WECO, 2014: 10 – translated from the Chinese). Still, the crisis in the Ukraine in 2014 and the boycott on Russian oil by Western Europe compelled Russian President Putin to commit to selling oil to China for 30 years at a price favourable to China.

Shopping for oil in democratic countries, such as Canada and Australia, which are far more stable politically than authoritarian states, is problematic. Investments by Chinese NOCs, which are viewed as tools of the Chinese Communist Party (CCP), are highly suspect, within these societies, within the government bureaucracy and within the governing elites (Chapters 5 and 6, this volume). Strengthening their NOCs would not necessarily yield more positive results for China in these countries. In the most positive case of 'greenfield investment', when compared to buying locally owned companies where China would be creating new jobs for these countries, barely half of the respondents in Canada or Australia see Chinese investments as a 'good' or a 'very good' development. When asked why they opposed such sales, 47 per cent of Australians worried that China 'might use ownership of Australian firms against us in conflict', and 37 per cent felt they 'could not trust China' (Hanson, 2013).

Assessing China's efforts at diversification

Has China actually diversified its sources of supply? Although China imports oil from over 50 countries (US Department of Defense, 2014), a good indicator of the real level of diversification is the percentage of crude oil that comes from China's top 10–12 importers, and any changes in that percentage over time. The larger the share imported from its top producers, the less diversified China's crude oil imports. A second measure assesses the variance of crude oil distribution *within* China's top 12 crude oil importers. A smaller variance suggests a more even distribution among the 12 major suppliers and less dependence on any one or two countries within this small group.

China's level of dependence has not improved, despite extraordinary efforts to develop new suppliers. The percentage of energy coming from the top 12 suppliers peaked in 2005 when 82.3 per cent of China's imports came from the top 12 suppliers. Thereafter, it declined in 2006 and 2007 to below 80 per cent, only to rise again to 82.2 per cent in 2008. After a decline to 78.7 per cent in 2010, greater dependence re-emerged in 2011 (Table 14.2).

The data on variance is more positive because, after peaking in 2008 when its supplies were more dependent on fewer countries, in 2011 its oil imports

Table 14.2 Diversification of China's energy supplies, 2003–2013 (top 12 oil suppliers)[1]

Year	Percentage	Variance
2003	75.6	0.30%
2004	73.3	0.28%
2005	78.5	0.29%
2006	80.2	0.31%
2007	81.2	0.27%
2008	85.5	0.35%
2009	82.9	0.33%
2010	82.3	0.27%
2011	85.0	0.24%
2012	83.7	0.28%
2013	85.5	0.26%

Sources: China Customs statistics, various years, from Tian (2013, 2014, 2010 and 2006).

Note:

1 Top twelve oil suppliers include Saudi Arabia, Angola, Russia, Iran, Oman, Iraq, Venezuela, Kuwait, Kazakhstan, UAE, Sudan and Yemen.

were more evenly diversified among its top 12 suppliers than at any time since 2005. Still, some projections suggest that China's dependence on Saudi Arabia will increase year on year, leaving it reliant on a regime that remains somewhat unstable (Chapter 4, this volume). Great efforts at diversification have therefore only yielded marginal, if any, improvements.

Assessing the US impact on China's oil supply

What do the cases in this book tell us about the ability or willingness of the US to influence China's relations with RRS? China's deep concerns are overstated. According to one researcher for a major Chinese NOC, Chinese firms do incorporate US interests into their bilateral energy strategies (personal communication, 2010), but there is little proof that the US tries to block China's access to oil in order to slow down or complicate China's rise.

In the case of Australia, a US ally, the US has refrained from commenting on enhanced Sino–Australian economic interaction; it speaks softly but firmly only when Australian leaders suggest waning support for the ANZUS (Australia, New Zealand, United States) treaty and Australia's commitment to support the US in the event of a conflict with China in East Asia. The US encourages closer Sino–Saudi energy ties as a means to wean China from its dependency on Iran. The Saudis also play a critical role in constructing China's SPR and working with Sinopec (and Exxon) on a downstream megaproject in Fujian Province. Even in Nigeria, which in the mid-2000s was a focus of US–China energy competition, there were only rumours that the West tried to exclude China. Recently, the US has dramatically cut its oil imports from Nigeria, leaving far fewer reasons for the US to engage with Lagos or worry about Chinese activity in that country. Chinese-accelerated imports of Nigerian crude should be seen in a positive light because such sales can stabilize Africa's largest country.

The US's 'pivot' or 'reorientation' to East Asia increases China's energy anxiety. China hopes to rely on gas or oil from the East or South China Sea, where China faces challenges from Japan, Vietnam, Malaysia, Indonesia and the Philippines. US support for these states complicates China's claims over the underwater resources of these nearby regions. In China's Defense White Paper of 2013, the Chinese government emphasizes the role of the PLAN (People's Liberation Army Navy) in providing security support for China's 'oil and gas exploitation' (IOSC, 2013). China cannot take any solace in President Obama's speech to West Point on 28 May 2014 when he said that 'regional aggression that goes unchecked – in southern Ukraine, the South China Sea, or anywhere else in the world – will ultimately impact our allies, and could draw in our military' (Obama, 2014). That speech also stated that 'the United States will use military force, unilaterally if necessary, when our core interests demand it', which includes 'when the security of our allies is in danger'. He also emphasized US efforts to strengthen and enforce international order, including getting China to resolve its maritime disputes under the Law of the Sea Convention, which he admitted the US Senate has refused to ratify.

The US pressures China not to buy petroleum from pariah states, such as Sudan and Iran, but, as Garver (Chapter 11, this volume) and Regler (Chapter 12, this volume) show, these efforts are due to moral concerns about genocide and nuclear proliferation, respectively, not about constraining China's access to oil. Similarly, as we saw in Canada, a US ally, members of former administrations (who probably work as lobbyists) and Congressmen do try to complicate Canada's energy ties with China (Chapter 6, this volume). One member of the US House of Representatives admitted that he blocked the sale to China of the US energy firm, Unocal, to slow down China's development. In the case of Venezuela, a state sitting in the US's backyard, the US did not challenge China's increased ties with that country in any major way; instead, they cut their oil imports from Venezuela and increased their reliance on domestic oil.

In fact, the US is no longer a direct competitor for China's energy in much of the world. Although in 2007, 57 per cent of Chinese college students surveyed saw the US as China's major competitor for oil, and 30 per cent saw Japan in this role (Zweig and Ye, 2008), today that concern should be minimal. Moreover, in 2005, there was an overlap in energy supplies from only a few parts of the world (Middle East and Africa) because the US relied predominantly on Mexico and Canada for its oil imports. If, however, China's energy requirement pushes it to North America, tensions could rise.

Threats to delivery and risks to supply lines

China's energy deliveries are at risk because of its dependence on SLOCs. The sea lanes that bring oil to China are vulnerable to terrorism and open to harassment by India or the US. In fact, in 2008, Dou (2008: 12–13) expressed concerns about India's naval base in the Andaman Islands, seeing India as capable of using their navy to block China's access to the Straits of Malacca. China's NOCs have therefore been actively building new pipelines from Russia, Kazakhstan and Myanmar, which, when operating at full capacity, will carry 1.1 million barrels per day of oil or 14 per cent of China's projected imports in 2015. China is actively building up its own oil tanker fleet (Tunsjo 2013), which could supply China if the US pressurizes other countries to halt supplies to China, in a manner reminiscent of US policy towards Japan before the Second World War.

Separate from diversification, other strategies can enhance China's energy security. Building an SPR is a strategy adopted by many countries. For China, this effort has involved collaboration with Saudi Arabia (Chapter 4, this volume).

Rather than forever relying on future US benevolence, the PLAN has used China's need for energy security to strengthen its position within the Chinese military by constantly stressing the vulnerability of its oil SLOCs (Chapter 1, this volume). In 2004, a specialist in the Chinese Institute of Contemporary International Relations (CICIR), reportedly quoted in Hong Kong *Wen Wei Po*, argued that China would face an energy crisis if its oil supply lines at sea were attacked. Therefore, 'the Chinese government stated that it has strategic interests over oil supply routes and would use naval force to control the shipping lanes to protect its strategic interests over oil supply routes' (*China Reform Monitor,* 2004).

China, however, will not be able to extricate itself from its dependence on the Straits of Malacca and the South China Sea as major SLOCs. According to the US Department of Defense,

> Given China's growing energy demand, new pipelines will only slightly alleviate China's maritime dependency on either the Strait of Malacca or the Strait of Hormuz. Despite China's efforts, the sheer volume of oil and liquefied natural gas that is imported to China from the Middle East and Africa will make strategic SLOCs increasingly important to China.
>
> (2014: 27)

Also, as Kazakh oil has come out of the ground, a major struggle has ensued over whether it will flow east or west (Chapter 9, this volume). Although China has invested heavily in a pipeline running to Xinjiang, the US has used its financial and fiscal capital to promote the Baku–Tiblisi–Ceyhan (BTC) pipeline, which will bring Kazakh oil to Europe, with the support of Turkey (Yergin, 2011).

Stabilizing prices and supply

China is concerned about the price fluctuations brought on by changes in the supply of oil or changes in the value of the US dollar. To resolve the first problem, China tries to establish long-term contracts at fixed prices, as they did in 2014 with Russia for the purchase of natural gas. An alternative strategy is to sign contracts in other currencies, such as euros, which fluctuates far less that the US dollar.[4]

China and the US share a common interest in stable global prices. Even if US oil majors benefit from higher prices, shocks to the global economy harm US interests in a stable economic order. In fact, according to Crooks (2014), the rise in world oil prices is being curbed by US production, which between 2005 and 2013 accounted for almost all of the increase in global supply. The US and China could therefore cooperate in establishing some mutual guarantee agreement whereby consumers commit to a stable level of oil purchases that would ensure suppliers a stable market. Production would fluctuate far less, mitigating drops in output due to declining demand. Proposed by the World Energy Group of CASS, this agreement could help China avoid price increases in the face of decreasing US global demand. The difficulties facing West Africa today in light of the US withdrawal as a consumer of its oil fits this situation because they will be forced to cut production, which will create price instability.

The shale revolution and China's energy security

A major transformation in global energy has come about due to the shale revolution, which has important implications for China. First, an energy-independent US will compete far less with China over sources of energy around the world. Second, many producer countries, such as those in West Africa, who had been selling oil to the US, will now see China as an important market.

However, it remains unclear if shale itself will become an immediate energy solution for China. According to WECO (2014: 6), 'there is no way for China to copy the success of shale gas in the US. And no fast development could result any time soon', despite China's wealth of shale deposits. Water, technology and geography complicate China's shale development.

The US can help China develop its own supplies because the US has the technological know-how to manage many of the difficulties involved in shale gas extraction. US companies Mitchell Energy and Devon Energy began the practice of hydraulic fracking for shale gas (Brennan, 2013). In 2013, Sinopec bought into Devon Energy in a joint venture in shale gas in Mississippi, which will speed up China's acquisition of expertise. Similarly, in 2010, CNPC and PetroChina purchased a 60 percent stake of an Oil Sands project in Alberta (CBC News, 2012), and also formed a joint venture with Canadian Encana to develop shale gas extraction in British Columbia (*Financial Times,* 2011). In 2014, CNOOC took over the Canadian firm Nexen, which is exploiting shale gas in British Columbia and in Poland. The US has also established bilateral agreements to develop Chinese shale oil and gas under the US–China Shale Gas Resource Initiative (Brennan, 2013). By helping China extract shale, the US may contribute in an important way to alleviating China's energy anxiety.

Improving global energy governance and rulemaking

Despite efforts to enhance ties with the developing world, China, as with other players, suffers due to the lack of good governance in the energy sector. During the conference for which the original papers in this volume were written, attendees felt that China, as a major consumer of oil, needed more transparency from its energy suppliers, even though it moves quite capably in the murky world of authoritarian regimes. As more and more oil comes from neutral states (van Guens, 2008), many of whom are non-transparent, China can benefit from strengthened energy norms at the regional and global level (Xu, 2014). As Table 14.3 and Figure 14.1 show, neutral and pariah states are far less transparent than the US's allies. In the case of Angola, therefore, even after China paid up-front to their so-called friends, the contracts were torn up and the deal cancelled.

According to Chinese analysts, China has been actively establishing new multilateral energy regimes, such as the Shanghai Cooperation Organization (SCO) in Central Asia. It has fared quite well in that region in terms of its energy

Table 14.3 Average transparency score of regime types, 2012

	Pariah	Neutral	US ally
Transparency score	18.7	25.8	47.8

Source: Corruption Perceptions Index. *Transparency International Index, 2012.*
Available online at: http://cpi.transparency.org/cpi2012.

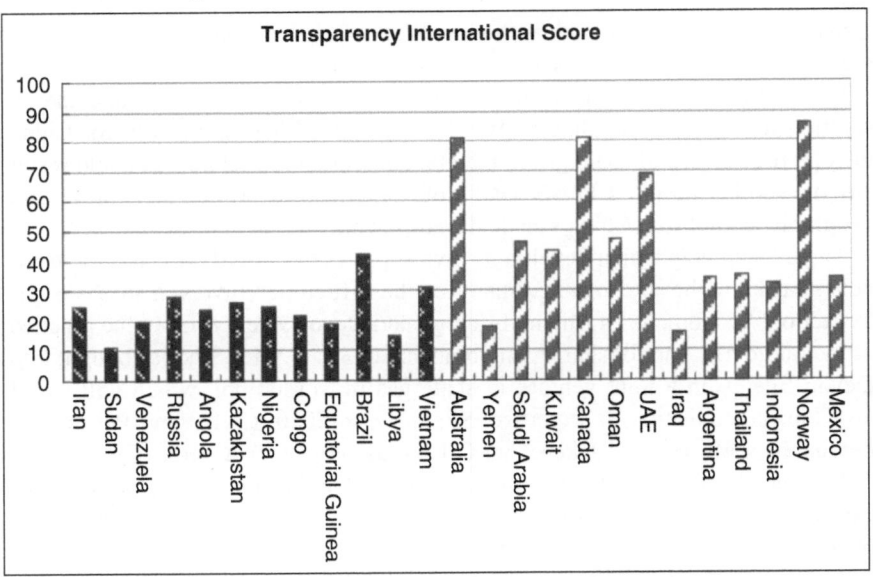

Figure 14.1 Transparency scores of energy rich states, by regime type, 2012.

Source: Corruption Perceptions Index. *Transparency International Index*, 2012.

ties, but given its economic weight in that part of the world, it remains unclear whether it is the norms or the economic muscle that explains this success. Moreover, only time will tell if Russia will abide by its long-term agreement of 2014. Nevertheless, these same Chinese energy analysts believe that China needs a stable global energy regime as it becomes more and more dependent on global supplies.

The US can play a role here as well. As the global hegemon, it has influence over the setting of rules in international organizations. Allowing China to enter the IEA would probably go a long way to relieve China of its energy anxiety, particularly its concern about energy shortages (Lieberthal and Herberg, 2006); however, China has shown a strong tendency to exclude the US from most new global grouping that it tries to establish.

The risk of war

The final risk that we need to consider would arise from a war between China and the US, which is generally unlikely (Kirschner, 2008), or one that involves one or more major energy producers. Such a war would immediately affect prices – instability in northern Iraq in June 2014 pushed up the price of West Texas crude by 4 percent (Crooks, 2014) – but is also likely to be highly disruptive to supplies coming from the Persian Gulf. In such an event, China would turn to its own tanker fleet to maintain shipments (Tunsjo, 2013). Under such conditions, the Chinese government would insist that its NOCs ship their equity oil back home, rather than sell it on the spot market; it would also draw on its SPR. In the case of such an eventuality, the US would be a major protagonist or closely allied with one or more of the combatants. China's ties to the US would be important because it would need to try rapidly to broker a ceasefire, if not an end to the war. Should the war involve combatants other than the US, China would need to work with the US to broker a ceasefire to ensure the continued flow of oil out of the Persian Gulf – assuming that it is still being pulled out of the ground.

Conclusion

As stated at the outset of this chapter, the role of the US in generating or ameliorating China's energy anxiety is a mixed one. On the one hand, China's expanding needs pushes it further out into a world largely influenced (if not dominated) by US power, and yet the US has rarely exercised that power to block China's rise (as the power transition theory would predict). When China's new energy sources are neither pariahs nor allies, the US remains neutral. A similar response has also occurred in places where one might have anticipated greater US agency, such as Canada or Venezuela.

The world energy situation, however, is in influx. Shale is forging new energy ties between the US and China, and opening new supplies for China, and as the US withdraws from some traditional markets, China is finding new supplies of oil. Such moves should ameliorate Chinese anxieties, and yet China cannot afford to be complacent. Increased domestic, social and industrial demands for oil will continue to propel China's NOCs and its government. An energy-hungry China will no doubt bump heads with the US in the future, but unlike US–China strategic competition, which reflects a more zero-sum game, the head bumping for the near future should be limited, even as the hand-holding expands.

Notes

1 The International Energy Agency (IEA) predicts lower demand growth at 1.9 per cent per year, but sees domestic energy production rising by only 1.4 per cent per year, making a greater supply gap than predicted by CASS [Chinese Academy of Social Sciences]' World Energy China Outlook (WECO, 2014).

2 A draft of this paper was presented at a seminar at the World Energy Division, Institute for World Economics and Politics, Chinese Academy of Social Sciences on 24 June 2014. A special thanks goes to its Director Xu Xiaojie, whose team gave me excellent comments. I also benefited from the comments of Øystein Tunsjo, whose insights, both in a seminar at the Norwegian Institute of Defense Studies, as well as in his current book, helped inform my framework.

3 'Energy production' refers to primary energy – petroleum (crude oil, natural gas liquids and oil from non-conventional sources), natural gas, solid fuels (coal, lignite, and other derived fuels), and combustible renewables and waste – and also primary electricity, all converted into oil equivalents.

4 This point was made by Chen Tansi, an MA student at the World Energy Division at CASS on 24 June 2014.

References

BP, 2014. *BP Statistical Review of World Energy, June 2014.* Available online at: www. bp.com/content/dam/bp/pdf/Energy-economics/statistical-review-2014/BP-statistical-review-of-world-energy-2014-full-report.pdf (accessed 27 March 2015).

Bradsher, Keith, 2014. China takes big risks in its quest for shale gas. *International New York Times,* 14 April, pp. 1, 15.

Brautigam, Deborah, 2009. *The dragon's gift: the real story of China in Africa.* Oxford, UK: Oxford University Press.

Brennan, Elliot, 2013. *Shale gas: the key in the US' Asia pivot?* Institute for Security and Development Policy, 8 March. Available online at: www.chinausfocus.com/energy-environment/shale-gas-the-key-in-the-us-asia-pivot/ (accessed 27 March 2015).

CBC News, 2012. *PetroChina buys entire Alberta oilsands project.* 3 January. Available online at: www.cbc.ca/news/business/petrochina-buys-entire-alberta-oilsands-project-1.1234007 (accessed 22 March 2015).

Chen, Shaofeng, 2008. Motivation behind China's foreign oil quest – a perspective from the Chinese government and the oil companies. *Journal of Chinese Political Science,* 13 (1): 79–103.

China Reform Monitor, 2004. Chinese officials vow use of force in oil-shipping lanes. *China Reform Monitor,* 527, 27 January.

China.org.cn, 2011. *Protest stops after power plant plan suspension.* 21 December. Available online at: www.china.org.cn/china/2011-12/21/content_24208700.htm (accessed 29 March 2015).

Crooks, Ed, 2014. US petroleum production hits 44-year high. *Financial Times,* 14 June. Available online at: www.ft.com/intl/cms/s/0/37afc468-f48f-11e3-a143-00144feabdc0. html#axzz34iuyaV2n (accessed 29 March 2015).

Dou, Chao, 2008. *Pojie shiyou kunju* [Analyzing and Explaining the Oil Predicament], *Jianzai wuqi* [Shipborne Weapons], 12: 10–13.

Downs, Erica, 2011. *Inside China, Inc.: China Development Bank's cross-border energy deals.* Brookings Institution, John L. Thornton China Center Monograph Series, No. 3, March. Washington, DC.

Economy, Elizabeth C. and Levi, Michael, 2014. *By all means necessary: How China's resource quest is changing the world.* New York, NY: Oxford University Press.

Faucon, Benoit and Swartz, Spencer, 2009. Africa pressures China's oil deals. *Wall Street Journal,* 30 September.

Financial Times, 2011. PetroChina in $5.4bn Canada gas buy. *Financial Times,* 11 February. Available online at: www.ft.com/intl/cms/s/0/94826968-34ab-11e0-9ebc-00144feabdc0.html#axzz3V5XzxN6i (accessed 22 March 2015).

Friedman, Thomas L., 2008. *Hot, flat, and crowded: why the world needs a green revolution – and how we can renew our global future.* London: Allen Lane.

Hanson, Fergus, 2013. *Australia and New Zealand in the world: public opinion and foreign policy.* Lowy Institute for International Policy.

Information Office of the State Council (IOSC), 2013. *The diversified employment of China's armed forces.* Available online at: http://eng.mod.gov.cn/Database/WhitePapers/index.htm (accessed 29 March 2015).

Kennedy, Andrew B., 2010. China's new energy–security debate. *Survival,* 52 (3): 137–58.

Kirshner, Jonathan, 2008. The consequences of China's economic rise for Sino–US relations: rivalry, political conflict and (not) war. *In:* Robert S. Ross and Feng Zhu, eds.

China's ascent: power, security, and the future of international politics. Ithaca, NY: Cornell University Press, pp. 238–59.

Lieberthal, Kenneth and Herberg, Mikkal, 2006. *China's search for energy security: implications for US policy.* National Bureau of Research, Vol. 17, No. 1, April.

Obama, 2014. President Obama's speech to West Point. Available online at: www.timesofisrael.com/full-text-of-barack-obamas-west-point-address/#ixzz346NmLsac (accessed 29 March 2015).

Odgaard, Ole and Delman, Jorgen, 2014. China's energy security and its challenges towards 2035. *Energy Policy,* 71: 107–17.

Patel, Prachi, 2012. China and South Africa pursue coal liquefaction. *Energy Quarterly,* 37, March. Available online at: www.mrs.org/bulletin (accessed 29 March 2015).

Reilly, Jill, 2013. China's smog has cut life expectancy by five-and-a-half years for those living in its polluted north. *Daily Mail,* 9 July. Available online at: www.dailymail.co.uk/news/article-2358689/Chinas-smog-cut-life-expectancy-half-years-living-polluted-north.html#ixzz38AteKi8I (accessed 29 March 2015).

Tammen, Ronald L., 2008. The Organski legacy: a fifty-year research program. *International Interactions,* 34.

Tian, Chunrong, 2006. *2005 Nian, Zhongguo shiyou tianranqi jinchukou zhuangkuang fenxi.* (Analysis of the situation of China's imports and exports of oil and natural gas) *Guoji shiyou jingji* (International energy economics), no. 3.

Tian, Chunrong, 2010. *2009 Nian, Zhongguo shiyou tianranqi jinchukou zhuangkuang fenxi.* (Analysis of the situation of China's imports and exports of oil and natural gas) *Guoji shiyou jingji* (International energy economics), no. 3.

Tian, Chunrong, 2013. *2012 Nian, Zhongguo shiyou tianranqi jinchukou zhuangkuang fenxi.* (Analysis of the situation of China's imports and exports of oil and natural gas) *Guoji shiyou jingji* (International energy economics), no. 3.

Tian, Chunrong, 2014. *2013 Nian, Zhongguo shiyou tianranqi jinchukou zhuangkuang fenxi.* (Analysis of the situation of China's imports and exports of oil and natural gas) *Guoji shiyou jingji* (International energy economics), no. 3.

Tunsjo, Øystein, 2013. *Security and profit in China's energy policy: hedging against risk.* New York, NY: Columbia University Press.

US Department of Defense, 2014. *Annual Report to Congress: Military and Security Developments Involving the People's Republic of China.* Washington, DC: Office of the Secretary of Defense.

van Geuns, Lucia, 2008. *China, Africa and the international oil market.* Clingendael International Energy Program. The Hague, Netherlands, 20 May.

World Energy China Outlook (WECO), 2014. *World Energy China Outlook* [Xu Xiaojie ed.]. Beijing: Chinese Academy of Social Sciences Press.

Xu, Xiaojie, 2014. Lecture by Xu Xiaojie, Hong Kong University of Science and Technology, 25 April 2014, Hong Kong.

Yergin, Daniel, 2011. *The quest: energy, security, and the remaking of the modern world.* New York, NY: Penguin Press.

Zweig, David and Shulan, Ye, 2008. A crisis is looming: China's energy challenge in the eyes of university students. *Journal of Contemporary China,* 17 (55): 273–96.

Index

Figures and tables are given in italics